CONTROL SYSTEMS

Analysis, Design, and Simulation

JOHN W. BREWER

Associate Professor
Department of Mechanical Engineering
University of California, Davis

Prentice-Hall, Inc., Englewood Cliffs, New Jersey

Library of Congress Cataloging in Publication Data

BREWER, JOHN W.
   Control systems: analysis, design, and simulation.

   Includes bibliographies.
   1. Automatic control.   2. Control theory.
3. System analysis.   I. Title.
TJ213.B64   629.8   73–19697
ISBN 0–13–171850–9

Dedicated to

M'lou,

Corrie,

Jeff

10 9 8 7 6 5 4 3 2 1

Printed in the United States of America

PRENTICE-HALL INTERNATIONAL, INC., *London*
PRENTICE-HALL OF AUSTRALIA, PTY. LTD., *Sydney*
PRENTICE-HALL OF CANADA, LTD., *Toronto*
PRENTICE-HALL OF INDIA PRIVATE LTD., *New Delhi*
PRENTICE-HALL OF JAPAN, INC., *Tokyo*

Contents

*Chapters and sections marked with an asterisk contain reference material and may be omitted during first reading of the text.

10 Feedback Control: Design Criteria and Elementary Methods for Stability Analysis   275

11 Design of Scalar Automatic Control Systems by the Root Locus Method   315

12 Design of Multivariable Control Systems by the Methods of Modal and Decoupling Control; The Use of Dynamic Observers* 365

13 Frequency Domain Analysis and Alternative Design Methods 387

Preface

It is the author's intention to integrate, into a form suitable for mechanical engineering undergraduates, an introductory treatment of

1. Physical system analysis,
2. Control system design, and
3. Sensitivity analysis.

It is the author's opinion that closed loop *insensitivity* to parameter change is the most important topic of feedback design theory.

A secondary objective is to prepare the student for modern control theory. State space concepts are introduced, and nonlinear regulators are discussed. The discussion of frequency domain centers on the Nyquist diagram because it is felt that the beginning student will, in this way, be prepared for advanced topics, such as Popov stability, if he thoroughly understands vector-loci techniques.

The outline for this text is prepared with the view that computers will play a major role in all aspects of modern engineering. For this reason, detailed instructions for construction of root locus diagrams, Bode diagrams, and Nyquist diagrams for complicated systems have been deleted. In place of such contruction rules, the author supplies discussion which will, it is hoped, increase the student's ability to extract all important information from these fundamental diagrams. Even at this date the construction of these diagrams is easily programmed on a computer (see Appendix C).

The author relied heavily on the contributions of Y. Takahashi, H.

Paynter, D. Karnopp, and R. Rosenberg in the theory of physical system analysis; R. Tomovic and I. Horowitz in sensitivity analysis; and W. Loscutoff in state space analysis. It is hoped that the text represents a true integration of the theory contributed by these researchers into a form suitable for the beginning student of systems theory.

There are several unique features in this text. The treatment of right half-plane zeros in the chapters on root locus and frequency domain techniques points up some interesting facts. State variable analysis is motivated with an introduction to simulation and synthesis techniques. Another unique feature is the use of frequency domain methods to show, for one simple example, the accuracy of various orders of truncated state variable models of a distributed parameter model. I think the users of this text will discover other innovations.

The text contains sufficient material for a one-year course on introductory systems theory even if the sections and chapters marked with an asterisk are omitted. The sections marked with an asterisk were included in order to provide the student with material that can aid his study of advanced topics.

A number of students made valuable corrections to and comments on the text. Ching Yung and John Seevers deserve special mention. Jeff Young bravely read the entire final draft and kindly concealed a good part of his amusement. Brian Crews assisted with figures.

H. Brandt and W. Giedt, Chairmen of the Department of Mechanical Engineering, supplied the encouragement to write this book. Acknowledgment is also made to typists Lois Wilson, Theda Strack, Winnie Letson, and Renee Dominguez; may their neurons unboggle and may they recover from their harrowing experience. A special thanks to Ruth Moorhead, who typed the final draft in such an excellent manner.

J.W.B.

Introduction: Feedback Control and Dynamic Systems Analysis

1

1.1 Feedback Control

To illustrate the concept of feedback control, the applied mathematician Norbert Wiener described a human being grasping an object.[13] The process is illustrated in Figure 1.1. The *desired hand position* is the position of the object to be grasped. The brain compares this position with the actual hand position, which is measured and *fed back* to the brain by the eye. The difference between the actual and desired position of the hand is the *error signal*, which, if not zero, causes another part of the brain (the *compensator*) to send appropriate signals to muscle cells. This *prime mover* converts available energy (stored chemical energy in this case) into useful work and moves the hand or *controlled object*. If all the elements of the feedback system are working properly, the hand will be moved in a manner so as to continually reduce the magnitude of the error signal until the hand is close enough to the object to grasp it.

The description of many other physiological functions in such a manner is quite popular, and, as a matter of fact, many different types of pathologies may be thought of as malfunctions of one of the elements of some physiological feedback system.[6] Wiener identified similar types of feedback loops in social organizations.[13] Allen describes the operation of feedback in economic systems.[1]

It is important to gain some insight into the necessity for feedback itself and for the compensator element within a feedback loop. In concise terms,

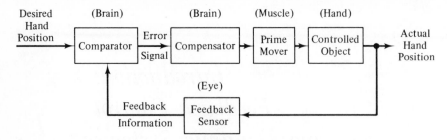

Figure 1.1. Elements of a feedback system; Wiener's classical example.

the feedback loop corrects for inaccuracies in the operation of individual loop elements and for the occurrence of disturbing signals. The eye, after all, provides only an estimate of the three-dimensional position, and the muscle contractions do not respond perfectly accurately and consistently to stimuli from the brain. However, the overall feedback system accuracy is much higher than the accuracy of either of these components because the compensator will always react to nonzero error and eventually force the hand to the correct position. Examples of disturbing signals are unforeseen forces applied to the hand from an external source (e.g., the result of the operation of a physiological feedback system of a coed) and physiological signals which affect the operation of the muscle. The effect of disturbance is an increase in the magnitude of the error signal, which, once again, is corrected by the operation of the feedback system.

A third function provided by the feedback system is compensation for *parameter variation*. Muscle fatigue could cause a great deal of change in the response of the prime mover in the feedback loop illustrated in Figure 1.1. A remarkable fact about feedback systems is that the variation in performance of the system is often far less than the associated variation of performance of any of the individual elements within the loop. Thus, one would expect that a great deal of muscle fatigue (parameter change) would occur before the performance of the feedback system, described by Wiener, is affected. It is little wonder that evolutionary processes provided the human body with many feedback loops.

The final concept to be gleaned from this simple example is the role of the compensator in the smooth operation of the feedback system. The hand, of course, has inertia and, once set in motion by the muscle prime mover, will continue to move until the muscles exert a braking or stopping force. If the braking force is exerted too late, the hand will overshoot its target. The resulting error will activate the feedback mechanism, which will tend to bring the hand back to the correct position, but a better-quality performance would be one in which the hand is brought smoothly to the desired position with little or no overshoot. No overshoot could be accomplished if the braking action were anticipatory, i.e., if the braking were applied slightly early. It is the function of the compensator to operate upon the feedback information and provide the anticipatory control actions.

The main objective of this text is to present the application of the feedback principles, described above, to the design of automatically controlled engineering systems. An example of such a system is the first radar-controlled antiaircraft gun, which Wiener helped develop during World War II. The development of this instrument is a significant event in the history of automatic control because, for the first time, mathematical techniques which had previously been used to design electrical feedback amplifiers were abstracted and applied to the design of a mechanical system. Abstract mathematical techniques will be strongly emphasized in this text.

In the radar-controlled weapon, the aircraft angular position is the desired angular position of the weapon and is measured by radar. The actual weapon position can be measured with a potentiometer and an electrical error signal is generated. The compensator is an electrical network which must be designed to provide the proper anticipation. The prime mover supplies the energy to drive the system and, in this case, is a hydraulic motor which is activated by an electrical signal from the compensator. The controlled object is the weapon, which has inertias and frictions which must be overcome by the prime mover in order to reduce the error. An important type of parameter variation is the extreme variation in the frictional properties of the grease-lubricated connections which is induced by temperature variations and/or contaminants. The feedback systems developed for such weapons are insensitive to such parameter variations.

1.2 Early Examples of Mechanical Feedback Systems[5]

While World War II marks the beginning of the development of mathematical automatic feedback control theory by Wiener and Bode and others, the actual use of feedback systems greatly precedes this event. Since several of the early feedback devices are still used today in some form or another, it is instructive to study the early applications of feedback.

The first example is the control of the volume flow rate of a fluid, which was accomplished by the Alexandrian Greek Ktesibios in the third century B.C.[5] The device is illustrated in Figure 1.2. The variable to be controlled is the flow rate at the outlet. The main purpose of this simple feedback system is to correct for large fluctuations or disturbances in the supply pressure. To understand the operation of the system, assume that some large, unforeseen increase occurs in the supply pressure. The inlet flow rate will increase, causing an error signal, i.e., a difference between inlet flow rate and outlet flow

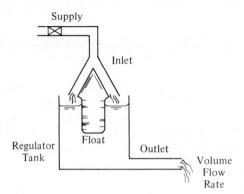

Figure 1.2. Flow control system of Ktesibios (third century B.C.).

rate. This error signal will ultimately cause the float to rise and restrict the inlet flow area. The system will then come to equilibrium as the inlet flow rate is decreased until it equals the outlet flow rate. Note that the system comes to equilibrium with a slightly increased head in the regulator tank and, since outlet flow rate is proportional to this head, with a slightly increased outlet flow rate. The fact that makes the system effective is that the inlet flow rate is strongly dependent on inlet flow area and, hence, on the vertical position of the float. Thus, negligibly small changes in outlet flow rate (and regulator tank head) are associated with large supply pressure disturbances. The main application of Ktesibios' invention was in the development of water clocks. Notice that no compensator or prime mover is included in Ktesibios' feedback loop.

The development of feedback sensors (see Figure 1.1) seems to have been the crucial step in the synthesis of early engineering feedback systems. Mayr describes many of these sensors.[5] For instance, the invention of the thermostat by the seventeenth-century Dutchman Cornelis Drebbel (working in England) led to the development of feedback temperature control systems. An improved eighteenth-century version of the thermostat is illustrated in Figure 1.3. The tube is immersed in the medium whose temperature is to be controlled. Because of differences between the coefficients of expansion of iron and lead, temperature changes in the medium induce corresponding changes in the displacement of the upper edge of the tube, x. The displacement, x, serves as *feedback information* (see Figure 1.1) in temperature feedback loops. The displacement, x, is input to a linkage which regulates the angular position of a damper, which in turn regulates the flow of gas to a combustion chamber (the prime mover). Large variations in the quantity and quality of fuel in the combustion chamber motivated the synthesis of these early temperature control systems.

Eighteenth- and nineteenth-century British millwrights were very inventive and provided many developments in feedback devices and synthesized several interesting feedback loops.[5] It is instructive to speculate on the reason

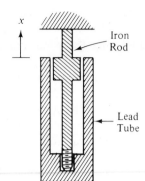

Figure 1.3. Early thermostat invented by Bonnemain (Paris, 1783).[5]

millwrights should be the source of important feedback inventions. The nature of the millwrights' prime mover was probably the source of motivation for the development of feedback control systems. Large fluctuations (disturbances) in the direction and strength of the wind would, if uncontrolled, lead to intolerably large variations in millstone velocity and, thereby, to large variations in the quality of the ground grain. The need to control the angular velocity of millstones was the mother of inventions which are used to this day. It might be noted, parenthetically, that millwrights had other control problems that were dealt with effectively, such as the maintenance of the attitude of the axis of the mainsails so as to transfer maximum power from the wind to the millstone and the control of the spacing between the grinding stones so as to ensure uniformity of ground meal.

As was the case with temperature control systems, the key invention for speed control was that of the feedback sensor. Millwrights called this sensor a centrifugal pendulum, and in later and modern applications, this device is called a *flyball governor*. In the general operation of the governor (see Figure 1.4) the slider displacement, x, is a monotonic function of shaft speed, ω,

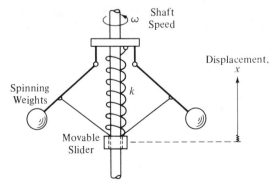

Figure 1.4. Flyball speed sensor patented by the British millwright, Thomas Mead (1787).[5]

and is the feedback information in a feedback loop. In the original application, the slider operated through a complex linkage to change the effective area of the mainsail facing the wind and thereby initiating the feedback control operation.

In 1788, James Watt learned of the invention and immediately incorporated a similar feedback loop on his steam engine. The flyball device was crucial to the development of the steam engine (and of the industrial revolution?) because of otherwise uncontrolled variations in power output from the steam prime mover. The variations are caused by variations in the fuel combustion process and by environmental disturbances which affect the steam temperature and pressure. Another source of disturbance in a steam engine is the application of a load which, if not for the feedback loop, would cause a reduction in shaft speed (*droop*). In the steam engine, the slider displacement automatically manipulates the steam valve.

The governor is still used in modern applications with modern prime movers and other devices of sophisticated design.[7]

1.3 Academic Background of the Modern Control Systems Engineer

As mentioned earlier, World War II marks the beginning of the mathematical and abstract theory of feedback synthesis and automatic control. A concerted effort is made throughout this text to provide the student with the mathematical background that will allow him to make use of this large body of mathematical theory in his design work. However, it is the author's opinion that the *mathematics* of control systems theory provides only half the academic background required of the well-prepared control systems engineer. The other half might be referred to as the *physics* of systems theory but has been called *dynamic systems theory*. The dynamic system theory developed dramatically during the 1960s.[3,4,8,11]

A background in dynamic systems theory is required of the control systems engineer because of the great diversity of applications of feedback principles. Chemical reactors, nuclear reactors, machine tools, aircraft, spacecraft, and temperature control units are only a small portion of the total number of modern engineering systems whose operation require well-designed feedback loops. Another measure of the diversity of feedback applications is a list of the types of prime movers used in such design. A greatly abbreviated list is jet engines, steam engines, electric motors, hydraulic motors, and pneumatic actuators. The mathematical techniques of control system design can be applied only after the systems engineer has mathematical models of his prime mover and feedback sensor and the devices that he proposes to use in his design. Dynamic systems theory is applied during the essential and fundamental modeling phase.

The goal of any presentation of a dynamic systems theory is to provide a single framework or formalism for the mathematical modeling of several different types of physical systems. In this text, the dynamic systems theory may be applied to the development of models of electrical networks, mechanical networks and drive trains, hydraulic networks, pneumatic networks, heat transfer networks, *and* devices which interface any two different types of networks (i.e., devices which convert hydraulic energy to mechanical energy or mechanical energy to electrical energy, etc.).

The first portion of this text is devoted to mathematics, dynamic systems theory, and computer simulation. The emphasis in the final portion of the book is on control system design. The material on control theory is, in turn, divided into two conceptual divisions: (1) linear control systems theory and *extended* linear control systems theory, and (2) a brief introduction to nonlinear control systems theory. Extended linear control systems theory refers to the techniques with which one applies linear control system design methods to nonlinear systems by ensuring that systems variables will remain in a limited range of values.

One can demonstrate the generality of the system methodology by studying academic examples from ecosystem theory[9,12] and mathematical social system theory.[1,2,10]

References

1. ALLEN, R. G. D., *Mathematical Economics*, Macmillan, New York, 1966.

2. BAILEY, N. T. J., *The Mathematical Theory of Epidemics*, Griffin, London, 1957.

3. CRANDALL, S., KARNOPP, D., et al., *Dynamics of Mechanical and Electromechanical Systems*, McGraw-Hill, New York, 1968.

4. KARNOPP, D., and ROSENBERG, R. C., *Analysis and Simulation of Multiport Systems*, M.I.T. Press, Cambridge, Mass., 1968.

5. MAYR, O., "The Origins of Feedback Control," *Scientific American*, Oct. 1970.

6. MILHORN, H. T., *The Application of Control Theory to Physiological Systems*, Saunders, Philadelphia, 1966.

7. OLDENBURGER, R., "Hydraulic Speed Governor with Major Governor Problems Solved," *A.S.M.E. Journal of Basic Engineering*, March 1965.

8. PAYNTER, H. M., *Analysis and Design of Engineering Systems*, M.I.T. Press, Cambridge, Mass., 1961.

9. PIELOU, E. C., *An Introduction to Mathematical Ecology*, Wiley, New York, 1969.

10. RICHARDSON, L. F., *Arms and Insecurity*, Boxwood Press, Pittsburgh, 1960.

11. TAKAHASHI, Y., RABINS, M. J., and AUSLANDER, D. M., *Control*, Addison-Wesley, Reading, Mass., 1970.

12. WATT, K. E. F., *Ecology and Resource Management*, McGraw-Hill, New York, 1968.

13. WIENER, N., *The Human Use of Human Beings*: *Cybernetics and Society*, Avon, New York, 1950.

Mathematical Preliminary: The Laplace Transform   *2*

Laplace transform theory is one of the more useful and general developments in the mathematics of system theory. The theory is most useful when applied to analysis of linear systems and allows the analyst to reduce calculus problems to problems in algebra. The theory developed below will be referred to in both the dynamic systems analysis and control system design portions of the text.

2.1 The Transformation Concept

Consider the problem of finding x^{10} with the value of x known. This problem can be solved in two ways, as is illustrated in Figure 2.1.

The indirect path represents a technique learned in high school and is usually the most accurate and efficient. The technique is made useful because the existence of logarithm tables facilitates the transformations and inverse transformations in the indirect path.

It is important to note that the logarithm transformation is defined for positive numbers only (called the *domain* of the transformation). Thus, the technique of the indirect path can be applied only to a proper subset of all possible real numbers.

The most important problem in dynamic systems analysis is to predict the outputs of a physical system when the inputs are known. The question is now asked, Does a transformation scheme exist to aid in the tasks of dynamic systems analysis? The answer is yes. In fact, several transformations

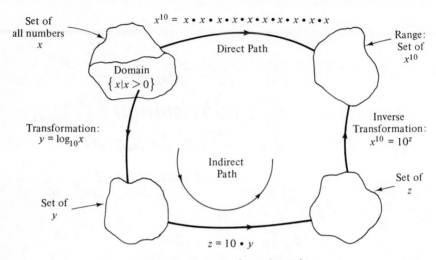

$$x^{10} = x \cdot x \cdot x \cdot x \cdot x \cdot x \cdot x \cdot x \cdot x \cdot x$$

Set of all numbers x

Direct Path

Range: Set of x^{10}

Domain $\{x \mid x > 0\}$

Transformation: $y = \log_{10} x$

Inverse Transformation: $x^{10} = 10^z$

Indirect Path

Set of z

Set of y

$$z = 10 \cdot y$$

Figure 2.1. Concept of transformation.

will do the job. For instance, the familiar Fourier transformation can be used; unfortunately, the domain for the Fourier transformation is "uninteresting." That is, too many important types of inputs do not belong to the domain of the Fourier transform.

The Laplace transform, on the other hand, has a very interesting domain

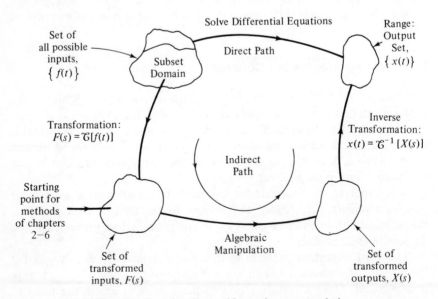

Solve Differential Equations

Set of all possible inputs, $\{f(t)\}$

Subset Domain

Direct Path

Range: Output Set, $\{x(t)\}$

Transformation: $F(s) = \mathcal{L}[f(t)]$

Inverse Transformation: $x(t) = \mathcal{L}^{-1}[X(s)]$

Indirect Path

Starting point for methods of chapters 2–6

Set of transformed inputs, $F(s)$

Algebraic Manipulation

Set of transformed outputs, $X(s)$

Figure 2.2. Transforming problems of systems analysis.

and the existence of tables of Laplace transforms facilitates the execution of the steps in the indirect path illustrated in Figure 2.2.

2.2 The Laplace Transform; Definition and Elementary Examples

The Laplace transform is used to transform functions of time to functions of a variable denoted by s which is chosen to be complex; i.e.,

$$s = \sigma + i\omega.$$

This choice is arbitrary[3] but proves to be very convenient in the development of dynamic systems theory. The transformation is defined by[4]

$$F(s) = \int_{0^-}^{\infty} e^{-st} f(t)\, dt. \tag{2.1}$$

Notice the lower limit of 0^-, which might be read "slightly before zero."

The Fourier transform is defined by (in one form)

$$F(s) = \int_{-\infty}^{\infty} e^{-st} f(t)\, dt; \quad s \text{ pure imaginary.} \tag{2.2}$$

The domains of the transformations differ, as is illustrated in Figure 2.3.

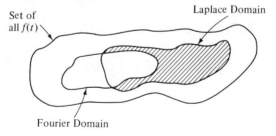

Figure 2.3. Illustration of the domains of integral transforms.

The domain of the Laplace transformation is described as follows: To be Laplace transformable, $f(t)$ must be

1. *Sectionally continuous*; i.e., $f(t)$ may have only a finite number of discontinuities in any finite interval; (2.3)

and

2. Of *exponential order*; i.e., for some real σ_0 and t_1, $e^{-\sigma_0 t} |f(t)|$ is bounded for all $t > t_1$. (2.4)

That region of the s plane for which $\text{Re}(s) > \sigma_0$ is called the *region of convergence*.

These conditions are sufficient conditions for the existence of the integral in Equation (2.1). It is important to note that, because of the definition of Equation (2.1), the Laplace transform of a function, $f(t)$, contains no information about that function for $t < 0^-$. For this reason, the inverse transformation (see Figure 2.2) will describe the output for $t \geq 0^-$; i.e., only for the so-called *causal* portion of the output function.

Following convention, the notation

$$F(s) = \mathcal{L}[f(t)] \quad \text{and} \quad f(t) = \mathcal{L}^{-1}[F(s)] \tag{2.5}$$

will be used.

Example 2.1

As an example consider the *unit step function* (see Figure 2.4):

$$u(t) = \begin{cases} 0, t < 0, \\ 1, t \geq 0. \end{cases} \tag{2.6}$$

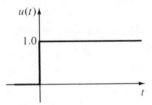

Figure 2.4. Unit step function.

This function satisfies Equations (2.3) and (2.4), and by Equation (2.1),

$$U(s) = \mathcal{L}[u(t)] = \int_{0^-}^{\infty} u(t)e^{-st}\, dt$$

$$= \int_0^{\infty} e^{-st}\, dt = -\frac{1}{s}e^{-st}\Big]_0^{\infty}$$

$$= -\frac{e^{-st}}{s}\Big|_{t=\infty} + \frac{1}{s}.$$

The first term is zero if $\sigma > 0$. To summarize,

$$U(s) = \mathcal{L}[u(t)] = \frac{1}{s} \tag{2.7}$$

for all s such that $\sigma > 0$.

* * *

One is not able to plot $U(s)$ vs. s in a very convenient manner because s is complex. It is very useful, however, to plot the *critical points*, i.e., points where $U(s)$ is zero (called *zeros*) and points where $U(s)$ is infinite (called *poles*), on the s plane.

For the unit step function, $U(s)$ has a single pole at the origin, and the pole zero diagram is illustrated in Figure 2.5. The analyst can solve a remarkably large number of problems by memorizing a small number of transforms. The transform of the unit step function is one that should be remembered, as should those derived in the next several examples.

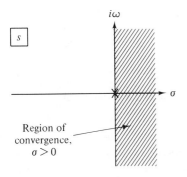

Figure 2.5. Pole zero diagram for the transform of the unit step function.

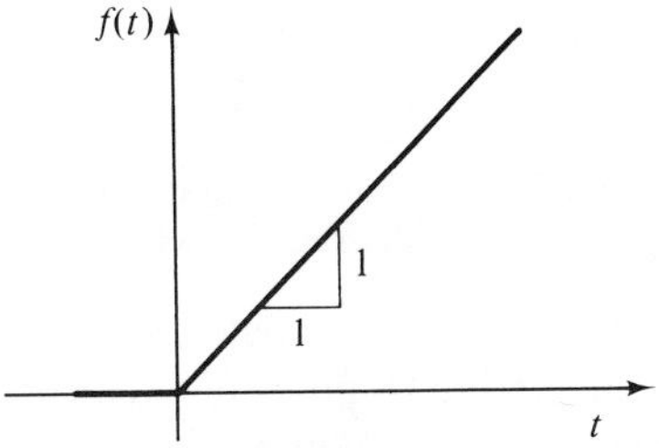

Figure 2.6. Ramp function.

Example 2.2

Consider $f(t) = t \cdot u(t)$. This function should not be confused with $f_0(t) = t$. $f(t)$ is called the *ramp* function (see Figure 2.6). From Equation (2.1), it follows that

$$F(s) = \int_{0^-}^{\infty} u(t)te^{-st}\,dt.$$

If the reader will refer to an elementary calculus text, he will find that condition (2.4) is also the condition for the interchange of limit operations; thus, the operator d/ds may be interchanged with the integral sign in the definition of the Laplace transformation. Hence,

$$F(s) = -\frac{d}{ds}\int_{0^-}^{\infty} u(t)e^{-st}\,dt = -\frac{d}{ds}U(s).$$

Use Equation (2.7) to find that

$$F(s) = -\frac{d}{ds}\left(\frac{1}{s}\right) = \frac{1}{s^2} \quad \text{if } \sigma > 0. \tag{2.8}$$

$F(s)$ has a double pole at the origin, as is illustrated in Figure 2.7.

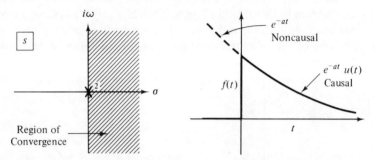

Figure 2.7. Pole zero diagram of the transform of the ramp function.   **Figure 2.8.** Exponential function.

The above two functions are commonly encountered physical signals. It is interesting to note that *neither* is Fourier transformable (they lie in the shaded region of Figure 2.3).

*   *   *

Example 2.3

Consider $f(t) = e^{-at}u(t)$ (which should not be confused with e^{-at}). See Figure 2.8. From Equation (2.1),

$$F(s) = \int_{0^-}^{\infty} u(t)e^{-at}e^{-st}\,dt = \left.\frac{-e^{-(s+a)t}}{s+a}\right|_0^{\infty} = \left.\frac{-e^{-(s+a)t}}{s+a}\right|_{t=\infty} + \frac{1}{s+a}.$$

The first term vanishes if $\sigma > -a$. Thus

$$F(s) = \frac{1}{s+a} \quad \text{for } \sigma > -a. \tag{2.9}$$

This transform turns out to be one of the most important for dynamic systems analysis, and its pole may be in the right half-plane if a is negative, in which case $f(t)$ is an exponential rise instead of an exponential decay. See Figure 2.9.

*   *   *

The transforms of trigonometric functions are used quite often. These transforms can be found by using Equation (2.9) and the fact that $s = \sigma + i\omega$

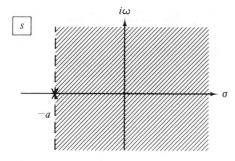

Figure 2.9. Pole zero diagram of the exponential function.

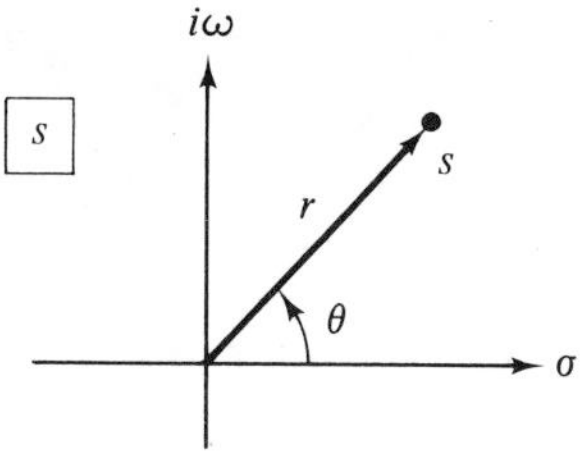

Figure 2.10. Complex s plane and the polar form.

can be written in polar form as

$$s = re^{i\theta}, \tag{2.10}$$

where

$$r = \sqrt{\sigma^2 + \omega^2} \tag{2.11}$$

and

$$\theta = \tan^{-1}\frac{\omega}{\sigma}. \tag{2.12}$$

(refer to Figure 2.10).

If the reader does not recall these facts, he will find Appendix B to be a useful reference. It is also shown, in this appendix, that

$$\boxed{\frac{1}{s} = \frac{e^{-i\theta}}{r}.} \tag{2.13}$$

It is strongly emphasized that Equations (2.10)–(2.13) are extremely important and will be referred to, time and time again, throughout the text. The reader is encouraged to spend the time necessary to completely understand these equations.

Another result derived in Appendix B is that

$$\boxed{e^{i\theta} = \cos\theta + i\sin\theta.} \tag{2.14}$$

It follows that

$$e^{-i\theta} = \cos\theta - i\sin\theta. \tag{2.15}$$

Combine the last two equations to find

$$\cos \theta = \frac{1}{2}[e^{i\theta} + e^{-i\theta}]; \tag{2.16}$$

$$\sin \theta = \frac{1}{2i}[e^{i\theta} - e^{-i\theta}]. \tag{2.17}$$

These equations will prove useful in the following example.

Example 2.4

Consider $f(t) = \cos(\omega t)u(t)$ (see Figure 2.11). It follows from Equation (2.16) that

$$f(t) = \left\{\frac{e^{i\omega t}}{2} + \frac{e^{-i\omega t}}{2}\right\}u(t).$$

Thus,

$$F(s) = \mathcal{L}[f(t)] = \tfrac{1}{2}\mathcal{L}[e^{i\omega t}u(t)] + \tfrac{1}{2}\mathcal{L}[e^{-i\omega t}u(t)].$$

Use Equation (2.9) to show that

$$F(s) = \frac{1}{2}\left\{\frac{1}{s - i\omega} + \frac{1}{s + i\omega}\right\} = \frac{s}{s^2 + \omega^2}. \tag{2.18}$$

In a similar manner, find that

$$\mathcal{L}[\sin(\omega t)u(t)] = \frac{\omega}{s^2 + \omega^2}. \tag{2.19a}$$

Equations (2.16) and (2.17) have been used in order to avoid the evaluation of the integral in Equation (2.1).

The pole zero diagram for the transform of Equation (2.19a) is illustrated in Figure 2.12.

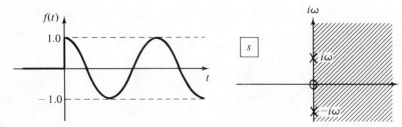

Figure 2.11. Cosine function.   Figure 2.12. Pole zero diagram of $\mathcal{L}[\cos(\omega t)u(t)]$.

* * *

The last example illustrates the usefulness of Equation (2.9) when deriving the Laplace transforms of functions more complex than the exponential function. This usefulness is further illustrated by deriving the Laplace transform of

$$f(t) = te^{-at}u(t).$$

First notice that

$$f(t) = -\frac{\partial}{\partial a} e^{-at}u(t).$$

Thus,

$$\mathcal{L}[f(t)] = -\frac{\partial}{\partial a} \int_{0^-}^{\infty} e^{-at}u(t)e^{-st}\, dt$$

$$= -\frac{\partial}{\partial a} \mathcal{L}[e^{-at}u(t)].$$

It follows from Equation (2.9) that

$$\mathcal{L}[te^{-at}u(t)] = \frac{1}{(s+a)^2}. \qquad (2.19b)$$

This transform has a *double* pole at $s = -a$, as is illustrated in Figure 2.13.

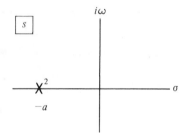

Figure 2.13. Pole zero diagram of $\mathcal{L}[te^{-at}u(t)]$.

A useful generalization of Equation (2.19b) is

$$\mathcal{L}\left[\frac{t^n e^{-at}u(t)}{n!}\right] = \frac{1}{(s+a)^{n+1}}. \qquad (2.19c)$$

As will be illustrated in Example 2.7 (Section 2.5), Equations (2.19b) and (2.19c) are useful when attempting to find the function of time associated with a given Laplace transform.

2.3   The Differentiation Theorem; Solution of Linear, Time-Invariant Differential Equations

Up to this point, only the Laplace transforms of commonly encountered time functions have been determined. The results of Laplace transformation theory which make the theory so valuable in dynamic systems analysis are outlined in this section. In particular, the solution of linear, ordinary differential equations with constant coefficients is discussed. The discussion begins with the following fundamental theorem.

Theorem

If

$$F(s) = \mathcal{L}[f(t)],$$

then

$$\mathcal{L}\left[\frac{df(t)}{dt}\right] = sF(s) - f(0^-). \tag{2.20}$$

Proof. As is indicated by Equation (2.1),

$$\mathcal{L}\left[\frac{df(t)}{dt}\right] = \int_{0^-}^{\infty} \frac{df(t)}{dt} e^{-st}\, dt.$$

To integrate by parts, let

$$w = e^{-st} \qquad\qquad v = f(t)$$

$$dw = -se^{-st}\, dt \qquad dv = \frac{df}{dt}\, dt.$$

Then, since $\int w\, dv = wv - \int v\, dw$, it follows that

$$\mathcal{L}\left[\frac{df(t)}{dt}\right] = e^{-st}f(t)\Big]_{0^-}^{\infty} - \int_{0^-}^{\infty} (-se^{-st})f(t)\, dt.$$

Assuming s lies in the region of convergence of $f(t)$, $e^{-st}f(t)\Big|_{t=\infty} = 0$ so that

$$\mathcal{L}\left[\frac{df(t)}{dt}\right] = -f(0^-) + s\int_{0^-}^{\infty} e^{-st}f(t)\, dt$$

$$= -f(0^-) + sF(s). \quad \text{Q.E.D.}$$

$$* \quad * \quad *$$

It is very important to distinguish between $f(0^-)$ (called the *initial condition*) and $f(0^+)$ (called the *initial value*). For example, for the unit step, the initial condition is 0.0 and the initial value is 1.0. The difference between these two quantities will be further illustrated in Section 2.6.

It is convenient to use the operator notation

$$Df = \frac{df}{dt}. \tag{2.21}$$

Then Equation (2.20) may be written

$$\mathcal{L}[Df] = sF(s) - f(0^-). \tag{2.22}$$

It follows immediately that

$$\mathcal{L}[D^2 f] = s\mathcal{L}[Df] - Df(0^-).$$

Apply Equation (2.22) again to find

$$\mathcal{L}[D^2 f] = s^2 F(s) - sf(0^-) - Df(0^-). \tag{2.23}$$

In general, it can be shown that

$$\mathcal{L}[D^n f] = s^n F(s) - \sum_{k=0}^{n-1} s^{n-1-k} D^k f(0^-). \tag{2.24}$$

The function $f(t)$ is called a *causal* function if

$$D^i f(0^-) = 0 \quad \text{for all } i \geq 0, \tag{2.25}$$

so that for causal functions, Equation (2.24) reduces to

$$\mathcal{L}[D^n f] = s^n F(s). \tag{2.26}$$

The above formulas can be applied to the solution of differential equations in the manner illustrated by the following example.

Example 2.5

The problem is to find the solution to the equation

$$Dx + \frac{1}{T}x = 0, \quad T = \text{constant},$$

with the initial condition

$$x(0^-) = 1.$$

Take the Laplace transform of the differential equation and take note of

Equation (2.22) to find that

$$sX(s) - x(0^-) + \frac{1}{T}X(s) = 0;$$

thus, solve for the unknown, $X(s)$, to find

$$X(s) = \frac{x(0^-)}{s + (1/T)} = \frac{1}{s + (1/T)}.$$

It follows from Equation (2.9) that

$$x(t) = e^{-t/T}u(t),$$

and the solution is complete. The reader should compare the above solution with the scheme indicated in Figure 2.2.

$$* \quad * \quad *$$

2.4 Dynamic Systems Analysis, the Transfer Function

One of the basic problems of dynamic systems analysis is to predict the output of a physical system when the input is known. The application of the material in the previous section to this problem will now be discussed.

A *linear* physical system is defined in this paragraph. Suppose the inputs $m_1(t)$, $m_2(t)$, $m_3(t) \cdots m_n(t)$ induce the responses $p_1(t)$, $p_2(t)$, $p_3(t) \cdots p_n(t)$, respectively. Suppose that the input $m_1(t) + m_2(t) + m_3(t) + \cdots + m_n(t)$ is applied, producing the response $p_0(t)$. A linear system satisfies the following criteria, which can be tested experimentally[2]:

1. *Principle of superposition*, i.e.,

$$p_0(t) = p_1(t) + p_2(t) + p_3(t) + \cdots + p_n(t); \qquad (2.27)$$

and

2. *Principle of homogeneity*, i.e.,

$$\text{The response to } \alpha m_1(t) \text{ is } \alpha p_1(t),$$
$$\text{the response to } \alpha m_2(t) \text{ is } \alpha p_2(t), \qquad (2.28)$$
$$\cdots \quad \text{the response to } \alpha m_n(t) \text{ is } \alpha p_n(t)$$

for any scalar α.

In this text, attention is devoted in the main to an important subclass of linear systems, namely, those systems for which the input and output are related by a linear differential equation with *constant* coefficients, for example, an equation of the form

$$D^2p + 2Dp + p = Dm + 2m. \tag{2.29}$$

The definition of some important terms will be illustrated by taking the Laplace transform of Equation (2.29).

It is assumed initially that the input, $m(t)$, is *causal*, because $t = 0$ is taken to be that instant when the input is first applied. The output, $p(t)$, will not, in general, be causal; i.e., $p(t)$ could have nonzero initial conditions (preloaded springs, charged capacitors, etc.). Thus the transform of Equation (2.29) is

$$s^2P(s) - sp(0^-) - Dp(0^-) + 2sP(s) - 2p(0^-) + P(s) = sM(s) + 2M(s).$$

Solve for the unknown, $P(s)$, to find

$$P(s) = \underbrace{\frac{sp(0^-) + Dp(0^-)}{s^2 + 2s + 1}}_{\substack{\text{Contribution to} \\ \text{response from} \\ \text{initial conditions}}} + \underbrace{\frac{s + 2}{s^2 + 2s + 1}M(s)}_{\substack{\text{Contribution} \\ \text{from input} \\ M(s)}}. \tag{2.30}$$

A system is said to be *initially quiescent* (hereafter called I.Q.) if $p(t)$ and all of its derivatives are zero slightly before the input is applied (i.e., at $t = 0^-$).

If the system of the present example were I.Q., then Equation (2.30) reduces to

$$P(s) = \frac{(s + 2)}{s^2 + 2s + 1}M(s). \tag{2.31}$$

The *transfer function* of a system is denoted $G(s)$ and is defined to be the ratio of the transformed output to the transformed input when the system is I.Q., i.e.,

$$G(s) \triangleq \frac{P(s)}{M(s)}. \tag{2.32}$$

For the present example

$$G(s) \triangleq \frac{P(s)}{M(s)} = \frac{s + 2}{s^2 + 2s + 1}.$$

Systems for which powers of s greater than, or equal to, 1 appear in the numerator of the transfer function are said to have *numerator dynamics*. Such systems are described by differential equations in which derivatives of the *input* appear on the right-hand side of the equation. As will be shown in the chapters on dynamic systems analysis, systems with numerator dynamics are the rule rather than the exception.

Interesting properties of the transfer function are illustrated by a consideration of the physical systems illustrated in Figure 2.14.

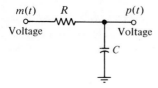

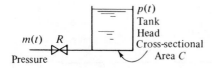

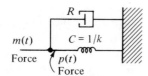

Figure 2.14. Three physical systems.

All three systems may be described by the same differential equation, namely,

$$RC\frac{dp}{dt} + p = m(t). \tag{2.33}$$

To find the transfer function, take the Laplace transform, assuming I.Q. and causal input, to find

$$RCsP(s) + P(s) = M(s).$$

Then, the transfer function

$$G(s) \triangleq \frac{P(s)}{M(s)} = \frac{1}{1 + RCs}. \tag{2.34}$$

Notice that the transfer function depends *only* on the physical constants R and C. The transfer function does not depend on the input or any initial conditions; for this reason the transfer function is often referred to as the *signature* of the system.

Equation (2.32) can be rewritten

$$P(s) = G(s)M(s). \tag{2.35}$$

This equation further illustrates the significance of the transfer function. Notice that the transfer function is always a factor of the output transform regardless of the input. One of the goals of dynamic systems analysis is the derivation of the transfer function of a physical system.

2.5 Introduction to Heavyside Expansion Techniques for Inverse Laplace Transformation

The inverse transformation step of the indirect path discussed in Section 2.1 can be performed in one of at least three ways:

1. *Table look-up*. This is the method implied in Section 2.1 and is the analog of inverting the logarithmic transformation. Extensive tables are available for this purpose.[3]

2. *Complex integration*. Using methods considered beyond the scope of this course, it can be shown that if

$$F(s) = \mathcal{L}[f(t)] \triangleq \int_{0-}^{\infty} f(t)e^{-st}\, dt,$$

then

$$f(t) = \mathcal{L}^{-1}[F(s)] = \frac{1}{2\pi i}\int_{c-i\infty}^{c+i\infty} e^{st}F(s)\, ds, \tag{2.36}$$

where c is any constant chosen so that the integral lies in the region of convergence.[3] This method will not be discussed in the present text.

3. *Heavyside expansion*. This method is a set of cleverly devised techniques wherein complex algebra and the handful of transforms derived in Section 2.2 (and summarized in Appendix A) are applied to the solution of a wide variety of problems. Important features of this method are outlined below. The treatment is introductory in nature but sufficient for present purposes. Many fine references are available which treat this method in more detail.[2,4,5] The method is best illustrated by examples. It will be instructive to consider a single class of systems in these examples, namely, those systems for which the output, $p(t)$, is related to the input, $m(t)$, by Equation (2.33) and for which the transfer function is identified in Equation (2.34). Particular reference will be made to the tank system.

Example 2.6

Assume, in this first example, that the input $m(t) = u(t)$, the unit step, and that the tank is partially filled initially so that

$$p(0^-) = \tfrac{1}{2}.$$

Equation (2.33) can be rewritten

$$RC\frac{dp}{dt} + p = u(t).$$

Take the Laplace transform to find

$$RCsP(s) - RC\left(\frac{1}{2}\right) + P(s) = \frac{1}{s}$$

or

$$P(s) = \underbrace{\frac{RC/2}{RCs + 1}}_{\substack{\text{Initial} \\ \text{condition} \\ \text{term}}} + \underbrace{\frac{1}{RCs + 1}}_{\substack{\text{Transfer} \\ \text{function}}} \underbrace{\frac{1}{s}}_{\substack{\text{Input} \\ \text{transform}}}.$$

Finally,

$$P(s) = \frac{\tfrac{1}{2}[s + (2/RC)]}{s[s + (1/RC)]}. \tag{2.37}$$

To invert this transform, attempt a *partial fraction expansion*, i.e.,

$$P(s) = \frac{A}{s} + \frac{B}{s + (1/RC)}. \tag{2.38}$$

A and *B* are unknown constants to be determined and are called *residues*. *A* is determined by using the following trick: Multiply Equation (2.38) by s to obtain

$$\frac{\tfrac{1}{2}[s + (2/RC)]}{[s + (1/RC)]} = A + \frac{sB}{[s + (1/RC)]},$$

and let s be zero to find

$$A = 1.$$

To find *B*, multiply Equation (2.38) by $s + (1/RC)$ to obtain

$$\frac{\tfrac{1}{2}[s + (2/RC)]}{s} = \frac{s + (1/RC)}{s}A + B.$$

Let $s = -1/RC$ to find

$$B = -\tfrac{1}{2}.$$

Substitute A and B back into Equation (2.38) to find

$$P(s) = \frac{1}{s} - \frac{\tfrac{1}{2}}{s + (1/RC)}.$$

It now follows from Equations (2.7) and (2.9) that

$$p(t) = u(t) - \tfrac{1}{2}e^{-t/RC}u(t),$$

so that the response is

$$p(t) = (1 - \tfrac{1}{2}e^{-t/RC})u(t). \tag{2.39}$$

The important point is that by the use of heavyside or partial fraction expansion, the transform (2.37) was reduced to a superposition of elementary transforms and is easily inverted term by term.

* * *

The heavyside technique illustrated above can be generalized as follows: Consider the transform

$$C(s) = \frac{A(s)}{B(s)} = \frac{s^m + a_{m-1}s^{m-1} + \cdots + a_0}{s^n + b_{n-1}s^{n-1} + \cdots + b_0},$$

where $A(s)$, $B(s)$ are polynomials in the Laplace variable, s. Assume that $A(s)$ has order at least one less than the order of $B(s)$, i.e.,

$$m \leq n - 1. \tag{2.40}$$

It is convenient to consider several distinct cases separately.

CASE OF REAL, DISTINCT POLES

Suppose $C(s)$ has the form

$$C(s) = \frac{A(s)}{(s + P_1)(s + P_2)\ldots(s + P_n)}, \tag{2.41}$$

where $A(s)$ is some polynomial and no two poles, P_i, are equal. Expand $C(s)$ as follows:

$$C(s) = \frac{R_1}{(s + P_1)} + \frac{R_2}{(s + P_2)} + \cdots + \frac{R_n}{(s + P_n)}. \tag{2.42}$$

The unknown residues, R_i, are to be determined. R_i is determined by multiplying Equation (2.42) by $s + P_i$ to obtain

$$\frac{A(s)}{(s + P_1)\cdots(s + P_{i-1})(s + P_{i+1})\cdots(s + P_n)} = \frac{(s + P_i)}{s + P_1}R_1 + \cdots$$

$$+ R_i + \cdots + \frac{(s + P_i)R_n}{s + P_n}.$$

Then set $s = -P_i$ to obtain

$$R_i = \frac{A(-P_i)}{(P_1 - P_i)\ldots(P_n - P_i)}. \tag{2.43}$$

Apply Equation (2.9) to the additive terms in Equation (2.42), and find that

$$c(t) = (R_1 e^{-P_1 t} + R_2 e^{-P_2 t} + \cdots + R_n e^{-P_n t})u(t). \tag{2.44}$$

Equation (2.43) can be written more conveniently as

$$\boxed{R_i = \{(s + P_i)C(s)\}\Big|_{s=-P_i}} \tag{2.45}$$

It is interesting to note that an exponential term appears in the solution (2.44) for every pole of $C(s)$. The numerator polynomial influences only the value of the residues.

CASE OF REAL, REPEATED POLES

When one or more poles are repeated, one proceeds as illustrated by the following example.

Example 2.7

Suppose that for the fluid flow system of Example 2.6, $RC = 1.0$ and the system is I.Q.; then from Equation (2.33) it follows that

$$sP(s) + P(s) = M(s)$$

or

$$P(s) = \underbrace{\frac{1}{s + 1}}_{\substack{\text{Transfer}\\\text{function}}} \underbrace{M(s)}_{\substack{\text{Input}\\\text{transform}}}.$$

The same result could have been obtained using Equation (2.32). Suppose $m(t) = u(t) - e^{-t}u(t)$ (see Figure 2.15); then

$$M(s) = \frac{1}{s} - \frac{1}{s + 1} = \frac{1}{s(s + 1)}$$

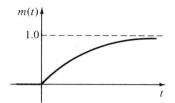

Figure 2.15. Input signal for Example 2.7.

and

$$P(s) = \frac{1}{s(s+1)^2}.$$

Notice that $P(s)$ has a double pole at $s = -1$. For this case, a useful procedure is to attempt an expansion of the form

$$P(s) = \frac{R_1}{s} + \frac{R_2}{s+1} + \frac{R_3}{(s+1)^2}.$$

R_1, R_3 are calculated in the same manner as before [Equation (2.45)], i.e.,

$$R_1 = sP(s)\Big|_{s=0} = 1$$

$$R_3 = (s+1)^2 P(s)\Big|_{s=-1} = -1.$$

To calculate R_2, another trick is used. Multiply $P(s)$ by $(s+1)^2$ to obtain

$$\frac{1}{s} = \frac{(s+1)^2 R_1}{s} + (s+1)R_2 + R_3.$$

To isolate R_2, *differentiate* with respect to s to find

$$-\frac{1}{s^2} = \left\{ \frac{2(s+1)}{s} - \frac{(s+1)^2}{s^2} \right\} R_1 + R_2,$$

and set $s = -1$ to find

$$R_2 = -1.$$

Finally, substitute for R_1, R_2, R_3 to find

$$P(s) = \frac{1}{s} - \frac{1}{s+1} - \frac{1}{(s+1)^2}$$

for which

$$p(t) = u(t) - e^{-t}u(t) - te^{-t}u(t)$$
$$= (1 - e^{-t} - te^{-t})u(t).$$

Symbolically, R_2 was found from the formula

$$R_2 = \left[\frac{d}{ds}\{(s + P_2)C(s)\}\right]_{s=-P_2} \tag{2.46}$$

* * *

The interested reader can easily find references which generalize Equation (2.46) for the case of higher-order *degeneracies* (i.e., repeated poles).[2, 4]

CASE OF COMPLEX POLES

Transforms with complex poles are also inverted by using Equations (2.45) and (2.46). For this case, the residues will be complex numbers. It is very convenient to use the polar form of complex numbers to calculate these residues. These manipulations are illustrated by the following example.

Example 2.8

For the I.Q. fluid flow system of Example 2.7, assume that the input pressure is

$$m(t) = \cos(\omega t)u(t) \tag{2.47}$$

(refer to Figure 2.16). As is indicated in Equation (2.18),

$$M(s) = \frac{s}{s^2 + \omega^2}.$$

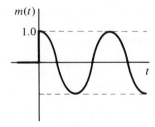

Figure 2.16. Input for Example 2.8.

Since the system is I.Q., it follows from Equation (2.32) that

$$P(s) = \frac{1}{1 + s}M(s) = \frac{s}{(s + 1)(s^2 + \omega^2)}.$$

Proceed as before to determine the heavyside expansion:

$$P(s) = \frac{s}{(s + 1)(s + i\omega)(s - i\omega)} = \frac{R_1}{s + 1} + \frac{R_2}{s + i\omega} + \frac{R_3}{s - i\omega}.$$

From Equation (2.45), it follows that

$$R_1 = (s+1)P(s)\Big|_{s=-1} = \frac{-1}{1+\omega^2},$$

$$R_2 = (s+i\omega)P(s)\Big|_{s=-i\omega} = \frac{-i\omega}{(-i\omega+1)(-2i\omega)} = \frac{1}{2(1-i\omega)},$$

$$R_3 = (s-i\omega)P(s)\Big|_{s=i\omega} = \frac{i\omega}{(i\omega+1)(2i\omega)} = \frac{1}{2(1+i\omega)}.$$

Notice that R_2 and R_3 are complex conjugates and use Equation (2.9) to show that

$$p(t) = \frac{-e^{-t}}{1+\omega^2}u(t) + \frac{1}{2(1-i\omega)}e^{-i\omega t}u(t) + \frac{1}{2(1+i\omega)}e^{i\omega t}u(t).$$

It follows from Equation (2.13) that

$$\frac{1}{1-i\omega} = \frac{e^{-i\phi}}{\sqrt{1+\omega^2}},$$

where

$$\phi = \tan^{-1}(-\omega). \tag{2.48a}$$

Derive a similar equation for $(1+i\omega)^{-1}$ and show that

$$p(t) = \left[\frac{-e^{-t}}{1+\omega^2} + \frac{1}{\sqrt{1+\omega^2}}\left(\frac{e^{-i(\omega t+\phi)} + e^{i(\omega t+\phi)}}{2}\right)\right]u(t).$$

Use Equation (2.16) to show that

$$p(t) = \left[\frac{-e^{-t}}{1+\omega^2} + \frac{1}{\sqrt{1+\omega^2}}\cos(\omega t+\phi)\right]u(t). \tag{2.48b}$$

$$*\quad *\quad *$$

Two facts which were illustrated in the above example are (1) the residues associated with complex conjugate poles are complex conjugates of one another, and (2) the solution, $p(t)$, must be real even when its transform has complex poles.

2.6 The Initial Value Theorem

The initial and final value theorems, which are to be discussed in the next two sections, have at least two very important practical applications:

1. To provide checks on the accuracy of mathematical manipulations with Laplace transformation.

2. To provide aids for hypothesizing the structure of dynamic systems models from an examination of operating records.

The initial value theorem is stated as follows:

Theorem

If

$$F(s) = \mathcal{L}[f(t)],$$

then

$$\lim_{s\to\infty} sF(s) = f(0^+). \tag{2.49}$$

$f(0^+)$ is called the *initial value*. Because $f(t)$ may be discontinuous at $t = 0$, the *initial value*, $f(0^+)$, is not equal to the *initial condition*, $f(0^-)$.

Proof. Start with Equation (2.20), which reads

$$\mathcal{L}\left[\frac{df}{dt}\right] = sF(s) - f(0^-).$$

But by definition,

$$\mathcal{L}\left[\frac{df}{dt}\right] = \int_{0^-}^{\infty} e^{-st}\frac{df}{dt}\, dt = \int_{0^-}^{0^+} \frac{df}{dt}\, dt + \int_{0^+}^{\infty} e^{-st}\frac{df}{dt}\, dt$$

$$= f(0^+) - f(0^-) + \int_{0^+}^{\infty} e^{-st}\frac{df}{dt}\, dt;$$

because f may be discontinuous at $t = 0$, the first two terms do *not* cancel. Thus,

$$sF(s) = f(0^+) + \int_{0^+}^{\infty} e^{-st}\frac{df}{dt}\, dt.$$

Taking the limit as $s \to \infty$,

$$\lim_{s\to\infty} sF(s) = f(0^+) + \lim_{s\to\infty} \int_{0^+}^{\infty} \frac{df}{dt} e^{-st}\, dt.$$

If the limit is taken so that $Re\{s\} \to +\infty$, then

$$\lim_{s\to\infty} e^{-st} = 0 \quad \text{for } 0^+ < t < \infty$$

so

$$\lim_{s\to\infty} sF(s) = f(0^+). \quad \text{Q.E.D.}$$

* * *

Example 2.9

To illustrate the use of the initial value theorem (and other things as well), consider the mechanical systems illustrated in Figures 2.17a and 2.17b.

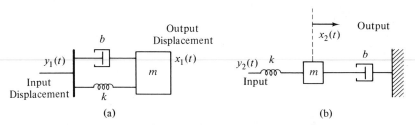

Figure 2.17. Spring-mass damped systems.

Assuming the systems are I.Q., it can be shown that the transfer functions

$$\frac{X_1(s)}{Y_1(s)} = \frac{k + bs}{k + bs + ms^2}, \qquad \frac{X_2(s)}{Y_2(s)} = \frac{k}{k + bs + ms^2}.$$

For a step input, $Y(s) = 1/s$,

$$X_1(s) = \frac{k + bs}{s(k + bs + ms^2)}, \qquad X_2(s) = \frac{k}{s(k + bs + ms^2)}.$$

It follows from Equation (2.49) that

$$x_1(0^+) = 0, \qquad\qquad x_2(0^+) = 0.$$

The interesting result is obtained when the initial values of the velocities are determined. Since the systems are I.Q., it follows from Equation (2.22) that the velocities are given by

$$V_1(s) = sX_1(s) = \frac{k + bs}{k + bs + ms^2}, \qquad V_2(s) = \frac{k}{k + bs + ms^2}.$$

From Equation (2.49), it follows that the initial velocities are

$$v_1(0^+) = \frac{b}{m}, \qquad v_2(0^+) = 0.$$

The system is I.Q.; thus, $v_1(0^-) = 0$. Hence for the first of the above systems a discontinuity in velocity occurs. The step responses for the above systems are illustrated in Figures 2.18a and 2.18b.

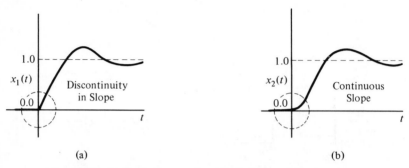

(a)   (b)

Figure 2.18. (a) Step response of system illustrated in Figure 2.17a, (b) step response of system illustrated in Figure 2.17b.

*   *   *

An experienced dynamic systems analyst gives considerable attention to initial values on operating records when attempting to develop empirical models for dynamic systems. Example 2.9 illustrates the reason.

2.7 The Final Value Theorem, Definition of *Gain*

The final value theorem is stated as follows:

Theorem

If $F(s) = \mathcal{L}[f(t)]$ and if the poles of $sF(s)$ lie *strictly* in the left half-plane, then

$$\lim_{s \to 0} sF(s) = f(\infty). \tag{2.50}$$

Proof. Begin with Equation (2.20), which reads

$$sF(s) - f(0^-) = \mathcal{L}\left[\frac{df}{dt}\right] \triangleq \int_{0^-}^{\infty} e^{-st}\frac{df}{dt}\, dt.$$

Thus,

$$\lim_{s \to 0} sF(s) = f(0^-) + \int_{0^-}^{\infty} \frac{df}{dt}\, dt$$

if the integral exists. Thus,

$$\lim_{s \to 0} sF(s) = f(0^-) + f(\infty) - f(0^-) = f(\infty)$$

provided $f(\infty)$ exists. $f(\infty)$ exists if $sF(s)$ has poles *strictly* in the left half-plane (i.e., the real parts of the poles < 0).   Q.E.D.

* * *

Example 2.10

Let $\dot{f}(t) = \cos \omega t \cdot u(t)$; then, as has been shown,

$$F(s) = \frac{s}{s^2 + \omega^2}.$$

An indiscriminate use of Equation (2.50) would lead to the conclusion that $f(\infty) = 0$. This ridiculous result was obtained because the poles of $sF(s)$ do not lie *strictly* in the left half plane in violation of the hypothesis of the final value theorem.

* * *

Example 2.11

For the mechanical system discussed in the previous section,

$$X_1(s) = \frac{k + bs}{s(k + bs + ms^2)}$$

It follows from Equation (2.50) that

$$x_1(\infty) = 1.0.$$

* * *

The *gain* of a system (usually denoted K) is defined as follows: Suppose a step change of magnitude a is made in the input $m(t)$; then define the gain in terms of the final value of the output, $p(\infty)$, by

$$K = \frac{p(\infty)}{a}. \tag{2.51}$$

It will be shown that for a system described by linear differential equations with constant coefficients, the gain is a constant and is independent of a. The gain is easily related to the transfer function. From the definition of a transfer function, $G(s)$, it follows that for an I.Q. system

$$P(s) = G(s)M(s).$$

If $m(t)$ is a step change of magnitude a, then

$$P(s) = \frac{aG(s)}{s}.$$

It follows from Equation (2.50) that

$$p(\infty) = a \lim_{s \to 0} G(s)$$

if $p(\infty)$ exists. Thus use Equation (2.51) to show

$$\boxed{K = \lim_{s \to 0} G(s).}\tag{2.52}$$

Thus, the gain of a system depends only on the transfer function.

Example 2.12

For the mechanical system of the previous section

$$\frac{X_1(s)}{Y_1(s)} = G_1(s) = \frac{k + bs}{k + bs + ms^2};$$

thus,

$$K = \lim_{s \to 0} \left[\frac{k + bs}{k + bs + ms^2} \right] = 1.0.$$

*   *   *

2.8 The Delta Function and the Convolution (Faltung) Integral

The *delta* or *impulse* function, denoted $\delta(t - T)$, is defined by

$$\int_a^b \delta(t - T) f(t) \, dt = \begin{cases} f(T), & \text{if } a < T < b, \\ 0, & \text{otherwise,} \end{cases}\tag{2.53}$$

for all $f(t)$. This function has played an extremely important role in systems analysis.

This definition is accepted and used in this text, but it should be pointed out that the above definition is a mathematical *absurdity*. If Equation (2.53) is to hold for all $f(t)$, it must hold for the functions illustrated in Figures 2.19a and 2.19b. By Equation (2.53)

$$\int_a^b \delta(t - T) f_1(t) \, dt = 0 \quad \text{and} \quad \int_a^b \delta(t - T) f_2(t) \, dt = 0.$$

But f_1, f_2 are strictly positive, thus

$$\delta(t - T) = 0 \quad \text{for } a \le t < T \text{ and for } T < t \le b.\tag{2.54}$$

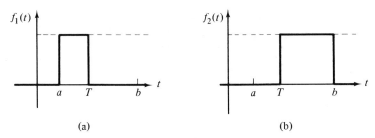

(a)   (b)

Figure 2.19. (a) and (b) are functions which vanish at T^- and T^+ respectively.

In fact, in some references the graph of Figure 2.20 is associated with $\delta(t - T)$. Equation (2.54) means that the integrand of Equation (2.53) is zero except at one point, T. But for *no* definition of the integral can such an integral be nonzero.

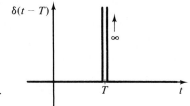

Figure 2.20. Graph of the delta function.

In some texts, it is seemingly shown that the delta function is the limit of a sequence of functions. In all such "proofs" an illegal interchange of limit and integral signs is made.

A rigorous treatment of delta functions is possible[7] but is considered well beyond the scope of the present discussion.

Consider the Laplace transformation of the delta function for which $T = 0$, i.e., $\delta(t)$. By definition

$$\mathcal{L}[\delta(t)] = \int_{0^-}^{\infty} \delta(t)e^{-st}\, dt.$$

From Equation (2.53)

$$\mathcal{L}[\delta(t)] = 1.0. \tag{2.55}$$

The above result leads to a physical interpretation of the inverse transform of the transfer function. By the definition of a transfer function, $G(s)$,

$$P(s) = G(s)M(s).$$

If $m(t)$ is the delta function, then by Equation (2.55), $M(s) = 1$ and $P(s) =$

$G(s)$. Thus, the response to the impulse input is

$$p(t) = \mathcal{L}^{-1}[G(s)] \triangleq g(t).$$ (2.56)

The inverse, $g(t)$, is called the *impulse response* for obvious reasons. The name *Green's function* is also associated with this quantity.

The significance of the impulse response to dynamic systems theory will be illustrated in an interpretation of the following theorem.

Convolution Theorem

If $F_1(s) = \mathcal{L}[f_1(t)]$ and $F_2(s) = \mathcal{L}[f_2(t)]$, then

$$\mathcal{L}\left[\int_0^t f_1(t - \xi)f_2(\xi)\,d\xi\right] = F_1(s)F_2(s).$$ (2.57)

Proof. By definition

$$\mathcal{L}\left[\int_0^t f_1(t - \xi)f_2(\xi)\,d\xi\right] = \int_{0^-}^{\infty} e^{-st}\int_0^t f_1(t - \xi)f_2(\xi)\,d\xi\,dt.$$

Interchange the order of integration and choose limits to observe the ordering

$$0^- \leq \xi \leq t \leq \infty$$

and find that the above equation reduces to

$$\int_{0^-}^{\infty} e^{-s\xi}f_2(\xi)\int_{\xi}^{\infty} e^{-s(t-\xi)}f_1(t - \xi)\,dt\,d\xi.$$

Let $\eta = t - \xi$ which implies $d\eta = dt$ in the second integral and show that

$$\mathcal{L}\left[\int_0^t f_1(t - \xi)f_2(\xi)\,d\xi\right] = \int_{0^-}^{\infty} e^{-s\xi}f_2(\xi)\int_{0^-}^{\infty} e^{-s\eta}f_1(\eta)\,d\eta\,d\xi$$

$$= F_1(s)\int_{0^-}^{\infty} e^{-s\xi}f_2(\xi)\,d\xi = F_1(s)F_2(s). \quad \text{Q.E.D.}$$

* * *

It follows from Equations (2.57) and (2.35) that

$$p(t) = \int_0^t g(t - \xi)m(\xi)\,d\xi,$$ (2.58)

where $g(t)$ is the impulse response defined in Equation (2.56). The impulse response and the integrand of Equation (2.58) are interpreted in Figures

2.21a and 2.21b. Equation (2.58) is a convenient formula, especially when $m(t)$ is known graphically only. The convolution integral (2.58) will be used to yield many useful theoretical results throughout the text.

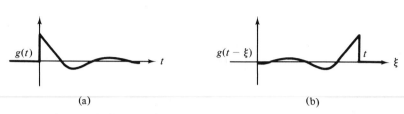

Figure 2.21. (a) Impulse response, (b) impulse response with inverted argument as in Equation (2.58).

2.9 The Integration and Shifting Theorems and Other Useful Results

A summary of several useful results is provided in this section.

Integration Theorem

$$\mathcal{L}\left[\int_{0-}^{t} f(\xi)\, d\xi\right] = \frac{1}{s}F(s) \tag{2.59}$$

if

$$\mathcal{L}[f(t)] = F(s).$$

Proof. By definition

$$\mathcal{L}\left[\int_{0-}^{t} f(\xi)\, d\xi\right] = \int_{0-}^{\infty} e^{-st} \int_{0-}^{t} f(\xi)\, d\xi\, dt.$$

Interchange the order of integration to find that

$$\mathcal{L}\left[\int_{0-}^{t} f(\xi)\, d\xi\right] = \int_{0-}^{\infty} f(\xi) \int_{\xi}^{\infty} e^{-st}\, dt\, d\xi$$

$$= \frac{1}{s}\int_{0-}^{\infty} f(\xi)e^{-s\xi}\, d\xi = \frac{1}{s}F(s). \quad \text{Q.E.D.}$$

* * *

This theorem will prove quite useful in the chapters on dynamic systems analysis.

Occasional use of the following theorem of Laplace transformation theory is made when transforming certain functions of time and in the analysis of some systems described by linear partial differential equations.

Shifting Theorem

If

$$\mathcal{L}[f(t)] \triangleq F(s),$$

then

$$\mathcal{L}[f(t - a)] = e^{-as}F(s). \tag{2.60}$$

Proof.

$$\mathcal{L}[f(t - a)] = \int_{0^-}^{\infty} e^{-st} f(t - a) \, dt = e^{-as} \int_{0^-}^{\infty} e^{-s(t-a)} f(t - a) \, dt.$$

Let

$$\zeta = t - a.$$

Then

$$\mathcal{L}[f(t - a)] = e^{-as} \int_{-a}^{\infty} e^{-s\zeta} f(\zeta) \, d\zeta.$$

But, the domain of Laplace transformation is such that

$$f(\zeta) = 0 \quad \text{for } \zeta < 0^-.$$

Thus,

$$\mathcal{L}[f(t - a)] = e^{-as} \int_{0^-}^{\infty} e^{-s\zeta} f(\zeta) \, d\zeta$$
$$= e^{-as}F(s). \quad \text{Q.E.D.}$$

* * *

Example 2.13

Find the Laplace transform of the delayed step illustrated in Figure 2.22. Use Equation (2.60) to show that

$$\mathcal{L}[u(t - \tau)] = e^{-\tau s}\mathcal{L}[u(t)].$$

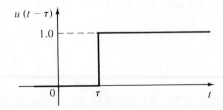

Figure 2.22. Delayed step function.

Thus,

$$\mathcal{L}[u(t-\tau)] = \frac{e^{-\tau s}}{s}. \tag{2.61}$$

* * *

Example 2.14

Find the Laplace transform of the two functions of time illustrated in Figures 2.23 and 2.24. Notice that $f_1(t)$ may be written

$$f_1(t) = t[u(t) - u(t-\tau)], \tag{2.62}$$

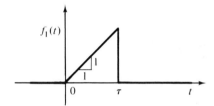

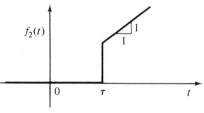

Figure 2.23. Truncated ramp.

Figure 2.24. Delayed and truncated ramp.

which can be rewritten as a superposition of functions of $t-\tau$ as follows:

$$f_1(t) = tu(t) - (t-\tau)u(t-\tau) - \tau u(t-\tau).$$

Use Equations (2.8) and (2.60) to show that

$$\mathcal{L}(f_1) = \frac{1}{s^2} - e^{-\tau s}\mathcal{L}[tu(t)] - \tau e^{-\tau s}\mathcal{L}[u(t)].$$

Finally,

$$\mathcal{L}[f_1(t)] = \frac{1}{s^2} - e^{-\tau s}\left[\frac{1}{s^2} + \frac{\tau}{s}\right]. \tag{2.63}$$

Notice that it is possible to write $f_2(t)$ as

$$f_2(t) = tu(t-\tau) = (t-\tau)u(t-\tau) + \tau u(t-\tau).$$

Use Equation (2.60) to show that

$$\mathcal{L}[f_2(t)] = e^{-\tau s}\mathcal{L}[tu(t)] + \tau e^{-\tau s}\mathcal{L}[u(t)].$$

Finally,

$$\mathcal{L}[f_2(t)] = e^{-\tau s}\left[\frac{1}{s^2} + \frac{\tau}{s}\right]. \tag{2.64}$$

* * *

It is useful to generalize a result derived and used in Example 2.2; if

$$F(s) = \mathcal{L}[f(t)] = \int_{0-}^{\infty} f(t)e^{-st}\, dt,$$

then

$$\frac{dF(s)}{ds} = -\int_{0-}^{\infty} tf(t)e^{-st}\, dt,$$

that is,

$$\mathcal{L}[t \cdot f(t)] = \frac{-d}{ds}\mathcal{L}[f(t)]. \tag{2.65}$$

It is easily shown that the generalization is

$$\mathcal{L}[t^n \cdot f(t)] = (-1)^n \frac{d^n}{ds^n}\mathcal{L}[f(t)]. \tag{2.66}$$

These equations are useful when applied to the inversion of Laplace transforms which have repeated poles. The heavyside expansions of such transforms contain terms of the form

$$F_n(s) = \frac{R_n}{(s + a)^n}, \tag{2.67}$$

where R_n is the residue. Notice that

$$F_n(s) = \frac{(-1)^n}{n!}\frac{d^n}{ds^n}\left[\frac{R_n}{s + a}\right]. \tag{2.68}$$

It follows from Equation (2.66) that

$$\mathcal{L}^{-1}[F_n(s)] = \frac{1}{n!}\mathcal{L}^{-1}\left[\frac{R_n}{s + a}\right]$$

or

$$\mathcal{L}^{-1}[F_n(s)] = \frac{1}{n!}t^n e^{-at}. \tag{2.69}$$

To this point, Laplace transformation theory has been applied to the solution of linear ordinary differential equations. The theory is also very useful in the solution of linear partial differential equations.[1,2,6] The crucial fact in the latter application is a formula with a rather innocent appearance. Consider the function $f(x, t)$, where x, t are independent variables; then,

$$\mathcal{L}\left\{\frac{\partial f}{\partial x}\right\} = \frac{\partial}{\partial x}\mathcal{L}\{f\}. \tag{2.70}$$

Condition (2.4) is the mathematical criterion for the interchange of differentiation and integration operations which is indicated in Equation (2.70). The application of this result to the solution of partial differential equations will be demonstrated by example.

Example 2.15

Consider the partial differential equation

$$\frac{\partial p}{\partial t} = -V\frac{\partial p}{\partial x} - kp; \quad V, k \text{ constants}, \tag{2.71}$$

with the initial condition

$$p(x, 0^-) = 0 \tag{2.72}$$

and boundary condition

$$p(0^-, t) = m(t), \quad t \geq 0, \tag{2.73}$$

where $m(t)$ is some function of time which will not be specified at present. Let

$$P(x, s) = \mathcal{L}\{p(x, t)\}. \tag{2.74}$$

It then follows from Equations (2.22) and (2.70) that

$$sP(x, s) - p(x, 0^-) = -V\frac{\partial P(x, s)}{\partial x} - kP(x, s).$$

Since no derivatives with respect to s appear in this equation, the partial derivatives with respect to x can be replaced with total derivatives. Use Equation (2.72) to show

$$\frac{dP(x, s)}{dx} + \frac{(k+s)}{V}P(x, s) = 0. \tag{2.75}$$

When the Laplace transform was applied to the solution of ordinary differential equations, the calculus problem was reduced to a problem in algebra (heavyside expansion). In the present case, the problem of solving a partial differential equation is reduced to a problem of solving an ordinary differential equation. Notice that since s and x are independent, s may be treated as a parameter in Equation (2.75), which, therefore, is a linear differential equation in the variable x. With a slight reorientation in thinking, this equation can be solved using Laplace transformation. Make the transformation $x \longrightarrow z$ using an integral definition similar to that of Equation (2.1), where z is some complex Laplace variable. Equation (2.75) becomes

$$zP(z, s) - P(0^-, s) + \frac{(k + s)}{V}P(z, s) = 0,$$

so that

$$P(z, s) = \frac{P(0^-, s)}{z + [(k + s)/V]}. \tag{2.76}$$

Use Equation (2.8) to make the inverse transformation $z \longrightarrow x$ and find

$$P(x, s) = P(0^-, s)e^{-(k+s)x/V}, \quad x > 0. \tag{2.77}$$

Take the Laplace transform of Equation (2.73) to show that

$$P(0^-, s) = M(s),$$

so that

$$P(x, s) = M(s)e^{-kx/V}e^{-sx/V}. \tag{2.78}$$

The inverse transformation, $s \longrightarrow t$, can be obtained using Equation (2.60) to find

$$p(x, t) = e^{-kx/V}m\left(t - \frac{x}{V}\right), \quad t > \frac{x}{V}. \tag{2.79}$$

The complete solution of Equation (2.71) has been obtained without specifying a form for $m(t)$.

*   *   *

2.10 Systems of Ordinary Differential Equations: State Variable Representations and Eigenvalue Problems*

Reference sections are included in many of the succeeding chapters which deal with the *state variable* representation of dynamic control systems. In

---

*The material in this section may be omitted during the first reading of the text.

this representation, the dynamics of a system are described by a set of coupled, ordinary differential equations which have the general form

$$\dot{x}_1 = a_{11}x_1 + a_{12}x_2 + \cdots + a_{1n}x_n + b_{11}u_1 + \cdots + b_{1m}u_m$$
$$\dot{x}_2 = a_{21}x_1 + a_{22}x_2 + \cdots + a_{2n}x_n + b_{21}u_1 + \cdots + b_{2m}u_m$$

$$\tag{2.80}$$

$$\dot{x}_n = a_{n1}x_1 + a_{n2}x_2 + \cdots + a_{nn}x_n + b_{n1}u_1 + \cdots + b_{nm}u_m.$$

The $u_k(t)$ may be thought of as a set of input functions. Use the compact matrix notation (reviewed in Appendix B) by denoting the array of a_{kp} by $\mathcal{C}$ and the array of b_{kp} by $\mathcal{B}$. The set of Equations (2.80) can then be written succinctly as

$$\dot{x} = \mathcal{C}x + \mathcal{B}u. \tag{2.81}$$

Consider the special case when $\mathcal{C}$ and $\mathcal{B}$ are arrays of constants. The application of Laplace transformation to the analysis of Equation (2.81) for this special case will be presented in this section. Apply the Laplace transform to Equations (2.80) and show that the results may be written in the form

$$sX(s) - x(0^-) = \mathcal{C}X(s) + \mathcal{B}U(s). \tag{2.82}$$

$X(s)$ is a vector of Laplace transforms of the $x_k(t)$; $U(s)$ is a vector of Laplace transforms of the $u_k(t)$; and $x(0^-)$ is a vector of initial conditions. Show that Equation (2.82) may be rewritten

$$\{s\mathcal{I} - \mathcal{C}\}X(s) = x(0^-) + \mathcal{B}U(s), \tag{2.83}$$

where $\mathcal{I}$ is the unit matrix. Thus, the unknown

$$X(s) = \Phi(s)x(0^-) + \mathcal{G}(s)U(s), \tag{2.84}$$

where

$$\Phi(s) = \{s\mathcal{I} - \mathcal{C}\}^{-1} \tag{2.85}$$

and the *transfer function matrix*

$$\mathcal{G}(s) = \Phi(s)\mathcal{B}. \tag{2.86}$$

Define the *fundamental solution matrix*

$$\phi(t) = \mathcal{L}^{-1}\{\Phi(s)\}. \tag{2.87}$$

Take the inverse transform of Equation (2.84), by using Equation (2.57) on the second term, to find

$$\mathbf{x}(t) = \boldsymbol{\phi}(t)\mathbf{x}(0^-) + \int_0^t \boldsymbol{\phi}(t - \xi)\boldsymbol{\mathfrak{B}}\mathbf{u}(\xi)\,d\xi. \tag{2.88}$$

It is instructive to investigate some of the properties of the transfer function matrix and of the fundamental solution matrix. Using the basic formula for an inverse matrix, derived in Appendix B, one finds that

$$\boldsymbol{\Phi}(s) = \frac{\text{adj}\{s\boldsymbol{\mathcal{I}} - \boldsymbol{\mathcal{C}}\}}{\det\{s\boldsymbol{\mathcal{I}} - \boldsymbol{\mathcal{C}}\}}, \tag{2.89}$$

where $\text{adj}\{s\boldsymbol{\mathcal{I}} - \boldsymbol{\mathcal{C}}\}$ is a matrix of cofactors. Since these cofactors are proportional to the determinant of $s\boldsymbol{\mathcal{I}} - \boldsymbol{\mathcal{C}}$, less one row and one column, they are $(n - 1)$st-order polynomials in s which determine the zeros of the elements of $\boldsymbol{\Phi}(s)$ and $\boldsymbol{\mathcal{G}}(s)$. The determinant of $s\boldsymbol{\mathcal{I}} - \boldsymbol{\mathcal{C}}$ is an nth-order polynomial in s. Notice that this determinant is the denominator polynomial of *all* elements of $\boldsymbol{\Phi}(s)$ and of $\boldsymbol{\mathcal{G}}(s)$. Thus, all of the transforms of these important matrices have the same poles, which are those values of s which satisfy the so-called *characteristic equation*

$$\det\{s\boldsymbol{\mathcal{I}} - \boldsymbol{\mathcal{C}}\} = 0. \tag{2.90}$$

Example 2.16

Consider the unforced system described by

$$\dot{x}_1 = x_2$$
$$\dot{x}_2 = -x_1.$$

It follows that

$$s\boldsymbol{\mathcal{I}} - \boldsymbol{\mathcal{C}} = \begin{bmatrix} s & 0 \\ 0 & s \end{bmatrix} - \begin{bmatrix} 0 & 1 \\ -1 & 0 \end{bmatrix} = \begin{bmatrix} s & -1 \\ 1 & s \end{bmatrix}.$$

The characteristic equation is

$$\det\begin{bmatrix} s & -1 \\ 1 & s \end{bmatrix} = s^2 + 1 = 0$$

and

$$\boldsymbol{\Phi}(s) = \frac{1}{s^2 + 1}\begin{bmatrix} s & 1 \\ -1 & s \end{bmatrix} = \begin{bmatrix} \dfrac{s}{s^2 + 1} & \dfrac{1}{s^2 + 1} \\ \dfrac{-1}{s^2 + 1} & \dfrac{s}{s^2 + 1} \end{bmatrix};$$

thus,

$$\phi(t) = \begin{bmatrix} \cos t & \sin t \\ -\sin t & \cos t \end{bmatrix}.$$

Finally, from Equation (2.88)

$$x_1(t) = x_1(0^-)\cos t + x_2(0^-)\sin t,$$
$$x_2(t) = -x_1(0^-)\sin t + x_2(0^-)\cos t.$$

* * *

An often encountered problem in analysis is, Find an *eigenvalue*, λ, and an *eigenvector*, e, of the matrix $\mathcal{C}$ which satisfy the equation

$$\mathcal{C}e = \lambda e. \tag{2.91}$$

This equation can be rewritten

$$\{\lambda \mathcal{I} - \mathcal{C}\}e = 0.$$

It is shown in Appendix B that a nontrivial vector e satisfies the above equation only if

$$\det\{\lambda \mathcal{I} - \mathcal{C}\} = 0. \tag{2.92}$$

When one compares Equations (2.90) and (2.92), it becomes apparent that the poles of the $\Phi(s)$ and $\mathcal{G}(s)$ are equal to the eigenvalues of $\mathcal{C}$.

The identity of poles and eigenvalues can be demonstrated another way; suppose $u = 0$ so that Equation (2.81) becomes

$$\dot{x} = \mathcal{C}x. \tag{2.93}$$

Attempt a solution of the form

$$x(t) = ce^{\alpha t}, \quad c = \text{constant}. \tag{2.94}$$

Substitute Equation (2.94) into Equation (2.93) to find

$$\alpha c e^{\alpha t} = \mathcal{C}c e^{\alpha t}$$

or

$$\mathcal{C}c = \alpha c. \tag{2.95}$$

Thus, the $x(t)$ of Equation (2.94) is a solution to Equation (2.93) only if α is an eigenvalue and c is an eigenvector of $\mathcal{C}$. Notice also that a term proportional to $e^{\alpha t}$ will appear in the solution to Equation (2.93) only if α is a pole of the elements of $\Phi(s)$ [see Equation (2.84)].

The above facts about eigenvalues and poles of Laplace transforms provide more than a simple, passing academic interest. In advanced references, it is useful to be able to derive facts about the polynomial equation

$$a_n s^n + a_{n-1} s^{n-1} + a_{n-2} s^{n-2} + \cdots + a_1 s + a_0 = 0. \qquad (2.96)$$

This equation arises when one attempts to find the poles or the zero of a Laplace transform. Clearly this equation can be associated with the differential equation

$$a_n \frac{d^n x}{dt^n} + a_{n-1} \frac{d^{n-1} x}{dt^{n-1}} + \cdots + a_1 \frac{dx}{dt} + a_0 x = 0. \qquad (2.97)$$

Define the set of variables $x_1, x_2, \ldots, x_n$ in the following way:

$$x_1 = x$$

$$x_k = \frac{d^{k-1} x}{dt^{k-1}}, \quad k = 2, 3, \ldots, n. \qquad (2.98)$$

Combine the above equations to show that

$$\dot{x}_1 = x_2,$$
$$\dot{x}_2 = x_3,$$
$$\cdot$$
$$\cdot \qquad \qquad (2.99)$$
$$\cdot$$
$$\dot{x}_{n-1} = x_n,$$
$$\dot{x}_n = -\frac{a_0 x_1}{a_n} - \frac{a_1 x_2}{a_n} - \cdots - \frac{a_{n-1} x_n}{a_n}.$$

These equations have the form of Equation (2.93), where

$$\mathcal{Q} = \begin{bmatrix} 0 & 1 & 0 & \cdots & 0 \\ 0 & 0 & 1 & \cdots & 0 \\ \cdot & & & & \cdot \\ \cdot & & & & \cdot \\ \cdot & & & & \cdot \\ 0 & 0 & 0 & \cdots & 1 \\ -\dfrac{a_0}{a_n} & -\dfrac{a_1}{a_n} & -\dfrac{a_2}{a_n} & \cdots & -\dfrac{a_{n-1}}{a_n} \end{bmatrix}. \qquad (2.100)$$

Notice that the eigenvalues of this matrix are the poles of the transform of Equation (2.97) and, hence, are the roots of Equation (2.96). For this reason, $\mathcal{Q}$ is said to be the *companion matrix* to the polynomial of Equation (2.96).

The important point is that any fact about the eigenvalues of the companion matrix applies to the roots of the associated polynomial. Often, it is easier to prove statements about the eigenvalues of matrices.

Another fact that has been demonstrated in the above development is that differential equations of the form of Equation (2.97) can always be converted to a set of simultaneous first-order differential equations of the form of Equation (2.99) by defining the so-called *phase space* variables as in Equation (2.98).

The final useful result to be presented is as follows: If λ is an eigenvalue of the companion matrix, then an eigenvector is

$$\mathbf{e} = \begin{bmatrix} 1 \\ \lambda \\ \lambda^2 \\ \cdot \\ \cdot \\ \cdot \\ \lambda^{n-1} \end{bmatrix}. \tag{2.101}$$

To prove this assertion, consider $\mathbf{v} = \mathbf{\alpha e}$; obviously

$$v_k = \lambda^k, \ 1 \leq k \leq n - 1, \tag{2.102}$$

and

$$v_n = \frac{-a_0}{a_n} + \frac{-a_1\lambda}{a_n} + \frac{-a_2\lambda^2}{a_n} + \cdots + \frac{-a_{n-1}\lambda^{n-1}}{a_n}.$$

But the characteristic equation of the companion matrix is Equation (2.96) (with s replaced by λ). Thus,

$$v_n = \lambda^n. \tag{2.103}$$

Combine Equations (2.102) and (2.103) to show

$$\mathbf{\alpha e} = \mathbf{v} = \begin{bmatrix} \lambda \\ \lambda^2 \\ \cdot \\ \cdot \\ \cdot \\ \lambda^n \end{bmatrix} = \lambda \begin{bmatrix} 1 \\ \lambda \\ \cdot \\ \cdot \\ \cdot \\ \lambda^{n-1} \end{bmatrix}$$

or

$$\mathbf{\alpha e} = \lambda \mathbf{e},$$

and assertion (2.101) is established.

References

1. CARSLAW, H. S., and JAEGAR, H. S., *Operational Methods in Applied Mathematics*, Dover, New York, 1963.

2. CHENG, D. K., *Analysis of Linear Systems*, Addison-Wesley, Reading, Mass., 1959.

3. CHURCHILL, R. V., *Operational Mathematics*, McGraw-Hill, New York, 1958.

4. LATHI, B. P., *Signals, Systems, and Communication*, Wiley, New York, 1965.

5. OGATA, K., *Modern Control Engineering*, Prentice-Hall, Englewood Cliffs, N.J., 1970.

6. PIPES, L. A., *Matrix Methods for Engineering*, Prentice-Hall, Englewood Cliffs, N.J., 1963.

7. ZADEH, L. A., and DESOER, C. A., *Linear System Theory: The State Space Approach*, McGraw-Hill, New York, 1963.

Problems

2.1. Use Equation (2.1) to determine the Laplace transforms of the two functions graphed in Figures 2.25 and 2.26. Repeat the exercise by using Equation (2.60).

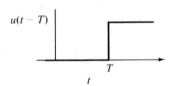

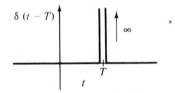

Figure 2.25. Delayed step.   **Figure 2.26.** Delayed impulse.

2.2. Derive Equation (2.8) from Equation (2.1).

2.3. Consider the physical systems described by Equations (2.33) and (2.34). Determine the system response, $p(t)$, when

$$RC = 1.0 \text{ sec}^{-1},$$
$$p(0^-) = 2.0,$$

and $m(t)$ is as graphed in Figure 2.27.

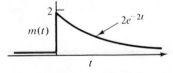

Figure 2.27. Input function.

2.4. Calculate all residues of the following Laplace transforms:

(a) $P(s) = \dfrac{s + 2}{s^2 + 7s + 12}$.

(b) $P(s) = \dfrac{2s + 2}{s^2 + 2s + 1}$.

(c) $P(s) = \dfrac{5}{s(0.5s^2 + s + 1)}$.

2.5. Use only the initial and final value theorems to sketch $p(t)$ when

(a) $P(s) = \dfrac{1}{s(s + 1)}$.

(b) $P(s) = \dfrac{s + 2}{s(s + 1)}$.

(c) $P(s) = \dfrac{s - 2}{s(s + 1)}$.

(d) $P(s) = \dfrac{2 - s}{s(s + 1)}$.

2.6. Find the Laplace transform of the *posi-cast function* (see Figure 2.28); use the shifting theorem.

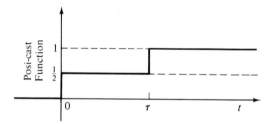

Figure 2.28. Posi-cast function.

2.7. The twin-*RC* network shown in Figure 2.29 may be described by the transfer function

$$\frac{E_0(s)}{E_i(s)} = \frac{s^2 + 1}{(s + \alpha)[s + (1/\alpha)]}$$

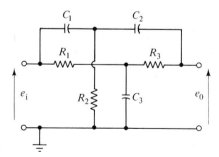

Figure 2.29. Twin-*RC* network.

for certain values of the resistors and capacitors. α is a constant. Find $e_0(t)$ when

$$e_i(t) = \sin t \cdot u(t).$$

What would be a practical application of such a network?

2.8. Use the Laplace transform to solve the following differential equations for $x(t)$, $y(t)$:

$$\frac{dy(t)}{dt} = ax(t), \quad x(0^-) = x_0, \; y(0^-) = y_0,$$

$$\frac{dx(t)}{dt} = by(t), \quad b, a, x_0, y_0 \text{ constants.}$$

2.9. Consider the hydraulic line shown in Figure 2.30. It can be shown that

$$\frac{\partial e}{\partial t} = \frac{-1}{C} \frac{\partial f}{\partial x},$$

$$\frac{\partial f}{\partial t} = \frac{-1}{L} \frac{\partial e}{\partial x}.$$

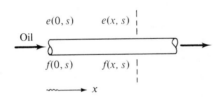

Figure 2.30.

C and L are physical constants. Find the matrix relating the vector

$$\begin{bmatrix} E(0, s) \\ F(0, s) \end{bmatrix}$$

to the vector

$$\begin{bmatrix} E(x, s) \\ F(x, s) \end{bmatrix}.$$

(*Hint:* Take the Laplace transform of the partial differential equations and then use the results of Problem 2.8.)

2.10. For the system shown in Figure 2.31, the elements are designed so that

$$\frac{dc}{dt} = r(t) - c(t - \tau).$$

The unavoidable delay τ occurs because the measuring device must be placed

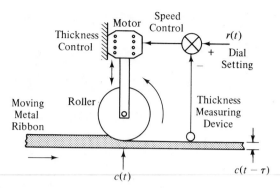

Figure 2.31.

downstream of the roller. Use the shifting theorem to find the transfer function

$$G(s) \triangleq \frac{C(s)}{R(s)}.$$

2.11. The transfer function of a dynamic system is given by

$$\frac{C(s)}{R(s)} = \frac{1}{s^2 + 1}.$$

Find $c(t)$ if $r(t)$ is the posi-cast function of Problem 2.6 ($\tau = \pi$). Sketch $c(t)$.

The Basic Concepts
of Dynamic Systems
Analysis: Electrical
and Magnetic Circuits

3

In this chapter, the discussion shifts from the mathematics of systems theory to the physics of systems theory. Methods which allow the analyst to derive formulas for the transfer functions of electrical and magnetic circuits will be introduced. In succeeding chapters, these same methods will be extended to the analysis of mechanical, hydraulic, and pneumatic networks.

The discussion will, in the main, center on networks of circuits of *passive* components. *Passive* components are those which can store, dissipate, and return stored energy but cannot supply *net* energy to another component or to a network of components.

The roles of energy and power concepts in dynamic systems analysis will be emphasized, as they have been in the basic references on the subject.[2,4,5] In this introductory presentation, no discussion of powerful graphical methods of systems analysis will be provided. The reader is encouraged to refer to the literature in order to learn these important graphical techniques[4,5,7] and thereby to gain depth in his understanding of the physics of system theory.

3.1 The Axioms of the Macroscopic Theory
of Passive Electrical Components;
The Generalized Concept of Impedance,
Power State, and Transfer Matrices[8]

In elementary courses, students learn the relationships between the voltage across and the current through the basic passive electrical elements.[3] These relationships are generally called *laws* (e.g., Ohm's law); however, in this text, the term *axiom* will be considered a more accurate term. The symbols for passive circuit elements are illustrated in Figure 3.1. R, L, and C are constants and have the units ohms, henries, and farads, respectively.

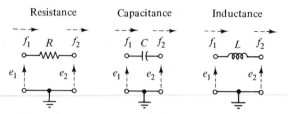

Figure 3.1. Passive electrical circuit elements. Voltage is denoted e and current f. The dashed arrows indicate the directions of positive current and positive voltage difference.

The axiomatic relations between the variables are as follows: For the resistor

$$f_1 = f_2,$$
$$e_1 = e_2 + Rf_2; \tag{3.1}$$

for the capacitance

$$f_1 = f_2,$$
$$e_1 = e_2 + \frac{1}{C} \int_0^t f_2(\xi)\, d\xi; \tag{3.2}$$

and for the inductance

$$f_1 = f_2,$$
$$e_1 = e_2 + \frac{L\, df_2}{dt}. \tag{3.3}$$

Define the *impedance*, denoted $Z(s)$, as the ratio of the voltage drop, $E_1 - E_2$, across such a circuit element to the current through the element after all variables have been transformed to the Laplace domain (assuming initial quiescence). Use the basic theorems derived in Chapter 2 to show that for the above components

$$Z(s) \triangleq \frac{E_1(s) - E_2(s)}{F_2(s)} = R \quad (resistance); \tag{3.4}$$

and

$$E_1(s) = E_2(s) + \frac{1}{Cs} F_2(s),$$

$$\therefore \quad Z(s) = \frac{1}{Cs} \quad (capacitance); \tag{3.5}$$

and

$$E_1(s) = E_2(s) + LsF_2(s),$$

$$\therefore \quad Z(s) = Ls \quad (inductance). \tag{3.6}$$

Notice that Equations (3.4)–(3.6) have the general form

$$E_1 = E_2 + Z(s)F_2,$$
$$F_1 = F_2 \tag{3.7}$$

when they are transformed to the Laplace domain. Equations (3.7) are the canonical form for the *series* impedances illustrated in Figure 3.1.

The three basic impedances can be connected in *shunt* in the manner indicated in Figure 3.2. For this configuration, $E_1 = E_2$, but $F_1 \neq F_2$ because

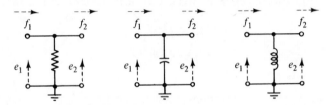

Figure 3.2. Basic components in shunt.

current now goes to ground. This current is E_2/Z so that canonical transformed equations for a shunted impedance are

$$E_1 = E_2,$$
$$F_1 = F_2 + \frac{E_2}{Z(s)}. \tag{3.8}$$

The basic equations (3.7) and (3.8) could be applied to the analysis of the dynamic system illustrated in Figure 3.3. Suppose that it is desired to determine the transfer function $F_1(s)/E_0(s)$. After recognizing series and shunted components, one could write a couplet of equations for each element and then perform the laborious algebraic task of eliminating all unwanted E and F variables until only a relation between E_0 and F_1 remains. This

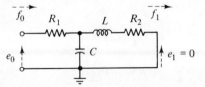

Figure 3.3. Ladder circuit.

procedure is greatly simplified by introducing matrix notation (reviewed in Appendix B).

To facilitate the introduction of matrix notation, the following definitions are made. The *power state* vector at any point in a ladder

$$\mathbf{P} = \begin{bmatrix} E \\ F \end{bmatrix}. \tag{3.9}$$

The name derives from the fact that the instantaneous power equals $e(t)f(t)$. Note, however, that the product $E(s)F(s)$ is not related to power for the simple reason that

$$\mathcal{L}[e(t)f(t)] \neq E(s)F(s). \tag{3.10}$$

Define the *series transfer matrix*

$$\mathfrak{I} = \begin{bmatrix} 1 & Z(s) \\ 0 & 1 \end{bmatrix} \tag{3.11}$$

and the *shunt transfer matrix*

$$\mathfrak{I} = \begin{bmatrix} 1 & 0 \\ Z^{-1}(s) & 1 \end{bmatrix}. \tag{3.12}$$

Notice that Equations (3.7) and (3.8) can be written

$$\mathbf{P}_1 = \mathfrak{I}\mathbf{P}_2. \tag{3.13}$$

Repeated application of Equation (3.13) and use of Equations (3.4)–(3.6) will produce the desired result. To illustrate this fact, consider the ladder illustrated in Figure 3.3. Repeated application of Equation (3.13) leads to

$$\begin{bmatrix} E_0 \\ F_0 \end{bmatrix} = \begin{bmatrix} 1 & R_1 \\ 0 & 1 \end{bmatrix}\begin{bmatrix} 1 & 0 \\ Cs & 1 \end{bmatrix}\begin{bmatrix} 1 & Ls \\ 0 & 1 \end{bmatrix}\begin{bmatrix} 1 & R_2 \\ 0 & 1 \end{bmatrix}\begin{bmatrix} 0 \\ F_1 \end{bmatrix}$$
$$= \begin{bmatrix} 1 + R_1Cs & R_1 \\ Cs & 1 \end{bmatrix}\begin{bmatrix} 1 & Ls + R_2 \\ 0 & 1 \end{bmatrix}\begin{bmatrix} 0 \\ F_1 \end{bmatrix}; \tag{3.14}$$

finally,

$$\begin{bmatrix} E_0 \\ F_0 \end{bmatrix} = \begin{bmatrix} \cdots & (1 + R_1Cs)(Ls + R_2) + R_1 \\ \cdots & \cdots \end{bmatrix}\begin{bmatrix} 0 \\ F_1 \end{bmatrix}. \tag{3.15}$$

The dots indicate matrix elements that need not be calculated because they do not enter the relationship between $E_0(s)$ and $F_1(s)$. It follows that

$$E_0(s) = \{(1 + R_1Cs)(Ls + R_2) + R_1\}F_1,$$

so that the desired transfer function

$$G(s) = \frac{F_1(s)}{E_0(s)} = \frac{1}{R_1CLs^2 + (L + R_1R_2C)s + R_1 + R_2}. \qquad (3.16)$$

The response of the system, $f_1(t)$, can be determined using the methods of Chapter 2 after the Laplace transform of the input, $E_0(s)$, has been determined.

The above procedure is applicable to the case when $e_0(t)$ can be arbitrarily varied with the aid of a voltage source. Suppose, for a second illustration of the procedure, that a current source is used so that $f_0(t)$ is the input. For this case, the transfer function $F_1(s)/F_0(s)$ is the goal of a dynamic systems analysis. The appropriate form of Equation (3.15) is

$$\begin{bmatrix} E_0 \\ F_0 \end{bmatrix} = \begin{bmatrix} \cdots & \cdots \\ \cdots & Cs(Ls + R_2) + 1 \end{bmatrix}\begin{bmatrix} 0 \\ F_1 \end{bmatrix}, \qquad (3.17)$$

so that

$$F_0 = \{Cs(Ls + R_2) + 1\}F_1 \qquad (3.18)$$

or

$$G(s) = \frac{F_1(s)}{F_0(s)} = \frac{1}{CLs^2 + R_2Cs + 1}. \qquad (3.19)$$

Notice that even though attention is focused on the same output variable, $f_1(t)$, in the same dynamic system, the form of the transfer function is greatly modified by a change in the type of input.

It is important to notice that the introduction of the concepts of impedance allows the analyst to determine transfer functions by performing algebraic manipulations with Laplace-transformed variables directly. The usual procedure of deriving differential equations and transforming them into Laplace domain is thereby circumvented.

3.2 Input Impedance, *Reflecting* or Inverse Transfer Matrices, Transfer Functions for Intermediate Ladder Variables

Equation (3.13) relates the two power states on either side of a series or shunted impedance in a ladder network. It follows that

$$\mathbf{P}_2 = \mathfrak{I}^{-1}\mathbf{P}_1. \qquad (3.20)$$

Using the technique summarized in Appendix B or, more simply, by manipulating Equations (3.7) and (3.8), one can show that for series transfer matrices

$$\mathcal{J}^{-1} = \begin{bmatrix} 1 & -Z(s) \\ 0 & 1 \end{bmatrix} \qquad (3.21)$$

and for shunt transfer matrices

$$\mathcal{J}^{-1} = \begin{bmatrix} 1 & 0 \\ -\dfrac{1}{Z(s)} & 1 \end{bmatrix}. \qquad (3.22)$$

As illustrated in the previous section, the analyst is able to work through a ladder network in the *forward* direction (i.e., in the direction selected for positive current) by stringing together a set of transfer matrix products. Knowledge of Equations (3.20)–(3.22) provides the analyst with the ability to move *backward* through the network (i.e., in the direction counter to that selected for positive current). This ability allows the analyst to determine formulas for the *input impedance*, Z_I, at any point in the network. The input impedance is defined by

$$Z_I(s) = \frac{E(s)}{F(s)}, \qquad (3.23)$$

where E and F are the voltage and current at the point in question.

The final fact that must be known in order to determine Z_I is that a ladder network must end in a series impedance or in a shunt impedance (see Figure 3.4). In the former case the final voltage is zero, while in the latter case the

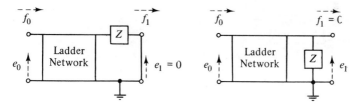

Figure 3.4. Two types of end conditions for a ladder network.

final current is zero, so that in any case one of the final power state variables is zero. The way in which the above facts are used to determine the input impedance is demonstrated by Examples 3.1 and 3.2.

Example 3.1

Determine the input impedance, E_0/F_0, at the input of the ladder network described in Figure 3.3. Notice that $E_1 = 0$ and work backward from this end condition by successive application of Equation (3.20) to find

$$\begin{bmatrix} 0 \\ F_1 \end{bmatrix} = \begin{bmatrix} 1 & -R_2 \\ 0 & 1 \end{bmatrix}\begin{bmatrix} 1 & -Ls \\ 0 & 1 \end{bmatrix}\begin{bmatrix} 1 & 0 \\ -Cs & 1 \end{bmatrix}\begin{bmatrix} 1 & -R_1 \\ 0 & 1 \end{bmatrix}\begin{bmatrix} E_0 \\ F_0 \end{bmatrix}.$$

After multiplication of the matrices, one finds

$$\begin{bmatrix} 0 \\ F_1 \end{bmatrix} = \begin{bmatrix} 1 + Cs(R_2 + Ls) & -R_1 - (R_2 + Ls)(1 + R_1Cs) \\ \cdots & \cdots \end{bmatrix}\begin{bmatrix} E_0 \\ F_0 \end{bmatrix},$$

where the dots denote matrix elements not needed to determine Z_I. The top equation is

$$0 = \{1 + Cs(R_2 + Ls)\}E_0 - \{R_1 + (R_2 + Ls)(1 + R_1Cs)\}F_0.$$

Notice that because of the zero end condition, only E_0 and F_0 appear in this equation. It follows immediately that

$$Z_I = \frac{E_0}{F_0} = \frac{R_1 + (R_2 + Ls)(1 + R_1C(s))}{1 + Cs(R_2 + Ls)}.$$

* * *

The significance of the input impedance is illustrated by the following problem: Suppose that the transfer function $E(s)/E_0(s)$ is to be determined, where $e(t) = \mathcal{L}^{-1}[E(s)]$ is the voltage at the intermediate position indicated in Figure 3.5. The analyst can work forward, with repeated application of

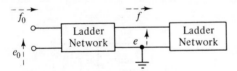

Figure 3.5. Ladder network wherein the output variable, $e(t)$, is at an intermediate position in the network.

Equation (3.13), to find

$$\begin{bmatrix} E_0(s) \\ F_0(s) \end{bmatrix} = \begin{bmatrix} T_{11}(s) & T_{12}(s) \\ T_{21}(s) & T_{22}(s) \end{bmatrix}\begin{bmatrix} E(s) \\ F(s) \end{bmatrix}. \tag{3.24}$$

The top equation reads

$$E_0(s) = T_{11}(s)E(s) + T_{12}(s)F(s). \tag{3.25}$$

This equation relates the output, $E(s)$, to the input, $E_0(s)$, but an extraneous power state variable, $F(s)$, is also present in the equation. However, Equation

(3.23) can be used to eliminate this variable; thus,

$$E_0(s) = T_{11}(s)E(s) + T_{12}(s)\frac{E(s)}{Z_i(s)}.$$

Thus, the transfer function

$$G(s) = \frac{E(s)}{E_0(s)} = \frac{Z_i(s)}{T_{11}(s)Z_i(s) + T_{12}(s)}. \tag{3.26}$$

The reader should be able to derive similar expressions for different definitions of input and output.

The presence of an extraneous variable in an expression such as Equation (3.25) can always be eliminated with the use of Equation (3.23) which quantifies the *loading effect*. The loading effect introduces the influence of that part of the network not intermediate between the input and the output into the transfer function dynamics.

Example 3.2

Denote the charge on the capacitor in the network illustrated in Figure 3.3 by $q(t)$ and its Laplace transform by $Q(s)$. Determine the transfer function $Q(s)/E_0(s)$.

Denote the power state variables just to the right of the capacitor by E, F. Since $CE = Q$, it follows that

$$\frac{Q(s)}{E_0(s)} = C\frac{E(s)}{E_0(s)}. \tag{i}$$

It is easily shown, by working forward, that

$$\begin{bmatrix} E_0 \\ F_0 \end{bmatrix} = \begin{bmatrix} 1 + R_1Cs & R_1 \\ \cdots & \cdots \end{bmatrix}\begin{bmatrix} E \\ F \end{bmatrix},$$

so that

$$E_0 = (1 + R_1Cs)E + R_1F. \tag{ii}$$

This expression is similar to Equation (3.25). Work backward to show that

$$\begin{bmatrix} 0 \\ F_1 \end{bmatrix} = \begin{bmatrix} 1 & -R_2 - Ls \\ \cdots & \cdots \end{bmatrix}\begin{bmatrix} E \\ F \end{bmatrix}.$$

Thus,

$$0 = E - (R_2 + Ls)F$$

or

$$F = \frac{1}{R_2 + Ls}E. \tag{iii}$$

Finally, combine Equations (i)–(iii) to find

$$\frac{Q(s)}{E_0(s)} = \frac{C(R_2 + Ls)}{(1 + R_1Cs)(R_2 + Ls) + R_1}.$$

$$* \quad * \quad *$$

3.3 Analysis of More General Types of Networks

In this section several methods for extending transfer matrix analysis to more general types of networks are presented. The simplest type of extension results from use of the simple rules of combining impedances, which are learned in elementary courses. If two impedances, Z_1 and Z_2, are in series, they may be replaced by the equivalent impedance

$$Z = Z_1 + Z_2. \tag{3.27}$$

If these impedances are in parallel, the equivalent impedance is defined by

$$\frac{1}{Z} = \frac{1}{Z_1} + \frac{1}{Z_2}. \tag{3.28}$$

The use of these relations is illustrated by example.

Example 3.3

Consider the network of Figure 3.6 and determine the transfer function

$$G(s) = \frac{E(s)}{F_0(s)}.$$

Notice that the capacitor and resistor, R_c, can be replaced with the single impedance

$$Z = \frac{1}{Cs} + R_c. \tag{3.29}$$

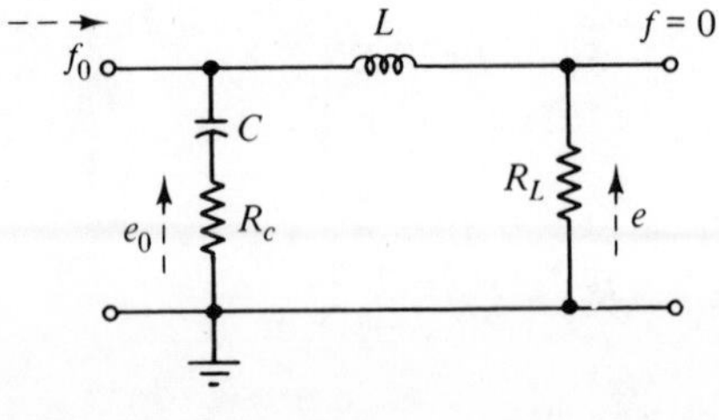

Figure 3.6. *R, L, C* network.

Thus,

$$\begin{bmatrix} E_0 \\ F_0 \end{bmatrix} = \begin{bmatrix} 1 & 0 \\ \dfrac{1}{Z} & 1 \end{bmatrix} \begin{bmatrix} 1 & Ls \\ 0 & 1 \end{bmatrix} \begin{bmatrix} 1 & 0 \\ \dfrac{1}{R_L} & 1 \end{bmatrix} \begin{bmatrix} E \\ 0 \end{bmatrix}$$

$$= \begin{bmatrix} \cdots & \cdots \\ \dfrac{1}{Z} + \dfrac{1}{R_L}\left\{ 1 + \dfrac{Ls}{Z} \right\} & \cdots \end{bmatrix} \begin{bmatrix} E \\ 0 \end{bmatrix}.$$

Thus,

$$F_0 = \left[\frac{R_L + Z + Ls}{R_L Z} \right] E.$$

Find

$$G(s) = \frac{E(s)}{F_0(s)} = \frac{R_L(1 + R_c Cs)}{1 + R_c Cs + Cs(R_L + Ls)}. \tag{3.30}$$

* * *

Example 3.4

The network indicated in Figure 3.7 is very important in feedback control systems applications. Find

$$G(s) = \frac{E(s)}{E_0(s)}.$$

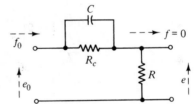

Figure 3.7. *Lead* compensator network.

It follows from Equation (3.28) that the R_c, C combination may be replaced with the impedance

$$Z = \frac{R_c}{1 + R_c Cs}.$$

Thus,

$$\begin{bmatrix} E_0 \\ F_0 \end{bmatrix} = \begin{bmatrix} 1 & Z \\ 0 & 1 \end{bmatrix} \begin{bmatrix} 1 & 0 \\ \dfrac{1}{R} & 1 \end{bmatrix} \begin{bmatrix} E \\ 0 \end{bmatrix}$$

$$= \begin{bmatrix} 1 + \dfrac{Z}{R} & \cdots \\ \cdots & \cdots \end{bmatrix} \begin{bmatrix} E \\ 0 \end{bmatrix}.$$

Thus,

$$E_0 = \left\{ 1 + \frac{R_C/R}{1 + R_C C s} \right\} E.$$

Show that

$$G(s) = \frac{E(s)}{E_0(s)} = \frac{R(1 + R_C C s)}{R + R_C + R R_C C s}. \tag{3.31}$$

* * *

One may also replace combinations of ladder networks with an equivalent ladder network.[6] An example of the equations of equivalent ladder networks is now provided. Consider the networks connected as in Figure 3.8. Let $\mathcal{F}$,

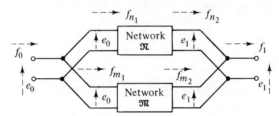

Figure 3.8. *Parallel* connection of networks.

$\mathcal{N}$, $\mathcal{M}$ be transfer matrices such that

$$\begin{bmatrix} E_0 \\ F_0 \end{bmatrix} = \mathcal{F} \begin{bmatrix} E_1 \\ F_1 \end{bmatrix}, \tag{3.32a}$$

$$\begin{bmatrix} E_0 \\ F_{n_1} \end{bmatrix} = \mathcal{N} \begin{bmatrix} E_1 \\ F_{n_2} \end{bmatrix}, \tag{3.32b}$$

and

$$\begin{bmatrix} E_0 \\ F_{m_1} \end{bmatrix} = \mathcal{M} \begin{bmatrix} E_1 \\ F_{m_2} \end{bmatrix}. \tag{3.32c}$$

The important task is to relate the elements of $\mathcal{F}$ to the elements of $\mathcal{M}$, $\mathcal{N}$. Write Equation (3.32b) in component form and then show that

$$F_{n_1} = \frac{N_{22}}{N_{12}} E_0 - \frac{1}{N_{12}} E_1 \tag{3.33a}$$

and

$$F_{n_2} = \frac{1}{N_{12}} E_0 - \frac{N_{11}}{N_{12}} E_1. \tag{3.33b}$$

(*Note:* Use det $\mathfrak{N} = 1$). Analogously,

$$F_{m_1} = \frac{M_{22}}{M_{12}} E_0 - \frac{1}{M_{12}} E_1, \tag{3.34a}$$

$$F_{m_2} = \frac{1}{M_{12}} E_0 - \frac{M_{11}}{M_{12}} E_1. \tag{3.34b}$$

Use the facts that

$$F_0 = F_{n_1} + F_{m_1} \tag{3.35}$$

$$F_1 = F_{n_2} + F_{m_2} \tag{3.36}$$

to show that

$$F_0 = \left\{\frac{N_{22}}{N_{12}} + \frac{M_{22}}{M_{12}}\right\} E_0 - \left\{\frac{1}{N_{12}} + \frac{1}{M_{12}}\right\} E_1, \tag{3.37a}$$

$$F_1 = \left\{\frac{1}{N_{12}} + \frac{1}{M_{12}}\right\} E_0 - \left\{\frac{N_{11}}{N_{12}} + \frac{M_{11}}{M_{12}}\right\} E_1. \tag{3.37b}$$

If these equations are compared with similar equations derived from Equation (3.32a), it can be shown (after a good deal of algebra) that

$$T_{11} = \frac{N_{11}M_{12} + M_{11}N_{12}}{N_{12} + M_{12}}, \qquad T_{12} = \frac{N_{12}M_{12}}{N_{12} + M_{12}}, \tag{3.38a}$$

$$T_{21} = \frac{(N_{11} - M_{11})(M_{22} - N_{22})}{N_{12} + M_{12}}, \qquad T_{22} = \frac{N_{12}M_{22} + M_{12}N_{22}}{N_{12} + M_{12}}. \tag{3.38b}$$

The use of these relations is now demonstrated by example.

Example 3.5

The network of Figure 3.9 has some interesting properties which the reader who has solved Problem 2.7 (see Figure 2.29) can affirm. Determine

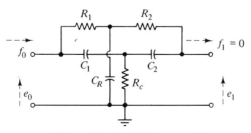

Figure 3.9. *Notch filter* network.

the transfer function $G(s) = E_1(s)/E_0(s)$. Begin by noting that some transfer matrix $\mathfrak{I}$ exists such that

$$\begin{bmatrix} E_0 \\ F_0 \end{bmatrix} = \begin{bmatrix} T_{11} & T_{12} \\ T_{21} & T_{22} \end{bmatrix} \begin{bmatrix} E_1 \\ 0 \end{bmatrix}.$$

Thus,

$$E_0 = T_{11} E_1$$

or

$$G(s) = \frac{E_1(s)}{E_0(s)} = \frac{1}{T_{11}(s)}. \tag{i}$$

Equation (3.38a) will be used to complete the solution of the problem. For the upper network, the transfer matrix

$$\begin{aligned} \mathfrak{N} &= \begin{bmatrix} 1 & R_1 \\ 0 & 1 \end{bmatrix} \begin{bmatrix} 1 & 0 \\ C_R s & 1 \end{bmatrix} \begin{bmatrix} 1 & R_2 \\ 0 & 1 \end{bmatrix} \\ &= \begin{bmatrix} 1 + R_1 C_R s & R_2(1 + R_1 C_R s) + R_1 \\ \cdots & \cdots \end{bmatrix}. \end{aligned} \tag{ii}$$

For the lower network, the transfer matrix

$$\begin{aligned} \mathfrak{M} &= \begin{bmatrix} 1 & \dfrac{1}{C_1 s} \\ 0 & 1 \end{bmatrix} \begin{bmatrix} 1 & 0 \\ \dfrac{1}{R_C} & 1 \end{bmatrix} \begin{bmatrix} 1 & \dfrac{1}{C_2 s} \\ 0 & 1 \end{bmatrix} \\ &= \begin{bmatrix} 1 + \dfrac{1}{R_C C_1 s} & \dfrac{1}{C_2 s}\left\{1 + \dfrac{1}{R_C C_1 s}\right\} + \dfrac{1}{C_1 s} \\ \cdots & \cdots \end{bmatrix}. \end{aligned} \tag{iii}$$

Combine Equations (i)–(iii) with Equation (3.38a) to show that

$$G(s) = \frac{1 + R_C(C_1 + C_2)s + R_C C_1 C_2 s^2\{R_1 + R_2(1 + R_1 C_R s)\}}{(1 + R_1 C_R s)(1 + R_C[C_1 + C_2]s) + C_2 s(1 + R_C C_1 s)\{R_1 + R_2(1 + R_1 C_R s)\}}.$$

An interesting choice of parameter values is

$$R_1 = R = R_2, \tag{3.39a}$$

$$C_1 = C = C_2, \tag{3.39b}$$

and

$$2R_C C = \frac{RC_R}{2}. \tag{3.40}$$

For these parameter values

$$G(s) = \frac{1 + 2R_CRC^2s^2}{1 + (RC_R + 2RC)s + 2R_CRC^2s^2} \tag{3.41}$$

because of the cancellation of a zero with a pole.

*   *   *

3.4 Transformers and Gyrators

There are two other types of passive elements that must be introduced in the discussion of dynamic systems analysis: the transformer and the gyrator. These elements will first be introduced by describing their transfer matrices. A discussion of the physical realizability of the elements is postponed to the section on magnetic circuits. For reasons that will become clear, the time domain transfer matrices will be defined.

The transformer transfer matrix satisfies

$$\begin{bmatrix} e_0(t) \\ f_0(t) \end{bmatrix} = \begin{bmatrix} n(t) & 0 \\ 0 & \dfrac{1}{n(t)} \end{bmatrix} \begin{bmatrix} e_1(t) \\ f_1(t) \end{bmatrix}. \tag{3.42}$$

Since the *transformer parameter*, $n(t)$, can vary with time, there is, in general, no Laplace domain equivalent of Equation (3.42). The power supplied to the first pair of terminals of the transformer is $e_0 f_0$, and the power supplied to the second pair of terminals of the transformer is $e_1 f_1$. Thus, the net power supplied by the transformer itself is

$$e_0 f_0 - e_1 f_1 = n(t)e_1 \frac{f_1}{n(t)} - e_1 f_1 = 0. \tag{3.43}$$

Thus, the transformer does not store or supply energy even on an instantaneous basis.

For the special, but not unusual, case when n is a constant

$$\begin{bmatrix} E_0(s) \\ F_0(s) \end{bmatrix} = \begin{bmatrix} n & 0 \\ 0 & \dfrac{1}{n} \end{bmatrix} \begin{bmatrix} E_1(s) \\ F_1(s) \end{bmatrix}. \tag{3.44}$$

For this case the analysis of circuits proceeds exactly as in previous sections.

The gyrator was first discussed by Tellegen.[9] The transfer matrix of this element satisfies

$$\begin{bmatrix} e_0(t) \\ f_0(t) \end{bmatrix} = \begin{bmatrix} 0 & g(t) \\ \dfrac{1}{g(t)} & 0 \end{bmatrix} \begin{bmatrix} e_1(t) \\ f_1(t) \end{bmatrix},$$

(3.45)

where $g(t)$ is the *gyrator parameter*. For the special case when g is a constant

$$\begin{bmatrix} E_0(s) \\ F_0(s) \end{bmatrix} = \begin{bmatrix} 0 & g \\ \dfrac{1}{g} & 0 \end{bmatrix} \begin{bmatrix} E_1(s) \\ F_1(s) \end{bmatrix}.$$

(3.46)

The circuit symbol commonly used for the gyrator is illustrated in Figure 3.10.

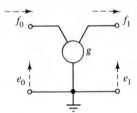

Figure 3.10. Symbol for the gyrator.

It is easily shown that for the transformer

$$\boldsymbol{\mathcal{T}}^{-1} = \begin{bmatrix} \dfrac{1}{n} & 0 \\ 0 & n \end{bmatrix}$$

(3.47)

and for the gyrator

$$\boldsymbol{\mathcal{T}}^{-1} = \begin{bmatrix} 0 & g \\ \dfrac{1}{g} & 0 \end{bmatrix}.$$

(3.48)

Example 3.6

Determine the input impedance of the circuit illustrated in Figure 3.11. Working backward,

$$\begin{bmatrix} E_1 \\ 0 \end{bmatrix} = \begin{bmatrix} 1 & 0 \\ -Cs & 1 \end{bmatrix} \begin{bmatrix} 0 & g \\ \dfrac{1}{g} & 0 \end{bmatrix} \begin{bmatrix} E_0 \\ F_0 \end{bmatrix} = \begin{bmatrix} \cdots & \cdots \\ \dfrac{1}{g} & -gCs \end{bmatrix} \begin{bmatrix} E_0 \\ F_0 \end{bmatrix}.$$

Thus,

$$0 = \frac{E_0}{g} - gCsF_0,$$

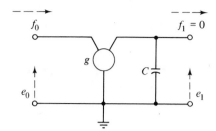

Figure 3.11. Gyrated impedance.

whence

$$Z_I = \frac{E_0}{F_0} = g^2 C s. \tag{3.49}$$

Thus, the inclusion of a gyrator has converted a capacitive impedance into an inductive impedance.

* * *

The above example illustrates one motivation for studying and synthesizing gyrators. Since gyrator constants can be fairly large ($\sim 10^8$ ohms), an extremely large inductive impedance can be realized.

The name *gyrator* is derived from the word gyroscope. As will be shown, there are many important mechanical and electromechanical devices described by gyrator transfer matrices.

3.5 Charge, Impulse, and the *Tetrahedron of State*[5]

Define the charge or displacement variable, q, by

$$q = \int f(t)\, dt \tag{3.50}$$

and the impulse variable, λ, by

$$\lambda = \int e(t)\, dt. \tag{3.51}$$

The four variables $e, f, q,$ and λ form what Paynter has called the *tetrahedron of state*.

Notice that the basic axioms [Equations (3.1)–(3.3)] can be rewritten

$$e_1 - e_2 = Rf_2, \tag{3.52}$$

$$q = c(e_1 - e_2), \tag{3.53}$$

and

$$\lambda = Lf_2, \tag{3.54}$$

where $\lambda = \int (e_1 - e_2)\, dt$. Thus, all of the basic axioms are *algebraic* equations when written in terms of variables of the tetrahedron of state.

In certain cases, the variables λ or q are the output variable rather than a voltage or a current. This fact is one reason for introducing these variables. Another reason is that nonlinear relations enter a dynamic systems analysis via generalizations of Equations (3.52)–(3.54). These generalizations must, therefore, be the starting point of nonlinear analysis. Symbolically, the generalization of the basic axioms are

$$e_1 - e_2 = \phi_r(f_2), \tag{3.55}$$

$$q = \phi_c(e_1 - e_2), \tag{3.56}$$

and

$$\lambda = \phi_L(f_2), \tag{3.57}$$

where ϕ_r, ϕ_c, and ϕ_L are nonlinear functions.

3.6 The Tetrahedron of State and the Basic Axioms of Magnetic Circuits[1,2]

In this section, the methods used to analyze electrical circuits in previous sections will be extended to the analysis of magnetic circuits. This procedure will be repeated in succeeding chapters for mechanical, hydraulic, and heat transfer networks. In all cases, the procedure will be the same: (1) define the tetrahedron of state; (2) state the basic axioms; and (3) describe the transformers and gyrators associated with the type of network being discussed.

In this text, the definition of the tetrahedron of state for magnetic circuits will conform with the unconventional definitions adopted by Buntenbach.[1] Define the charge (displacement) variable to be

$$q_m(t) = \text{magnetic flux.} \tag{3.58}$$

It follows from Equation (3.50) that the *flow* variable, $f(t)$, is given by

$$f_m(t) = \text{time rate of change of magnetic flux.} \tag{3.59}$$

The definitions of $e_m(t)$ and $\lambda_m(t)$ are not quite as straightforward. The *magnetomotive force*, $e_m(t)$, is defined by

$$f_m(t)e_m(t) = p_m(t), \qquad (3.60)$$

where $p_m(t)$ is the instantaneous power in the magnetic circuit. The impulse, $\lambda_m(t)$ is defined as in Equation (3.51).

Before the basic axioms are presented, it will be useful to explore the nature of the interconnection of electrical and magnetic circuits, illustrated in Figure 3.12. *Faraday's law*[3] (axiom) states that the voltage, $e(t)$, equals

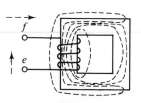

Figure 3.12. Electrical circuit interfaced to a magnetic circuit by a coil with N turns. The dashed lines are magnetic field lines.

the rate of change of flux linking the circuit, or, in terms of the variables defined above,

$$e(t) = Nf_m(t). \qquad (3.61)$$

Consider the ideal case when the electrical power is converted, without dissipation, to the magnetic circuit. It follows from Equation (3.60) that

$$f_m(t)e_m(t) = f(t)e(t). \qquad (3.62)$$

Substitute Faraday's law to find

$$f_m(t)e_m(t) = f(t)Nf_m(t),$$

so that current is simply related to magnetomotive force by

$$f(t) = \frac{1}{N}e_m(t).$$

This last equation can be summarized along with Equation (3.61) in the matrix equation

$$\begin{bmatrix} e(t) \\ f(t) \end{bmatrix} = \begin{bmatrix} 0 & N \\ \frac{1}{N} & 0 \end{bmatrix} \begin{bmatrix} e_m(t) \\ f_m(t) \end{bmatrix}. \qquad (3.63)$$

Thus, the interconnection is described by a transfer matrix of the gyrator type with the gyrator constant equal to the number of turns of the coil.

A conceptual difficulty associated with stating the axioms of magnetic circuits is that the magnetic properties of the circuit material are distributed throughout the circuit and not conveniently lumped in discrete elements as in most electric circuits. For this reason, magnetic circuits are properly described by partial differential equations[1] rather than by ordinary differential equations. Discussion of *distributed parameter* systems (systems described by partial differential equations) is delayed until Chapter 4 and axioms of *lumped* magnetic circuit parameters, which may be used in the synthesis of approximate models, are discussed in the present chapter.

The flux, q, in a magnetic circuit is a monotone increasing function of the applied magnetomotive force (hence, to the electrical current). For a limited range, this function is linear so that

$$q_m(t) = Pe_m(t). \tag{3.64}$$

The proportionality constant, P, is called *permeance* and can be related to geometrical and magnetic properties of the circuit material.[1] Compare Equations (3.64) and (3.53) and conclude that the above axiom describes a magnetic capacitor. Differentiate Equation (3.64) to find

$$\frac{de_m(t)}{dt} = \frac{1}{P}f_m(t).$$

In the Laplace domain

$$sE_m(s) = \frac{1}{P}F_m(s),$$

so that the magnetic impedance

$$Z_m(s) = \frac{E_m(s)}{F_m(s)} = \frac{1}{Ps}. \tag{3.65}$$

Example 3.7

Determine the electrical input impedance for the electromagnetic system illustrated in Figure 3.12. Assume that the magnetic circuit is described completely by Equation (3.64). Use Equations (3.63) and (3.65). Work backward to find

$$\begin{bmatrix} E_m \\ 0 \end{bmatrix} = \begin{bmatrix} 1 & 0 \\ -Ps & 1 \end{bmatrix} \begin{bmatrix} 0 & N \\ \dfrac{1}{N} & 0 \end{bmatrix} \begin{bmatrix} E \\ F \end{bmatrix}$$

$$= \begin{bmatrix} \cdots & \cdots \\ \dfrac{1}{N} & -NPs \end{bmatrix} \begin{bmatrix} E \\ F \end{bmatrix}.$$

Thus,

$$0 = \frac{E}{N} - NPsF$$

or

$$Z(s) = \frac{E(s)}{F(s)} = N^2 Ps, \tag{3.66}$$

which is an inductive impedance for which

$$L = N^2 P. \tag{3.67}$$

* * *

If the flux is varying in a magnetic circuit, electrical *eddy currents* will be induced in any neighboring electrical conductors.[3] From *Lenz's rule*[3] it follows that these eddy currents in turn generate magnetic fields which oppose those that generate the currents. The net effect is a loss or drop in magneto-motive force. This drop is a monotone function of the rate of change of flux. For a limited range, the function is linear so that

$$e_{m_2}(t) - e_{m_1}(t) = R_m f_m(t). \tag{3.68}$$

The proportionality constant, R_m, is called *magnetic resistance*.

It should not be difficult for the reader to demonstrate that this rule defines a resistive impedance

$$Z(s) = R_m. \tag{3.69}$$

The energy dissipation associated with this resistance is, of course, a result of the eddy currents.

Example 3.8

For the system illustrated in Figure 3.13, determine the transfer function $E_1(s)/F_0(s)$. Begin by "lumping" the magnetic circuit as a series resistance, R_m, and a shunt capacitor, P. Then

$$\begin{bmatrix} E_0 \\ F_0 \end{bmatrix} = \begin{bmatrix} 0 & N_1 \\ \frac{1}{N_1} & 0 \end{bmatrix} \begin{bmatrix} 1 & R_m \\ 0 & 1 \end{bmatrix} \begin{bmatrix} 1 & 0 \\ Ps & 1 \end{bmatrix} \begin{bmatrix} 0 & N_2 \\ \frac{1}{N_2} & 0 \end{bmatrix} \begin{bmatrix} E_1 \\ 0 \end{bmatrix},$$

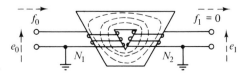

Figure 3.13. Electromagnetic system.

so that

$$\begin{bmatrix} E_0 \\ F_0 \end{bmatrix} = \begin{bmatrix} \cdots & \cdots \\ \dfrac{R_m}{N_1 N_2} & \cdots \end{bmatrix} \begin{bmatrix} E_1 \\ 0 \end{bmatrix}.$$

Thus,

$$F_0 = \frac{R_m}{N_1 N_2} E_1$$

or

$$\frac{E_1(s)}{F_0(s)} = \frac{N_1 N_2}{R_m}.$$

* * *

References

1. BUNTENBACH, R. W., "Improved Circuit Models for Inductors Wound on Dissipative Magnetic Cores," in *Conference Report, Second Asilomar Conference on Circuits and Systems*, I.E.E.E., Asilomar, Calif., Oct. 1968.

2. CRANDALL, S., KARNOPP, D., et al., *Dynamics of Mechanical and Electromechanical Systems*, McGraw-Hill, New York, 1968.

3. FEYNMAN, R., LEIGHTON, R., and SANDS, M., *The Feynman Lectures on Physics*, Vol. II: *Mainly Electromagnetism and Matter*, Addison-Wesley, Reading, Mass., 1964.

4. KARNOPP, D., and ROSENBERG, R., *Analysis and Simulation of Multiport Systems*, M.I.T. Press, Cambridge, Mass., 1968.

5. PAYNTER, H. M., *Analysis and Design of Engineering Systems*, M.I.T. Press, Cambridge, Mass., 1961.

6. PIPES, L. A., *Matrix Methods for Engineering*, Prentice-Hall, Englewood Cliffs, N.J., 1963.

7. SHEARER, J., MURPHY, A., and RICHARDSON, H., *Introduction to System Dynamics*, Addison-Wesley, Reading, Mass., 1967.

8. TAKAHASHI, Y., RABINS, M. J., and AUSLANDER, D. M., *Control*, Addison-Wesley, Reading, Mass., 1970.

9. TELLEGEN, B. D. H., "The Gyrator, a New Electric Circuit Element," *Philips Research Reports*, *3*, No. 2, Apr. 1948, pp. 81–101.

Problems

3.1. Show that the determinant of all transfer matrices introduced in this chapter is 1.0.

3.2. Prove the validity of Equations (3.21) and (3.22).

3.3. Derive an equation similar to Equation (3.26) for the transfer function $F(s)/E_0(s)$.

3.4. For the circuit illustrated in Figure 3.3, determine the transfer function $F(s)/F_0(s)$, where $F(s)$ is the current through the inductor.

3.5. Determine the input impedance, $E_0(s)/F_0(s)$, for the circuit illustrated in Figure 3.6 for the special case

$$R_C C = \frac{L}{R_L}. \tag{3.70}$$

Repeat the derivation, but assume in addition that

$$R_L = R_C. \tag{3.71}$$

3.6. Find the transfer function $E_1(s)/E_0(s)$ for the network illustrated in Figure 3.14. Assume that $F_1 = 0$.

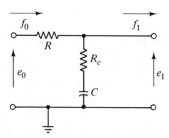

Figure 3.14. *Lag* compensator.

3.7. Find $e_1(t)$ when $e_0(t) = u(t)$, $e_1(0^-) = 0$, and

$$R_C = R = 10^6 \text{ ohms,}$$
$$C = 10^{-6} \text{ farad}$$

for both the lead compensator (Figure 3.7, Example 3.3) and the lag compensator (Problem 3.6).

3.8. Determine the transfer function, $G(s) = E_1(s)/E_0(s)$ for the circuit illustrated in Figure 3.15.

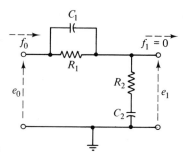

Figure 3.15. *Lag-lead* compensator.

3.9. Consider the measurement of acceleration using piezoelectricity (Figure 3.16).

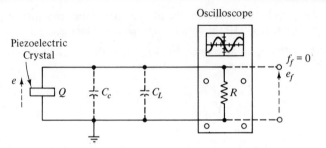

Figure 3.16. Acceleration measurement equipment.

The crystal is mounted on a structural member whose acceleration is to be measured. You may assume that the *charge* generated by the crystal

$$q(t) = ka(t);$$

k is a constant and a is the acceleration. The crystal has some inherent capacitance C_C. The crystal is connected to an oscilloscope with resistive input impedance, R, by a shielded two-conductor wire which has inherent capacitance C_L. Find the transfer function $E_f(s)/A(s)$.

3.10. In the previous example (Problem 3.9), suppose that $A(s) = 1/s$. Find $e_f(t)$. Plot $[(C_C + C_L)/k]e_f$ vs. $t/R(C_L + C_C)$. Is the system illustrated in Figure 3.16 a wise choice for measurement of constant acceleration?

3.11. The *dual circuit* of a given circuit is defined as a circuit whose *currents* are equal to the voltages of the original circuit at *all* instants of time. The same is true of the voltages in the dual circuit and the *currents* in the given circuit. Examination of the transfer matrix leads directly to a definition of dual elements of a passive ladder. For instance, consider the *series* resistor (Figure 3.1) for which the transfer matrix equation is

$$\begin{bmatrix} E_1 \\ F_1 \end{bmatrix} = \begin{bmatrix} 1 & R \\ 0 & 1 \end{bmatrix} \begin{bmatrix} E_2 \\ F_2 \end{bmatrix}.$$

The corresponding element in a dual circuit must, therefore, satisfy the relationship

$$\begin{bmatrix} E_{1d} \\ F_{1d} \end{bmatrix} = \begin{bmatrix} 1 & 0 \\ R & 1 \end{bmatrix} \begin{bmatrix} E_{2d} \\ F_{2d} \end{bmatrix}.$$

Thus the dual element of a series resistor, R, is a shunt resistor of magnitude $1/R$.

(a) Find dual elements for shunted *and* series resistors, capacitors, and inductors.

(b) Find the duals for series transformers and gyrators.

(c) Then construct the dual of the circuit illustrated in Figure 3.3.

3.12. Consider the circuit illustrated in Figure 3.17. Derive a differential equation for the charge $q_1(t)$ on the capacitor C_1 in this circuit. Solve the equation for

$$C_1 = C_2 = 10^{-6} \text{ farad,}$$
$$g = 10^8 \text{ ohms,}$$
$$q_1(0^-) = 10^{-6} \text{ coulomb,}$$
$$q_2(0^-) = 0.$$

Caution: Notice that, for the sign convention adopted in Figure 3.17,

$$-C_2 e_2 = \int f_2(t) \, dt.$$

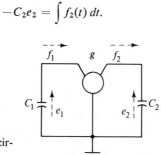

Figure 3.17. Unforced *gyroscope* circuit.

3.13. Assume that the magnetic circuit within the system illustrated in Figure 3.13 has permeance but negligible resistance. Show that an equivalent all-electrical system is an inductance and a transformer with transformer constant $n = N_1/N_2$.

Mechanical Network Systems: The Lumping Approximation and Rotary Machine Systems

4

The analysis techniques introduced in this chapter are useful in the analysis of instruments, rotary machines and other mechanical devices in which the kinematic relationships are not complex.

The first task to complete in order to be able to apply the techniques of Chapter 3 to mechanical systems is to define the e, f, λ, and q variables (the tetrahedron of state). In this text, the *effort* variable

$$e \triangleq \text{Force}, \qquad (4.1a)$$

and the *flow* variable

$$f \triangleq \text{Velocity}. \qquad (4.1b)$$

These definitions are arbitrary and are not universally accepted.[8] The exploration of the ramifications of different assignments of e and f variables will not be discussed here because of space limitations. As before, the impulse and displacement variables are defined by

$$\lambda = \int e(t)\, dt, \qquad (4.2)$$

$$q = x = \int f(t)\, dt. \qquad (4.3)$$

The more conventional symbol, x, will be used in the remainder of the text to denote mechanical displacement.

4.1 The Elementary Axioms for Networks of Mechanical Elements in Translation

The procedure to be followed is to define Laplace domain impedances and transfer matrices for the passive elements of mechanical systems. First, consider inertia (Figure 4.1). *Note: e_2 is the force exerted by the mass. e_1 is*

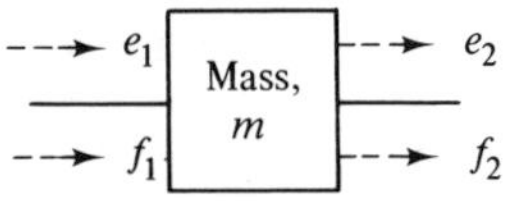

Figure 4.1. Inertia.

the force applied *to* the mass element. The dashed lines indicate sign convention. Notice that

$$f_1 = f_2. \qquad (4.4)$$

Notice that a reaction force equal and opposite to e_2 is applied to the mass and write Newton's second law in the Laplace domain to find

$$E_1 = E_2 + msF_2 \qquad (4.5)$$

(since sF_2 is the Laplace transform of the acceleration). Therefore, the impedance is given by

$$Z(s) = \frac{E_1 - E_2}{F_2} = ms. \qquad (4.6)$$

The transfer matrix equation implied by Equations (4.4) and (4.5) is

$$\mathbf{P}_1 = \begin{bmatrix} 1 & ms \\ 0 & 1 \end{bmatrix} \mathbf{P}_2. \qquad (4.7)$$

Thus, mass in a mechanical system is analogous to inductance in an electrical system. This analogy is a result of the definition of Equations (4.1a) and (4.1b).

An important point is that there is no mechanical analogy to a shunted inductance. The reader should verify that Equation (4.7) also applies to the inertia element illustrated in Figure 4.2. A beginning student often will make

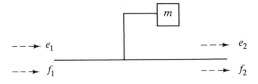

Figure 4.2. Inertia element which might, mistakenly, be taken to be analogous to a shunt inductance.

the mistake of describing such an element with a matrix for a shunted inductance. A mistake will never be made if one asks himself, What variable drops across the element? E drop implies *series* impedance and F drop implies *shunt* impedance. This exercise need not be performed with electric circuit elements because, in that case, it is always clear whether an element is in series or shunt.

The next element is a *massless* spring for which there is an e drop (Figure 4.3); thus, this spring is described as a *series* impedance. k is a proportionality

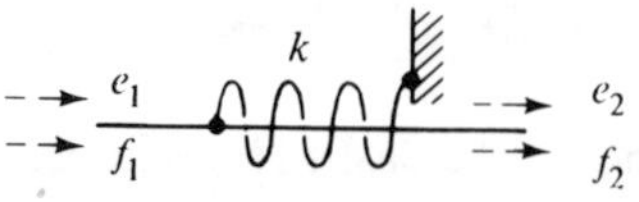

Figure **4.3.** Hooke's law: series spring. Note the convention described in Figure 4.1.

constant in Hooke's law:

$$e_1 - e_2 = k \int_0^t f(\xi)\, d\xi.$$

In the Laplace domain

$$E_1 = E_2 + \frac{k}{s} F_2,$$

so that the series transfer matrix equation is

$$\begin{bmatrix} E_1 \\ F_1 \end{bmatrix} = \begin{bmatrix} 1 & \dfrac{k}{s} \\ 0 & 1 \end{bmatrix} \begin{bmatrix} E_2 \\ F_2 \end{bmatrix}. \tag{4.8}$$

The impedance is

$$Z = \frac{k}{s}, \tag{4.9}$$

which is analogous to a capacitor of capacitance $1/k$.

Consider the arrangement in Figure 4.4. For a massless spring Newton's second law reads

$$e_1 = e_2.$$

However,

$$f_1 \neq f_2;$$

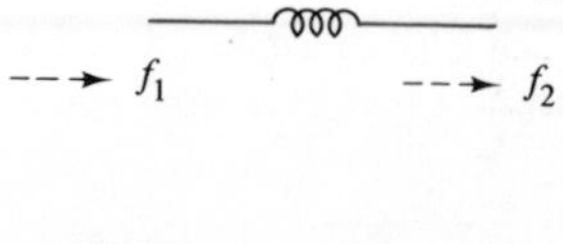

Figure **4.4.** Shunted spring.

thus, an f drop implies *shunt* impedance. The appropriate transfer matrix is

$$
\mathfrak{J} = \begin{bmatrix} 1 & 0 \\ \dfrac{1}{Z} & 1 \end{bmatrix} = \begin{bmatrix} 1 & 0 \\ \dfrac{s}{k} & 1 \end{bmatrix}. \tag{4.10}
$$

The last element to be considered in this section is the dashpot (Figure 4.5). It will be shown in the next chapter that the volume flow in the pipe is

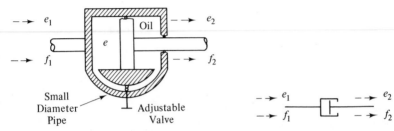

Figure 4.5. Dashpot. **Figure 4.6.** Dashpot symbol.

given by

$$
f_p = \frac{A^2 e}{b}. \tag{4.11}
$$

A is the area of the piston, b is a constant, and e is the oil pressure in the cylinder. b is called the dashpot constant and is a function of the valve opening. The symbol for a dashpot is demonstrated in Figure 4.6. If mass is neglected, the second law reads

$$
e_1 = e_2, \tag{4.12}
$$

which indicates that the dashpot is a shunted impedance.

From the principle of the conservation of mass, it follows that

$$
Af_1 = Af_2 + f_p = Af_2 + \frac{A^2 e}{b},
$$

where A is the piston area. Thus,

$$
f_1 = f_2 + \frac{Ae}{b} = f_2 + \frac{e_2}{b}. \tag{4.13}
$$

Thus, the matrix equation relating power states is

$$
\begin{bmatrix} E_1 \\ F_1 \end{bmatrix} = \begin{bmatrix} 1 & 0 \\ \dfrac{1}{b} & 1 \end{bmatrix} \begin{bmatrix} E_2 \\ F_2 \end{bmatrix}. \tag{4.14}
$$

Obviously, the dashpot is analogous to a resistor of resistance

$$R = b. \tag{4.15}$$

The series arrangement is illustrated in Figure 4.7. For this arrangement

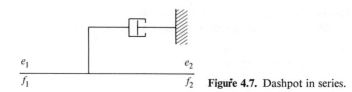

e_1 e_2

f_1 f_2 **Figure 4.7.** Dashpot in series.

Table 4.1

ELEMENTARY IMPEDANCES OF MECHANICAL TRANSLATION NETWORKS

Series resistance	*Series capacitance*	*Series inductance*
$\mathfrak{I} = \begin{pmatrix} 1 & R \\ 0 & 1 \end{pmatrix}$	$\mathfrak{I} = \begin{pmatrix} 1 & \dfrac{1}{Cs} \\ 0 & 1 \end{pmatrix}$	$\mathfrak{I} = \begin{pmatrix} 1 & Ls \\ 0 & 1 \end{pmatrix}$

Shunt resistance	*Shunt capacitance*	*Shunt inductance*
$\mathfrak{I} = \begin{pmatrix} 1 & 0 \\ \dfrac{1}{R} & 1 \end{pmatrix}$	$\mathfrak{I} = \begin{pmatrix} 1 & 0 \\ Cs & 1 \end{pmatrix}$	$\mathfrak{I} = \begin{pmatrix} 1 & 0 \\ \dfrac{1}{Ls} & 1 \end{pmatrix}$

$$f_1 = f_2,$$
$$e_1 = e_2 + bf_2,$$
$$\begin{bmatrix} E_1 \\ F_1 \end{bmatrix} = \begin{bmatrix} 1 & b \\ 0 & 1 \end{bmatrix} \begin{bmatrix} E_2 \\ F_2 \end{bmatrix}. \tag{4.16}$$

The material presented in this section is summarized in Table 4.1.

Example 4.1

For the network illustrated in Figure 4.8, find

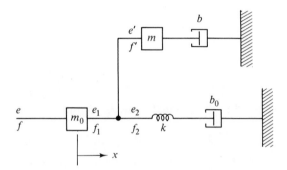

Figure 4.8. Network of translated mechanical elements.

$$\frac{X(s)}{E(s)}.$$

The network will first be reduced to the equivalent network illustrated in Figure 4.9 by finding an appropriate $Z_0(s)$. Since

$$f' = f_1 = f_2,$$

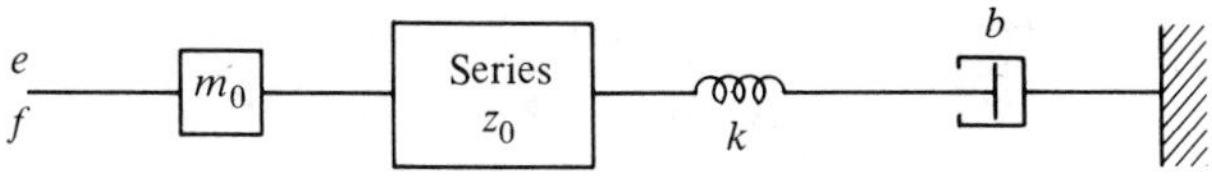

Figure 4.9. Network equivalent to that illustrated in Figure 4.8.

it follows that Z_0 is a series impedance.

To find Z_0, notice that $e' = e_1 - e_2$; thus,

$$Z_0 \triangleq \frac{E_1 - E_2}{F_2} = \frac{E'}{F'};$$

i.e., Z_0 is an input impedance. The upper network is redrawn in Figure 4.10.

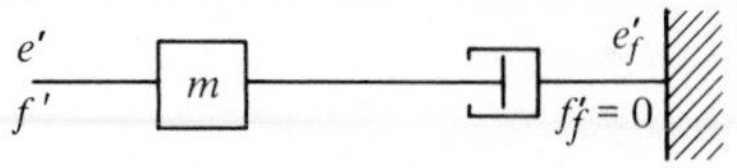

Figure 4.10. Upper network of Figure 4.8.

Notice $f'_f = 0$ and work backward:

$$\begin{bmatrix} E'_f \\ 0 \end{bmatrix} = \begin{bmatrix} 1 & 0 \\ -\dfrac{1}{b} & 1 \end{bmatrix} \begin{bmatrix} 1 & -ms \\ 0 & 1 \end{bmatrix} \begin{bmatrix} E' \\ F' \end{bmatrix}$$

$$= \begin{bmatrix} 1 & -ms \\ \dfrac{1}{b} & 1 + \dfrac{ms}{b} \end{bmatrix} \begin{bmatrix} E' \\ F' \end{bmatrix}.$$

From the bottom equation, it follows that

$$Z_0 = \frac{E'}{F'} = b + ms. \tag{i}$$

Now work backward in the original network to find

$$\begin{bmatrix} E_f \\ 0 \end{bmatrix} = \begin{bmatrix} 1 & 0 \\ -\dfrac{1}{b_0} & 1 \end{bmatrix} \begin{bmatrix} 1 & 0 \\ -\dfrac{s}{k} & 1 \end{bmatrix} \begin{bmatrix} 1 & -Z_0 \\ 0 & 1 \end{bmatrix} \begin{bmatrix} 1 & -m_0 s \\ 0 & 1 \end{bmatrix} \begin{bmatrix} E \\ F \end{bmatrix}$$

$$= \begin{bmatrix} 1 & 0 \\ -\left(\dfrac{1}{b_0} + \dfrac{s}{k}\right) & 1 \end{bmatrix} \begin{bmatrix} 1 & -(m_0 s + Z_0) \\ 0 & 1 \end{bmatrix} \begin{bmatrix} E \\ F \end{bmatrix}$$

$$= \begin{bmatrix} \cdots & \cdots \\ -\left(\dfrac{1}{b_0} + \dfrac{s}{k}\right) & 1 + \left(\dfrac{1}{b_0} + \dfrac{s}{k}\right)(m_0 s + Z_0) \end{bmatrix} \begin{bmatrix} E \\ F \end{bmatrix}.$$

Thus,

$$\frac{F(s)}{E(s)} = \frac{[(1/b_0) + (s/k)]}{1 + [(1/b_0) + (s/k)](m_0 s + Z_0)}.$$

It follows from Equation (4.3) that

$$X(s) = \frac{1}{s} F(s),$$

so that

$$\frac{X(s)}{E(s)} = \frac{[1 + (b_0/k)s]}{s[b_0 + [1 + (b_0/k)s][b + (m + m_0)s]]}.$$

* * *

4.2   Friction in Mechanical Networks

There are two distinct types of friction in mechanical engineering systems: the *dry* or *coulomb* friction and *viscous* friction. The friction force acts on an inertial element so as to inhibit the motion of the element (see Figure 4.11 for the case $e_1 > e_2$). If the mass element is on the verge of slipping, the coulomb friction force is given by

$$e_f = \gamma_0 \, mg; \tag{4.17}$$

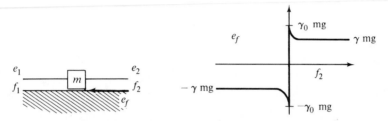

Figure 4.11. Coulomb friction (stationary dry surface).

Figure 4.12. Coulomb or *dry* friction.

γ_0 is the coefficient of *static* friction. If the mass slips,

$$e_f = \gamma \, mg; \tag{4.18}$$

γ is called the coefficient of *sliding* friction. e_f may be plotted as a function of f_2 in the manner indicated in Figure 4.12.

If the interface between the block and plane is lubricated, the friction will be *viscous*. For viscous friction, the friction force is given by

$$e_f = \frac{A\mu}{d} f_2, \tag{4.19}$$

where μ is the viscosity coefficient, d is the separation between the two surfaces, and A is the lubricated area of the inertial element. It will be left as an exercise for the reader to show that Equation (4.19) is a consequence of Newton's law of viscosity:

$$\tau = \mu \frac{du}{dy}; \tag{4.20}$$

τ is the shear stress in the lubricating medium. (*Hint:* Assume a constant velocity gradient and that the lubricant near a surface has the same velocity as that of the surface.)

It is not difficult to show that viscous friction may be described by the transfer matrix

$$\mathfrak{J} = \begin{bmatrix} 1 & \dfrac{A\mu}{d} \\ 0 & 1 \end{bmatrix}, \tag{4.20a}$$

which may be placed on either side of the inertia transfer matrix in the matrix equation relating $\mathbf{P}_1$ to $\mathbf{P}_2$.

4.3 A Special Case in Mechanical Networks

The configuration of dashpots and springs illustrated in Figure 4.13 should be treated as a special case. Since dashpots and springs are assumed massless, it follows from the second law that $e_1 = e_2$, which implies that the entire configuration is a shunt impedance. Applying Newton's second law to the boundary indicated by the dashed line, it follows that

$$e_2 = e^{(1)} + e^{(2)} + \cdots + e^{(n)},$$

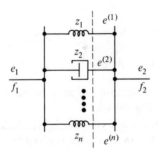

Figure 4.13. Special case.

or, in the Laplace domain,

$$E_2 = E^{(1)} + E^{(2)} + \cdots + E^{(n)}$$
$$= (F_1 - F_2)Z_1 + (F_1 - F_2)Z_2 + \cdots + (F_1 - F_2)Z_n,$$

whence

$$F_1 = F_2 + \frac{1}{\sum Z_i}E_2.$$

Thus, the transfer matrix for the special configuration is

$$\mathfrak{J} = \begin{bmatrix} 1 & 0 \\ \dfrac{1}{\sum Z_i} & 1 \end{bmatrix}. \tag{4.21}$$

Note: Don't confuse $1/\sum Z_i$ with $\sum (1/Z_i)$; that is, the above is *not* an equivalent impedance for several impedances in parallel.

The electrical analogy of the mechanical system illustrated in Figure 4.13 is indicated in Figure 4.14.

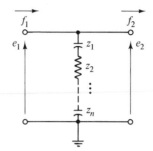

Figure 4.14. Electrical analogy of the mechanical element illustrated in Figure 4.13.

Example 4.2

A *lumped* model of a vehicle is illustrated in Figure 4.15. Determine $x(t)$ when $y(t)$ is as shown. Thus, $y(t) = y_0 u(t)$ or

$$Y(s) = \frac{y_0}{s}. \tag{i}$$

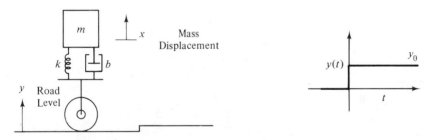

Figure 4.15. Model of an automobile. The tire is modeled as perfectly rigid and the unsprung mass is ignored.

The first task is to determine the transfer function $G(s) = X(s)/Y(s)$. The illustration is redrawn in Figure 4.16. It follows from Equation (4.21) that

$$\begin{bmatrix} E \\ F \end{bmatrix} = \begin{bmatrix} 1 & 0 \\ \dfrac{1}{b + (k/s)} & 1 \end{bmatrix} \begin{bmatrix} 1 & ms \\ 0 & 1 \end{bmatrix} \begin{bmatrix} 0 \\ F_f \end{bmatrix}$$

or

$$\begin{bmatrix} E \\ F \end{bmatrix} = \begin{bmatrix} 1 & ms \\ \dfrac{1}{b + (k/s)} & 1 + \dfrac{ms}{b + (k/s)} \end{bmatrix} \begin{bmatrix} 0 \\ F_f \end{bmatrix}.$$

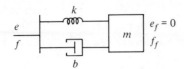

Figure 4.16. Vehicle suspension as a mechanical network.

Therefore,

$$F = \frac{b + (k/s) + ms}{b + (k/s)} F_f. \tag{ii}$$

Notice that

$$X(s) = \frac{1}{s} F_f(s) \tag{iii}$$

and that

$$Y(s) = \frac{1}{s} F(s). \tag{iv}$$

Combine Equations (ii)–(iv) to find

$$\frac{X(s)}{Y(s)} = \frac{k + bs}{k + bs + ms^2}. \tag{4.22}$$

Combine Equations (i) and (4.22) to find

$$X(s) = \frac{y_0(k + bs)}{s(k + bs + ms^2)} = \frac{y_0 b(s + \alpha)}{s[(s + a)^2 + \omega^2]},$$

where

$$\alpha = \frac{k}{b}, \quad a = \frac{b}{2m}, \quad \omega^2 = \frac{k}{m} - a^2. \tag{4.23}$$

Assume that

$$\omega^2 > 0 \quad (\textit{underdamped} \text{ condition}); \tag{4.24}$$

then

$$x(t) = \frac{by_0}{m}\left[\frac{\alpha}{a^2 + \omega^2} + \frac{1}{\omega}\sqrt{\frac{(\alpha - a)^2 + \omega^2}{a^2 + \omega^2}}\, e^{-at} \sin(\omega t + \phi)\right],$$

where

$$\phi = \tan^{-1}\frac{\omega}{\alpha - a} - \tan^{-1}\frac{\omega}{-a}.$$

* * *

4.4 Levers as Transformers

Now that the basic axioms have been reviewed, we shall turn our attention to the transformers of mechanical networks. Consider the lever illustrated in Figure 4.17. Take note of the sign convention, define θ to be the angle that the lever makes with the vertical, and take moments about the pivot to show that

$$e_1 a \cos \theta = e_2 b \cos \theta.$$

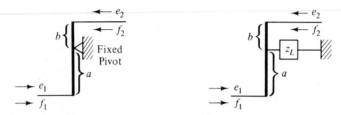

Figure 4.17. Lever transformer. Note the sign convention.

Figure 4.18. Lever with fulcrum impedance.

Thus,

$$e_1 = \frac{b}{a} e_2. \tag{4.25}$$

It is left as an exercise for the reader to show that

$$f_1 = \frac{a}{b} f_2. \tag{4.26}$$

It follows that if one defines

$$n_L = \frac{b}{a}, \tag{4.27}$$

then the lever is described by the transfer matrix equation

$$\mathbf{P}_1 = \begin{bmatrix} E_1 \\ F_1 \end{bmatrix} = \begin{bmatrix} n_L & 0 \\ 0 & \frac{1}{n_L} \end{bmatrix} \begin{bmatrix} E_2 \\ F_2 \end{bmatrix}. \tag{4.28}$$

Thus, *a lever is analogous to a transformer.*

Other types of lever configurations are used. For the configuration illustrated in Figure 4.18, it is not difficult to show that the following matrix equation applies:

$$\begin{bmatrix} E_1 \\ F_1 \end{bmatrix} = \begin{bmatrix} \dfrac{b}{a} & 0 \\ \dfrac{(a+b)^2}{abZ_L} & \dfrac{a}{b} \end{bmatrix} \begin{bmatrix} E_2 \\ F_2 \end{bmatrix}. \tag{4.29}$$

The configuration illustrated in Figure 4.19 is described by Equation (4.28) if the sign convention is taken as indicated.

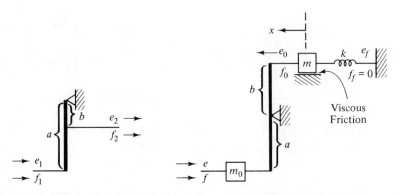

Figure 4.19. Another lever configuration. Compare the sign convention with that in Figure 4.17.

Figure 4.20. Mechanical network with a lever component.

Example 4.3

Determine the transfer function $X(s)/E(s)$ for the system illustrated in Figure 4.20. Define $n_L = b/a$, $b_f = \mu A/d$. Work forward to find

$$\begin{bmatrix} E \\ F \end{bmatrix} = \begin{bmatrix} 1 & m_0 s \\ 0 & 1 \end{bmatrix} \begin{bmatrix} n_L & 0 \\ 0 & \dfrac{1}{n_L} \end{bmatrix} \begin{bmatrix} E_0 \\ F_0 \end{bmatrix}$$

$$= \begin{bmatrix} n_L & \dfrac{m_0 s}{n_L} \\ 0 & \dfrac{1}{n_L} \end{bmatrix} \begin{bmatrix} E_0 \\ F_0 \end{bmatrix}.$$

Thus,

$$E = n_L E_0 + \frac{m_0 s}{n_L} F_0. \tag{i}$$

E_0 is an extraneous power state variable the presence of which indicates a loading effect. Work backward to find

$$\begin{bmatrix} E_f \\ 0 \end{bmatrix} = \begin{bmatrix} 1 & 0 \\ -\dfrac{s}{k} & 1 \end{bmatrix} \begin{bmatrix} 1 & -ms \\ 0 & 1 \end{bmatrix} \begin{bmatrix} 1 & -b_f \\ 0 & 1 \end{bmatrix} \begin{bmatrix} E_0 \\ F_0 \end{bmatrix}$$

or

$$\begin{bmatrix} E_f \\ 0 \end{bmatrix} = \begin{bmatrix} \cdots & \cdots \\ -\dfrac{s}{k} & 1 + \dfrac{s}{k}(ms + b_f) \end{bmatrix} \begin{bmatrix} E_0 \\ F_0 \end{bmatrix}.$$

Thus, the bottom equation may be used to show that

$$E_0 = \frac{1 + (s/k)(ms + b_f)}{s/k} F_0. \tag{ii}$$

Combine Equations (i) and (ii) to find that

$$E = \left[n_L \frac{(ms^2 + b_f s + k)}{s} + \frac{m_0 s}{n_L} \right] F_0,$$

so that

$$\frac{F_0}{E} = \frac{n_L s}{(m_0 + n_L^2 m)s^2 + n_L^2 b_f s + n_L^2 k}.$$

Since

$$X(s) = \frac{1}{s} F_0(s),$$

it follows that

$$\frac{X(s)}{E(s)} = \frac{n_L}{(m_0 + n_L^2 m)s^2 + n_L^2 b_f s + n_L^2 k}. \tag{4.30}$$

$$* \quad * \quad *$$

4.5 Approximate Models of the Passive Elements of Rotary Machines and Vehicle Drive Trains

Shafts, bearings, and gears are examples of passive elements in rotary mechanical systems. Such systems may also be analyzed using the transfer matrix approach.[6] The technique can be quite complete; however, for present purposes only an approximate procedure will be introduced. The nature of the approximation will become clear in the succeeding discussion.

Begin the definition of the tetrahedron of state with the assignment of effort and flow variables:

$$e(t) = \text{torque},$$
$$f(t) = \text{angular velocity},$$
(4.31)

so that $q(t)$ is angular displacement and $\lambda(t)$ is angular impulse.

Begin the discussion of rotary impedances with the rotary inertia illustrated in Figure 4.21. Denote the *polar moment of inertia* by J and show that

$$E_1 = E_2 + JsF_2,$$
$$F_1 = F_2.$$

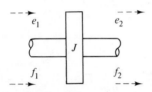

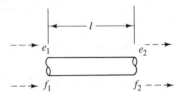

Figure 4.21. Rotational inertia. The sign convention indicates positive directions of vector quantities (recall the *right-hand* rule).

Figure 4.22. Idealized elastic shaft (massless).

The corresponding transfer matrix equation is

$$\mathbf{P}_1 = \begin{bmatrix} E_1 \\ F_1 \end{bmatrix} = \begin{bmatrix} 1 & Js \\ 0 & 1 \end{bmatrix} \begin{bmatrix} E_2 \\ F_2 \end{bmatrix}.$$
(4.32)

The impedance is given by

$$Z = Js.$$
(4.33)

Thus, as in a mechanical translation network, inertia is analogous to an electrical inductance in series.

Consider the shaft illustrated in Figure 4.22. The model of the shaft is idealized by ignoring inertia and considering only the elastic properties of the shaft. Since inertia is ignored, $e_1 = e_2$, which implies that the shaft is a shunted impedance.

Denote

$$d = \text{shaft diameter},$$
$$G = \text{shear modulus},$$
$$q = \text{angle of twist}.$$
(4.34)

Refer to any elementary text[9] and find that

$$q = \int (f_1 - f_2)\, dt = \frac{32l}{\pi G d^4} e_2.$$

Thus, in the Laplace domain,

$$F_1 = F_2 + \frac{32l}{\pi G d^4} s E_2.$$

Therefore, the approximate transfer matrix equation is

$$\begin{bmatrix} E_1 \\ F_1 \end{bmatrix} = \begin{bmatrix} 1 & 0 \\ \dfrac{32ls}{\pi G d^4} & 1 \end{bmatrix} \begin{bmatrix} E_2 \\ F_2 \end{bmatrix}.$$ (4.35)

This equation is suggestive of the analogy to a shunted capacitance. It follows that the impedance

$$\frac{1}{Z} = Cs = \frac{32l}{\pi G d^4} s.$$

It is important to note that a real shaft has both inertia and elasticity, which are distributed parameters.

The *journal* bearing, illustrated in Figure 4.23, is another passive element. This type of bearing is simply a hole in the housing through which the

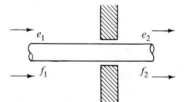

Figure 4.23. Journal bearing (viscous friction).

journal (or shaft) passes with small clearance. The clearance is filled with oil, which may or may not be pressurized.[9] Using Newton's law of viscosity [Equation (4.20)], assuming a constant velocity gradient between journal and housing, and ignoring eccentricity, it can be shown that

$$e_1 = e_2 + \frac{2\pi r^3 l \mu}{c} f_2,$$ (4.36)

where

$$r = \text{shaft radius},$$
$$c = \text{radial clearance},$$
$$\mu = \text{lubricant viscosity},$$
$$l = \text{bearing length}.$$

Equation (4.36) is known as *Petroff's law*.[9] The appropriate matrix equation is

$$\begin{bmatrix} E_1 \\ F_1 \end{bmatrix} = \begin{bmatrix} 1 & \dfrac{2\pi r^3 l \mu}{c} \\ 0 & 1 \end{bmatrix} = \begin{bmatrix} E_2 \\ F_2 \end{bmatrix}. \tag{4.37}$$

Making the obvious analogy with a series resistor, the impedance for a journal bearing is

$$Z = R = \frac{2\pi r^3 l \mu}{c}.$$

Petroff's law is not exact. A more exact expression for the resistive impedance of a journal bearing can be found by referring to a more advanced text on mechanical design or lubrication.

Roller or ball bearings (the so-called *antifriction* bearings) are more accurately described by the coulomb (dry) friction type of relation than by viscous friction relations.[9]

Finally, consider the gear set illustrated in Figure 4.24. Note the sign

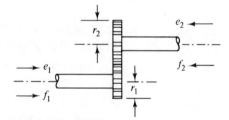

Figure 4.24. Gear set.

convention; ignore the inertias of the gears and show that

$$e_1 = \frac{r_1}{r_2} e_2, \tag{4.38}$$

where r_1 is the radius of the pitch circle of the driving (pinion) gear and r_2 is the radius of the pitch circle of the driven gear.[9] It is left as an exercise for the reader to show that

$$f_1 = \frac{r_2}{r_1} f_2. \tag{4.39}$$

Denote

$$n = \frac{r_1}{r_2}, \tag{4.40}$$

so that, in the Laplace domain, Equations (4.38) and (4.39) take the appearance of the transformer-type matrix equation

$$\begin{bmatrix} E_1 \\ F_1 \end{bmatrix} = \begin{bmatrix} n & 0 \\ 0 & \dfrac{1}{n} \end{bmatrix} \begin{bmatrix} E_2 \\ F_2 \end{bmatrix}. \tag{4.41}$$

It can be shown that n is also equal to the ratio of the number of teeth on the pinion and driven gear.

In a more complete analysis of the components of rotational systems, one finds that the power state is not simply described by the two variables e and f.[6,9] Other *degrees of freedom* are important; for instance, the components translate in a direction perpendicular to the axial. These translations are usually of small amplitude and vibratory in nature. The gear set is an example of a component which transfers considerable energy between rotational and translational power states. The reader is encouraged to refer to the literature for a discussion of these considerations as well as for a discussion of the important gyroscopic effects in rotary machinery.[6]

Example 4.4

Denote the maximum shear stress at the point indicated in Figure 4.25 by $\sigma(t)$ and its Laplace transform by $\Sigma(s)$. Determine the transfer function

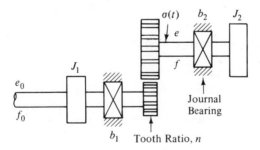

Figure 4.25. Drive train.

$\Sigma(s)/E_0(s)$. First, it is easily shown that[9]

$$\sigma(t) = \frac{re(t)}{I}, \tag{i}$$

where

$$\begin{aligned} r &= \text{shaft radius,} \\ I &= \text{moment of inertia.} \end{aligned} \tag{4.42}$$

Work forward to the output power state:

$$\begin{bmatrix} E_0 \\ F_0 \end{bmatrix} = \begin{bmatrix} 1 & J_1 s \\ 0 & 1 \end{bmatrix}\begin{bmatrix} 1 & b_1 \\ 0 & 1 \end{bmatrix}\begin{bmatrix} n & 0 \\ 0 & \dfrac{1}{n} \end{bmatrix}\begin{bmatrix} E \\ F \end{bmatrix}$$

$$= \begin{bmatrix} 1 & b_1 + J_1 s \\ 0 & 1 \end{bmatrix}\begin{bmatrix} n & 0 \\ 0 & \dfrac{1}{n} \end{bmatrix}\begin{bmatrix} E \\ F \end{bmatrix} = \begin{bmatrix} n & \dfrac{(b_1 + J_1 s)}{n} \\ \cdots & \cdots \end{bmatrix}\begin{bmatrix} E \\ F \end{bmatrix}.$$

Thus,

$$E_0 = nE + \frac{(b_1 + J_1 s)}{n} F. \tag{ii}$$

F is an extraneous power state variable which indicates a loading effect and which can be eliminated by working backward:

$$\begin{bmatrix} 0 \\ F_f \end{bmatrix} = \begin{bmatrix} 1 & -J_2 \\ 0 & 1 \end{bmatrix}\begin{bmatrix} 1 & -b_2 \\ 0 & 1 \end{bmatrix}\begin{bmatrix} E \\ F \end{bmatrix} = \begin{bmatrix} 1 & -(b_2 + J_2 s) \\ \cdots & \cdots \end{bmatrix}\begin{bmatrix} E \\ F \end{bmatrix}.$$

From the top equation it follows that

$$F = \frac{1}{b_2 + J_2 s} E. \tag{iii}$$

Combine Equations (i)–(iii) to find that

$$\frac{\Sigma(s)}{E_0(s)} = \frac{(nr/I)(b_1 + J_2 s)}{n^2 b_2 + b_1 + (n^2 J_2 + J_1)s}. \tag{4.43}$$

It follows from Equation (2.52) that the gain

$$k_\sigma = \frac{\sigma(\infty)}{e_0(\infty)} = \frac{nrb_1}{I(b_1 + n^2 b_2)}. \tag{4.44}$$

* * *

4.6　Euler's Equations, Modulated Gyrators, and Gyroscopes

Consider the rotation of a rigid body and the coordinate systems indicated in Figure 4.26. The Euler equations[1,2,4] are relationships between effort and flow variables defined in the 1-2-z coordinate system. These basic

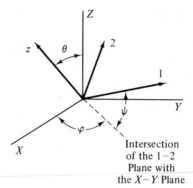

Figure 4.26. Coordinate systems for the rotation of a rigid body. The X-Y-Z frame is an inertial reference system and the 1-2-z frame is rigidly attached to the rotating body.

equations form the starting point for the discussion of this section and are

$$I_1\dot{f}_1 = (I_2 - I_z)f_zf_2 + e_1,$$
$$I_2\dot{f}_2 = (I_z - I_1)f_zf_1 - e_2, \qquad (4.45)$$
$$I_z\dot{f}_z = (I_1 - I_2)f_1f_2 + e_z.$$

The constants I_k represent moments of inertia about the k axis, and the positive directions for the vector efforts and flow are taken to be in the positive 1-2-z directions. The torque e_2 is taken to be the torque exerted *by* the rotating body, while the other torques act *on* the rotating body.

The succeeding discussion could be conducted on a more general level,[3, 5] but, for present purposes, it is sufficient to consider the special case of a *symmetric body* for which

$$I_1 = I = I_2. \qquad (4.46)$$

For this case, the Euler equations reduce to

$$I\dot{f}_1 = -g(t)f_2 + e_1,$$
$$I\dot{f}_2 = g(t)f_1 - e_2, \qquad (4.47)$$
$$\dot{g} = \left\{\frac{I_z - I}{I_z}\right\}e_z,$$

where, by definition,

$$g(t) = \{I_z - I\}f_z. \qquad (4.48)$$

A few further specialized cases illustrate the nature of the above equations. In some important instances, the inertial accelerations, $I\dot{f}_1$ and $I\dot{f}_2$, are negligible so that the Euler equations can be put in the transfer matrix form

$$
\begin{bmatrix} e_1 \\ f_1 \end{bmatrix} = \begin{bmatrix} 0 & g(t) \\ \dfrac{1}{g(t)} & 0 \end{bmatrix} \begin{bmatrix} e_2 \\ f_2 \end{bmatrix}.
\tag{4.49}
$$

And one notes the characteristic gyrator transfer matrix. Notice that, in general, g is time varying. Further notice that

$$
e_1 f_1 = g(t) f_2 \frac{e_2}{g(t)} = e_2 f_2.
\tag{4.49a}
$$

Thus, regardless of the value of $g(t)$, energy flows from the 1-axis power state to the 2-axis power state without any energy flow to or from the z-axis power state. It follows from Equations (4.47) and (4.49) that even though the z-axis power state does not exchange energy with the 1- or 2-axis power states, it does *modulate* the rate of flow of energy exchange between the latter two power states. Paynter has named the mechanical component described by Equation (4.49) a *modulated gyrator*. Other types of modulated gyrators will be discussed in succeeding chapters.

Another distinct case is that for which f_z (hence, g) is constant. For this case, one may take the Laplace transform of Equations (4.47) to find

$$
sIF_1 = -gF_2 + E_1,
$$
$$
sIF_2 = gF_1 - E_2,
$$

or

$$
\begin{bmatrix} E_1 \\ F_1 \end{bmatrix} = \begin{bmatrix} \dfrac{(Is)}{g} & \dfrac{(Is)^2}{g} + g \\ \dfrac{1}{g} & \dfrac{Is}{g} \end{bmatrix} \begin{bmatrix} E_2 \\ F_2 \end{bmatrix},
\tag{4.49b}
$$

which is the result one would have obtained if he had related the power states by the transfer matrix equation

$$
\begin{bmatrix} E_1 \\ F_1 \end{bmatrix} = \begin{bmatrix} 1 & Is \\ 0 & 1 \end{bmatrix} \begin{bmatrix} 0 & g \\ \dfrac{1}{g} & 0 \end{bmatrix} \begin{bmatrix} 1 & Is \\ 0 & 1 \end{bmatrix} \begin{bmatrix} E_2 \\ F_2 \end{bmatrix}.
\tag{4.49c}
$$

A *gyroscope* is an example of a device which may be described by the above equations.

Notice that the gyrator transfer matrix appears in all of the above descriptions.

Example 4.5

For the gyroscope, assume $e_1 = 0 = e_2$ and $f_1(0^-) = A$ radians/sec. Determine $f_1(t)$. It follows from Equation (4.47) that

$$I\dot{f_1} = -gf_2,$$
$$I\dot{f_2} = gf_1,$$

where, for this case, g is a constant. Take the Laplace transform to find

$$IsF_1 - IA = -gF_2,$$
$$IsF_2 = gF_1;$$

thus,

$$F_1(s) = \frac{As}{s^2 + (g^2/I^2)}.$$

Finally,

$$f_1(t) = A \cos\left\{\frac{g}{I}t\right\}. \tag{4.50}$$

A must be small so that g will be a constant. (Why?)

* * *

4.7 The Lumping Approximation and Distributed Parameter Systems

Many dynamic systems are more accurately described by partial differential equations than by ordinary differential equations. However, it is often useful, because of the resulting simplification of mathematical manipulations, to approximate the description of such systems with ordinary, constant coefficient differential equations. This is known as *lumping* because the distributed parameters of inertia, elasticity, and damping are assumed to be lumped or concentrated at certain points.

As an example, consider the solid bar subjected to a force or displacement input and illustrated in Figure 4.27. The elastic and inertia parameters are distributed throughout the bar so that the dynamic description of the bar is more accurately provided by a partial differential equation. Several alternative lumped models of the bar are indicated in Figure 4.28. It is left as an

Figure 4.27. Solid bar with distributed inertial, elastic, and friction properties.

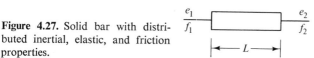

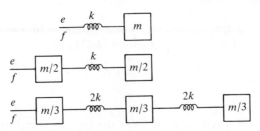

Figure 4.28. Various lumped models for the solid bar.

exercise to show that

$$k = \frac{\mathcal{E}A}{L}. \tag{4.51}$$

$\mathcal{E}$ is the elastic modulus, A is the cross-sectional area, and L is the length of the bar.

Which, or indeed if any, lumped model is appropriate depends on the type and accuracy of information desired. It is obvious that such lumped models can be analyzed using the techniques of passive network analysis discussed previously.

An alternative procedure is to deal directly with the partial differential equations. As will now be shown, a transfer matrix also results from this procedure. A differential element of the solid bar is indicated in Figure 4.29.

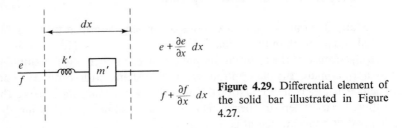

Figure 4.29. Differential element of the solid bar illustrated in Figure 4.27.

Define new parameters:

$$\frac{1}{k'} = \frac{1}{\mathcal{E}A} \quad \left(\frac{1}{k} \text{ per unit length}\right), \tag{4.52}$$

$$m' = \frac{m}{L} \quad \text{(mass per unit length)}. \tag{4.53}$$

Then, in the Laplace domain, the power states on either side of the element are related by

$$
\begin{bmatrix} E \\ F \end{bmatrix} = \begin{bmatrix} 1 & 0 \\ \dfrac{s\,dx}{k'} & 1 \end{bmatrix} \begin{bmatrix} 1 & m's\,dx \\ 0 & 1 \end{bmatrix} \begin{bmatrix} E + \dfrac{\partial E}{\partial x}\,dx \\ F + \dfrac{\partial F}{\partial x}\,dx \end{bmatrix}
$$

$$
= \begin{bmatrix} 1 & m's\,dx \\ \dfrac{s}{k'}\,dx & 1 + \dfrac{m'}{k'}s^2\,(dx)^2 \end{bmatrix} \begin{bmatrix} E + \dfrac{\partial E}{\partial x}\,dx \\ F + \dfrac{\partial F}{\partial x}\,dx \end{bmatrix}. \tag{4.54}
$$

Thus, from the first of Equations (4.54)

$$
E = E + \frac{\partial E}{\partial x}\,dx + m'sF\,dx + m's\frac{\partial F}{\partial x}(dx)^2,
$$

or, neglecting terms of $(dx)^2$,

$$
\frac{\partial E}{\partial x} = -m'sF. \tag{4.55}
$$

Also, from the second of Equations (4.54)

$$
F = \frac{s}{k'}\,dx\,E + \frac{s}{k'}\frac{\partial E}{\partial x}(dx)^2 + F + \frac{\partial F}{\partial x}\,dx + 0((dx)^2),
$$

where $0((dx)^2)$ is read "terms of order $(dx)^2$." Thus,

$$
\frac{\partial F}{\partial x} = -\frac{s}{k'}E. \tag{4.56}
$$

It follows from the solution of Problem 2.9 that the power states are related by

$$
\begin{bmatrix} E_1 \\ F_1 \end{bmatrix} = \begin{bmatrix} \cosh\sqrt{\dfrac{m'}{k'}}sL & Z_0\sinh\sqrt{\dfrac{m'}{k'}}sL \\ \dfrac{1}{Z_0}\sinh\sqrt{\dfrac{m'}{k'}}sL & \cosh\sqrt{\dfrac{m'}{k'}}sL \end{bmatrix} \begin{bmatrix} E_2 \\ F_2 \end{bmatrix}, \tag{4.57}
$$

where

$$
Z_0 = \sqrt{m'k'}. \tag{4.58}
$$

Notice that transcendental functions of s appear in this distributed parameter transfer matrix.

Table 4.2 contains a comparison of analysis methods discussed to this point.

Table 4.2

COMPARISON OF THE TYPES OF MODELS

Model	Comment
1. Lumped parameter transfer matrix	Linear model; transfer function is the ratio of polynomials in s.
2. Distributed parameter transfer matrix	Linear, transfer function is a transcendental function of s. More accurate than (1), but more difficult to manipulate.
3. Full equations of elasticity	Nonlinear partial differential equations. Difficult to solve. Provide most accurate description.

Example 4.6

Consider the high-speed camera illustrated in Figure 4.30.[7] Find the transfer function $F_2(s)/F_1(s)$. It is assumed that $f_1(t)$ will be of the general

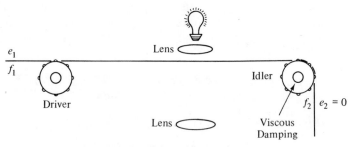

Figure 4.30. High-speed camera.

form illustrated in Figure 4.31. It follows from Equation (4.57) that

$$\begin{bmatrix} E_1 \\ F_1 \end{bmatrix} = \begin{bmatrix} \cosh\sqrt{\frac{m'}{k'}}sL & Z_0\sinh\sqrt{\frac{m'}{k'}}sL \\ \frac{1}{Z_0}\sinh\sqrt{\frac{m'}{k'}}sL & \cosh\sqrt{\frac{m'}{k'}}sL \end{bmatrix}\begin{bmatrix} 1 & b \\ 0 & 1 \end{bmatrix}\begin{bmatrix} 0 \\ F_2 \end{bmatrix}$$

$$= \begin{bmatrix} \cdots & \cdots \\ \frac{1}{Z_0}\sinh\sqrt{\frac{m'}{k'}}sL & \cosh\sqrt{\frac{m'}{k'}}sL + \frac{b}{Z_0}\sinh\sqrt{\frac{m'}{k'}}sL \end{bmatrix}\begin{bmatrix} 0 \\ F_2 \end{bmatrix}.$$

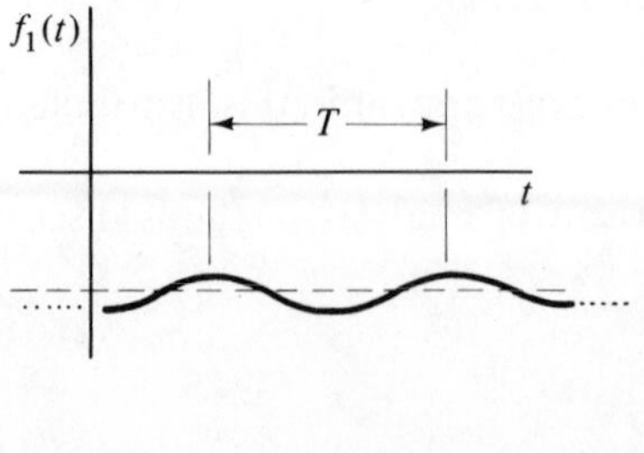

Figure 4.31. Typical input to film.

Thus,

$$\frac{F_2(s)}{F_1(s)} = \frac{1}{\cosh\sqrt{m'/k'}\,sL + (b/Z_0)\sinh\sqrt{m'k'}\,sL}.$$

Suppose that the idler is designed so that

$$b = Z_0 \triangleq \sqrt{m'k'}. \tag{4.59}$$

Then, since

$$\cosh x = \frac{1}{2}(e^x + e^{-x}); \quad \sinh x = \frac{1}{2}(e^x - e^{-x}),$$

it follows that

$$\frac{F_2(s)}{F_1(s)} = \frac{1}{e^{\sqrt{m'k'}\,Ls}}$$

or

$$F_2(s) = e^{-\sqrt{m'k'}\,Ls}F_1(s). \tag{4.60}$$

Use the shifting theorem (Chapter 2) to show that

$$f_2(t) = f_1\left(t - \sqrt{\frac{m'}{k'}}L\right)u\left(t - \sqrt{\frac{m'}{k'}}L\right). \tag{4.61}$$

Thus, if the length, L, is chosen such that

$$\sqrt{\frac{m'}{k'}}L = nT, \tag{4.62}$$

where n is some integer, the film will run smoothly.

$$* \quad * \quad *$$

4.8 The State Variable Approach to Dynamic Systems Analysis*

The transfer matrix approach to dynamic systems analysis provides several significant advantages:

1. All mathematical manipulations are algebraic in nature.
2. Loading effects are made obvious and are easily dealt with.
3. Distributed parameter components are easily described and incorporated in to this method of analysis (see the previous section).

* The material in this section may be omitted during the first reading of the text.

The disadvantages of transfer matrix methods are

1. Nonlinear components are not easily described and analyzed by this method.

2. Only ladder-like networks are easily analyzed.

3. Some of the newer simulation and design procedures are applied to mathematical models stated in the form of systems of simultaneous first-order differential equations (the so-called *state variable formalism*) rather than in the form of transfer functions.

It will be shown in Chapter 8 that a transfer function model can always be converted to a state variable model (it was shown in Chapter 2, Section 2.10, that state variable models can always be converted to transfer function models). It will be shown in the present section that a state variable model can be the direct result of a dynamic systems analysis procedure, which is an alternative to transfer matrix procedures. This alternative analysis procedure is performed in the time domain and can be applied equally well to nonlinear systems as in the next chapter.

The analyst begins by identifying all of the energy storage components in the systems. These devices will either be capacitors (springs) or inductances (inertias). As was indicated in Chapter 3, the capacitive devices are all described by differential equations of the form

$$C \frac{de_k}{dt} = f_{1k} - f_{2k}, \qquad (4.63)$$

and inductive devices are described by differential equations of the form

$$L \frac{df_i}{dt} = e_{1i} - e_{2i}. \qquad (4.64)$$

The efforts, e_k, associated with capacitors (springs) and the flows, f_i, associated with inductances (inertias) are called *state variables*. The analyst completes his task by eliminating all flows and efforts from the right-hand sides of Equations (4.63) and (4.64) which are not state variables *or* sources and inputs. This final task is often performed using relationships for efforts and flows associated with resistors (viscous friction), gyrators, and transformers. The elimination of extraneous variables can be a cumbersome task which is greatly facilitated by sophisticated graphical techniques[3, 5, 8] which will not be presented in the present text.

The state variable analysis technique will now be illustrated with a few simple examples.

Example 4.7

Consider the mechanical network illustrated in Figure 4.32. The model is somewhat suggestive of the dynamics of a machine tool. Assume that the velocity f_0 is the input to the system. Begin the analysis by noting that three

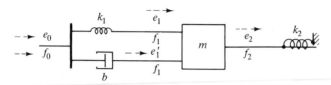

Figure 4.32. Mechanical network with three energy storage components.

energy storage components (hence, three state variables) are present in this system. For this case, Equations (4.63) and (4.64) become

$$\frac{1}{k_1}\frac{de_1}{dt} = f_0 - f_1,$$

$$\frac{1}{k_2}\frac{de_2}{dt} = f_2, \tag{4.65}$$

$$m\frac{df_2}{dt} = e_1 + e_1' - e_2.$$

The variables f_1 and e_1' are neither state variables (e_1, e_2, f_2) nor inputs (f_0) and must be eliminated from Equation (4.65) in order to complete the analysis. First notice that

$$f_1 = f_2. \tag{i}$$

From the defining equation of a dashpot, if follows that

$$e_1' = b(f_0 - f_1)$$
$$= b(f_0 - f_2). \tag{ii}$$

Equations (i) and (ii) relate the extraneous variables to state variables and inputs; combine them with Equations (4.65) to show that

$$
\begin{bmatrix} \dot{e}_1 \\ \dot{e}_2 \\ \dot{f}_2 \end{bmatrix} =
\begin{bmatrix} 0 & 0 & -k_1 \\ 0 & 0 & k_2 \\ \dfrac{1}{m} & -\dfrac{1}{m} & -\dfrac{b}{m} \end{bmatrix}
\begin{bmatrix} e_1 \\ e_2 \\ f_2 \end{bmatrix} +
\begin{bmatrix} k_1 \\ 0 \\ \dfrac{b}{m} \end{bmatrix} f_0. \tag{4.66}
$$

* * *

Example 4.8

For the system of Example 4.7, assume that e_0, rather than f_0, is the input. For this case, the list of extraneous variables on the right-hand sides of Equations (4.65) is f_0, f_1, and e_1'. As before

$$f_1 = f_2. \tag{i}$$

Since the dashpot and spring are massless,

$$e_1 + e_1' = e_0$$

or

$$e_1' = e_0 - e_1. \tag{ii}$$

The dashpot relation

$$e_1' = b(f_0 - f_1)$$

can be used to show

$$f_0 = \frac{1}{b}e_1' + f_1$$

$$= \frac{1}{b}(e_0 - e_1) + f_1. \tag{iii}$$

Combine Equations (i)–(iii) with Equations (4.65) to show that

$$\begin{bmatrix} \dot{e}_1 \\ \dot{e}_2 \\ \dot{f}_2 \end{bmatrix} = \begin{bmatrix} -\dfrac{k_1}{b} & 0 & 0 \\ 0 & 0 & k_2 \\ 0 & -\dfrac{1}{m} & 0 \end{bmatrix} \begin{bmatrix} e_1 \\ e_2 \\ f_2 \end{bmatrix} + \begin{bmatrix} \dfrac{k_1}{b} \\ 0 \\ \dfrac{1}{m} \end{bmatrix} e_0. \tag{4.67}$$

* * *

Example 4.9

The circuit of Example 3.2 is redrawn in Figure 4.33. Determine the state variable equations when f_0 is the input. First notice the two storage devices for which Equations (4.63) and (4.64) become

$$C\frac{de_c}{dt} = f_c,$$

$$L\frac{df_L}{dt} = e_0 - e_L. \tag{4.68}$$

All of the variables on the right-hand side of these equations must be elimi-

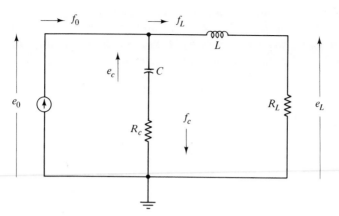

Figure 4.33. *R, L, C* circuit.

nated. First notice that

$$f_0 = f_c + f_L,$$

so that

$$f_c = f_0 - f_L. \qquad \text{(i)}$$

Then notice that

$$e_0 = e_c + R_c f_c$$
$$= e_c + R_c(f_0 - f_L). \qquad \text{(ii)}$$

Finally,

$$e_L = R_L f_L. \qquad \text{(iii)}$$

Combine Equations (i)–(iii) with Equations (4.68) to find

$$\begin{bmatrix} \dot{e}_c \\ \dot{f}_L \end{bmatrix} = \begin{bmatrix} 0 & -\dfrac{1}{C} \\ \dfrac{1}{L} & -\dfrac{(R_c + R_L)}{L} \end{bmatrix} \begin{bmatrix} e_c \\ f_L \end{bmatrix} + \begin{bmatrix} \dfrac{1}{c} \\ \dfrac{R_c}{L} \end{bmatrix} f_0. \qquad (4.69)$$

* * *

Another type of differential equation analysis, which is used quite often in mechanical engineering, results in the formulation of a set of simultaneous *second-order* differential equations. In matrix notation, these equations have the form

$$\mathfrak{M}\ddot{\mathbf{x}} + \mathfrak{B}\dot{\mathbf{x}} + \mathfrak{K}\mathbf{x} = \mathbf{e} + \mathfrak{K}_0\mathbf{u}, \qquad (4.70)$$

where **x** is the vector of all independent absolute displacements, **e** is the vector of input forces, and **u** is the vector of input displacements. $\mathfrak{M}$ is the *mass* matrix, $\mathfrak{B}$ is the *damping* matrix, and $\mathfrak{K}$ and $\mathfrak{K}_0$ are *elastic* matrices. The equations are derived by applying Newton's second axiom to the masses associated with the independent displacements.

*   *   *

Example 4.10

Determine $\mathfrak{M}$, $\mathfrak{B}$, and $\mathfrak{K}$ for the system illustrated in Figure 4.32. First note that the independent displacements are x_0 and x_2. No mass is associated with x_0 so the second axiom takes the degenerate form

$$b(\dot{x}_0 - \dot{x}_2) + k_1(x_0 - x_2) = e_0. \tag{i}$$

For x_2, the second axiom reads

$$m\ddot{x}_2 + b(\dot{x}_2 - \dot{x}_0) + k_1(x_2 - x_0) + k_2 x_2 = 0. \tag{ii}$$

Take

$$\mathbf{x} = \begin{bmatrix} x_0 \\ x_2 \end{bmatrix}, \qquad \mathbf{e} = \begin{bmatrix} e_0 \\ 0 \end{bmatrix},$$

and show, using Equations (i) and (ii), that

$$\mathfrak{M} = \begin{bmatrix} 0 & 0 \\ 0 & m \end{bmatrix}, \qquad \mathfrak{B} = \begin{bmatrix} b & -b \\ -b & b \end{bmatrix},$$

and

$$\mathfrak{K} = \begin{bmatrix} k_1 & -k_1 \\ -k_1 & k_1 + k_2 \end{bmatrix}.$$

*   *   *

References

1. CRANDALL, S., KARNOPP, D., et al., *Dynamics of Mechanical and Electromechanical Systems*, McGraw-Hill, New York, 1968.

2. FEYNMAN, R., LEIGHTON, R., and SANDS, M., *The Feynman Lectures on Physics*, Vol. I: *Mainly Mechanics, Radiation, and Heat*, Addison-Wesley, Reading, Mass., 1963.

3. KARNOPP, D., and ROSENBERG, R., *Analysis and Simulation of Multiport Systems*, M.I.T. Press, Cambridge, Mass., 1968.

4. MERIAM, J. E., *Dynamics*, Wiley, New York, 1966.

5. PAYNTER, H. M., *Analysis and Design of Engineering Systems*, M.I.T. Press, 1961.

6. PESTEL, E., and LECKIE, F., *Matrix Methods in Elastomechanics*, McGraw-Hill, New York, 1963.

7. PIPES, L. A., *Matrix Methods for Engineering*, Prentice-Hall, Englewood Cliffs, N.J., 1963.

8. SHEARER, J., MURPHY, A., and RICHARDSON, H., *Introduction to System Dynamics*, Addison-Wesley, Reading, Mass., 1967.

9. SHIGLEY, J. E., *Mechanical Engineering Design*, McGraw-Hill, New York, 1963.

Problems

4.1. Prove Equation (4.26) two ways:
(a) By using the usual relationship between tangential and angular velocities.
(b) By assuming that no power is stored or dissipated in the lever.

4.2. Prove the validity of Equation (4.29).

4.3. Prove the validity of Equation (4.19).

4.4. Prove the validity of Equation (4.36).

4.5. Prove the validity of Equations (4.38) and (4.39).

4.6. Prove the validity of Equation (4.51).

4.7. Does every passive electric circuit have a passive mechanical analog?

4.8. Consider the hydraulic device illustrated in Figure 4.34. All three pistons have the same area. Show that if the compressibility and inertia of the oil are

Figure 4.34. Three-piston hydromechanical device.

negligible, then

$$e_1 = e_2 = e_3$$

and

$$f_1 = f_2 + f_3. \tag{4.71}$$

Then sketch a hydromechanical analog of the circuit illustrated in Figure 3.9. Could a purely passive mechanical analog be constructed? For what purpose could the hydromechanical analog be used? (Refer to Problem 2.7.)

4.9. Sketch a mechanical analog of the lag compensator illustrated in Figure 3.14. (*Hint:* Refer to Figures 4.13 and 4.14.)

4.10. Sketch a mechanical analog of the lead compensator illustrated in Figure 3.7.

4.11. Determine the transfer function $X_2(s)/X_0(s)$ for the system illustrated in Figure 4.32.

4.12. $y(t) = u(t)(1 - 2e^{-t})$ is the displacement at the point shown in Figure 4.35. What is the displacement $x(t)$?

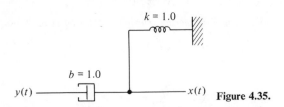

Figure 4.35.

4.13. Find the transfer function relating the output velocity of the first mass, J_1, to a torque input, e (Figure 4.36). The bearings are journal bearings and the sections of shafting have negligible *windup* (elastic deflection).

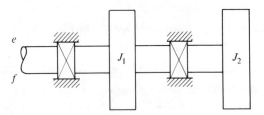

Figure 4.36.

4.14. For the gyroscope, assume $e_2 = 0$ and determine the input impedance $E_1(s)/F_1(s)$. The form of this impedance suggests what analogies?

4.15. The spring k, in the *rate gyroscope* device illustrated in Figure 4.37, exerts a torque proportional to the angular displacement, q_2. Assume that the inertial accelerations $I\dot{f}_1$ and $I\dot{f}_2$ are negligible and that g is a constant. Determine the transfer function $Q_2(s)/F_1(s)$.

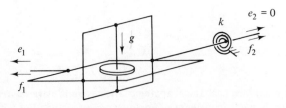

Figure 4.37. Rate gyroscope.

4.16. Repeat the solution of Problem 4.15 but assume that the inertial accelerations are no longer negligible.

Hydraulic and Pneumatic Networks: Linearization and Extended Linear Analysis 5

The analysis methods described in previous chapters are extended to the analysis of networks of passive hydraulic devices. Active hydraulic devices (pumps, motors, *servo-valves*) are used as prime movers for many important engineering systems. The prime movers will be discussed in Chapter 7. The techniques described in this chapter may also be applied to the dynamic flow of any fluid for which the *hydraulic approximation* is valid.

The analysis of networks of passive pneumatic devices is also described in this chapter. Pneumatic devices are used as prime movers and as information transmission networks in engineering systems.[2] Pneumatic networks are often substituted for electrical networks for reasons of safety and reliability (simplicity). Obviously, such a substitution can be made only if the relatively sluggish response of pneumatic networks is of little consequence.

5.1 Control Volume Analysis, the Hydraulic Approximation, and the Tetrahedron of State

Denote some important property of a fixed quantity of fluid (e.g., mass, momentum, or energy) by N and define the distribution function, η, by

$$N = \iiint \eta \rho \, dv, \qquad (5.1)$$

where ρ denotes the mass density, dv denotes a volume element, and the integration is carried out over the entire volume of the fixed quantity of the fluid (hereafter referred to as a *slug*).

A *control volume* is a reference volume which in this text is taken to be fixed in space; the surface of the control volume is called the *control surface*. In succeeding sections, it will be convenient to assume that the control volume is contained within a hydraulic network device in the manner indicated in Figure 5.1.

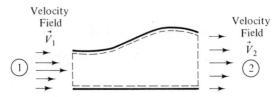

Figure 5.1. Stationary control volume—instantaneously coincident with a slug of fluid. $\vec{V}$ denotes fluid velocity.

Consider the instant when the slug is superimposed with the control volume. A remarkable equation which relates the properties of the slug to the physical phenomena occurring within and about the control volume is[4]

$$\left.\frac{dN}{dt}\right|_{\text{slug}} = \oiint_{\substack{\text{control}\\\text{surface}}} \eta(\rho\vec{V}\cdot d\vec{A}) + \frac{\partial}{\partial t} \iiint_{\substack{\text{control}\\\text{volume}}} \eta\rho \, dv. \tag{5.2}$$

$d\mathbf{A}$ is an elemental vector area of the control surface. A rough interpretation of this equation, which is sufficient for present purposes, is as follows: The first term on the right-hand side of Equation (5.2) is the *net* flux of N through the control surface and the second term is the *net* production of N within the control volume.

The *Navier-Stokes* equations are a set of partial differential equations which can be derived from Equation (5.2) by applying it to a differential control volume and by letting N be, successively, mass, linear momentum, and energy. These partial differential equations may then be solved to obtain accurate solutions for the velocity field and the spatial distribution of mass. For many important cases, a less accurate but useful procedure may be employed. This alternative procedure will be discussed in this text.

Two of the important approximations used in the alternative method are the assumed distributions of density and velocity on the control surface. These ad hoc distributions are often guesses based on experience and/or are selected for convenience. After this step is taken, one no longer needs to consider the full Navier-Stokes equations. A typical set of assumed distributions are that density and velocity are constant across cross section ① and across cross section ②.

To illustrate another commonly used approximation, the particular case when N is the total energy of the slug of fluid will be considered in some detail. Since the slug is a fixed quantity of fluid, one may correctly apply the first axiom of thermodynamics to it to find

$$\frac{dN}{dt}\bigg|_{\text{slug}} = \frac{dQ}{dt} - \frac{dW_s}{dt} + \oiint_{\substack{\text{control} \\ \text{surface}}} \sigma \vec{V} \cdot d\vec{A}, \tag{5.3}$$

where σ is the normal stress in the fluid acting on the control surface. The first term on the right-hand side of Equation (5.3) is the rate of heat supplied to the slug; the second term is the *shaft* power supplied by the slug; and the third term is the power supplied *to* the slug by the surrounding fluid. Also,

$$\eta = \frac{V^2}{2} + gZ + i, \tag{5.4a}$$

where g is the acceleration due to gravity, Z is the vertical distance above some reference in a gravitational field, and i is the specific internal thermal energy. Obviously, the energy terms on the right-hand side of Equation (5.4a) are specific kinetic, potential, and internal thermal energies. Combine Equations (5.2)–(5.4a) to find

$$\frac{\partial}{\partial t} \iiint_{\substack{\text{control} \\ \text{volume}}} \eta \rho \, dv = \frac{dQ}{dt} - \frac{dW_s}{dt} - \oiint_{\substack{\text{control} \\ \text{surface}}} \left\{ \frac{V^2}{2} + gZ + i - \frac{\sigma}{\rho} \right\} \rho \vec{V} \cdot d\vec{A}$$

or

$$\frac{dN}{dt}\bigg|_{\substack{\text{control} \\ \text{volume}}} = \frac{dQ}{dt} - \frac{dW_s}{dt} - \oiint_{\substack{\text{control} \\ \text{surface}}} \left\{ \frac{V^2}{2} + gZ + i - \frac{\sigma}{\rho} \right\} \rho \vec{V} \cdot d\vec{A}. \tag{5.4b}$$

The control surface is composed of two distinct portions: that portion through which fluid flows (e.g., cross sections ① and ② in Figure 5.1) and the remaining portion. For the remaining portion $\vec{V} \cdot d\vec{A} = 0$ so that the surface integral in Equation (5.4a) need be determined only for that portion of the control surface through which fluid flows.

The *hydraulic approximation* is defined by the following list of assumptions:

1. $V^2/2$, gZ, and i are negligible when compared to $-(\sigma/\rho)$.
2. The normal stress, σ, is equal to the negative of the pressure, which is denoted $+e$. 　(5.5)
3. The pressure, e, is a constant over a cross-sectional area of an inlet or of an outlet.

The surface integral in Equation (5.4b) may be interpreted as the power supplied to the control volume by other parts of the network. Consider the specific case of inlet cross section ① in Figure 5.1 and assume that the hydraulic approximation applies to find that the power

$$\iint_1 \sigma \vec{V} \cdot d\vec{A} = e_1 f_1, \qquad (5.6)$$

where, by definition, the volume flow rate

$$f_1 = - \iint_1 \vec{V} \cdot d\vec{A}. \qquad (5.7)$$

Taking a cue from Equation (5.6), the effort variable is taken to be pressure and the flow variable is taken to be volume flow rate. Following the precedent of foregoing chapters,[3]

$$q = \int f(t)\, dt,$$

$$\lambda = \int e(t)\, dt. \qquad (5.8)$$

The sign convention adopted is such that at an outlet

$$f_2 = \iint_2 \vec{V} \cdot d\vec{A} \qquad (5.9)$$

so that $e_2 f_2$ is the power supplied *by* the control volume to the downstream portion of the network. For a flow for which the hydraulic approximation is valid, Equation (5.4b) becomes

$$\left. \frac{dN}{dt} \right|_{\text{control volume}} = e_1 f_1 - e_2 f_2 + \frac{dQ}{dt} - \frac{dW_s}{dt}, \qquad (5.10)$$

where N is the total energy of the control volume.

5.2 The Hydraulic Impedances

Now that the tetrahedron of state has been defined, the impedances of commonly encountered hydraulic network components can be determined. First consider the fluid storage tank illustrated in Figure 5.2. Notice that

$$e_1 = e_2 = h\gamma, \qquad (5.11)$$

where γ is the fluid density (pounds per cubic foot). Thus the tank is a shunt

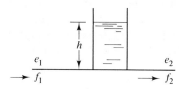

Figure 5.2. Storage or *surge* tank.

Figure 5.3. Another tank configuration which provides the same impedance as does the tank illustrated in Figure 5.2.

impedance. Let A denote the cross-sectional area of the tank and assume, for the time being, that

$$A(h) = \text{Constant.} \tag{5.12}$$

Notice that the tank head and the volume flow rates are related by the simple equation

$$\frac{1}{A} \int_0^t [f_1(\xi) - f_2(\xi)] \, d\xi = h(t), \quad \xi = \text{dummy time variable.}$$

Take Laplace transforms and use Equation (5.11) to show that

$$\frac{1}{As}[F_1(s) - F_2(s)] = \frac{E_2(s)}{\gamma}.$$

Thus,

$$E_1 = E_2,$$

$$F_1 = \frac{As E_2}{\gamma} + F_2,$$

so the transfer matrix is

$$\mathfrak{T} = \begin{bmatrix} 1 & 0 \\ \dfrac{As}{\gamma} & 1 \end{bmatrix}. \tag{5.13}$$

Thus, a tank is analogous to a shunted capacitance with

$$C = \frac{A}{\gamma}. \tag{5.14a}$$

A tank *always* acts like a shunt impedance since the tanks illustrated in both Figures 5.2 and 5.3 are described by the transfer matrix of Equation (5.13). One is reminded of the impedances associated with inertia in mechanical systems.

The next type of impedance to be determined is associated with the inertia of the fluid. As will be shown, the determination of the impedance is straightforward, but the interpretation will be counterintuitive and interesting. Consider the section of pipe illustrated in Figure 5.4. It is desired to

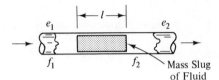

Figure 5.4. Fluid *inertance*. Constant cross-sectional area.

determine the inertia properties of this pipe section (i.e., of a control volume), but Newton's second law may be applied, correctly, only to a fluid slug. Fortunately, Equation (5.2) provides a connecting link between the second law and the inertia of the cross section. Denote the total mass of the fluid slug (hence, of the control volume or pipe section) by M. Take $\vec{\mathbf{N}} = M\vec{\mathbf{V}}$ so that $\vec{\boldsymbol{\eta}} = \vec{\mathbf{V}}$. It will be left as an exercise for the reader to show that if one neglects fluid compressibility and the frictional force between pipe and fluid, Equation (5.2) becomes

$$e_1 A - e_2 A = 0 + \rho A l \frac{dV}{dt}, \tag{5.14b}$$

where A is the cross-sectional area of the tank. Since the volume flow rate

$$f_1 = f_2 = AV, \tag{5.15}$$

the above equation can be rewritten

$$e_1 = e_2 + \frac{\rho l}{A}\frac{df_2}{dt}. \tag{5.16}$$

Rewrite Equations (5.15) and (5.16) in the Laplace domain to find

$$\begin{bmatrix} E_1 \\ F_1 \end{bmatrix} = \begin{bmatrix} 1 & Ls \\ 0 & 1 \end{bmatrix}\begin{bmatrix} E_2 \\ F_2 \end{bmatrix}, \tag{5.17}$$

where the *inertance*

$$L = \frac{\rho l}{A} = \frac{\gamma l}{gA}. \tag{5.18}$$

A counterintuitive result is that this inertia parameter *decreases* with *increasing* cross-sectional area A. Said another way, pipe sections with relatively small cross-sectional areas are likely to provide relatively high inertia. Close examination of Equations (5.14b) and (5.15) will provide an understanding of this effect (notice that pressure and volume flow rate must not be confused with force and velocity). Obviously, inertance is analogous to a series inductance.

The final general type of impedance is, of course, the resistive type. Consider the pipe section illustrated in Figure 5.5. Neglect compressibility and inertia. Then

$$f_1 = f_2. \tag{5.19}$$

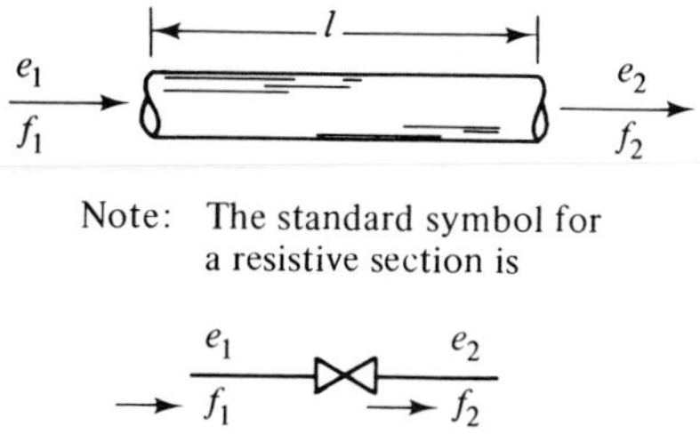

Note: The standard symbol for a resistive section is

Figure 5.5. Hydraulic resistance.

Using dimensional analysis[4] it can be shown that

$$e_1 - e_2 = \frac{\alpha \gamma l}{2dg} V^2, \tag{5.20}$$

where d is the pipe diameter and α, called the *friction factor*, is a dimensionless parameter which must be determined by experiment. It also can be shown that α must be a function of Reynolds number and the dimensionless ratio

$$\beta = \epsilon/d, \tag{5.21}$$

where ϵ is the characteristic *roughness* of the pipe. The parameter ϵ must also be determined experimentally. The original experiments were conducted by Nikuradze and the results were similar to those summarized in Figures 5.6a and 5.6b.

In the laminar range, i.e., for

$$R_e < 2300, \tag{5.22}$$

the experimental results agree with the theoretical equation[4]

$$\alpha = \frac{64}{R_e}. \tag{5.23}$$

In the turbulent regime ($R_e > 2300$), the approximation

$$\alpha \simeq K_\beta, \tag{5.24}$$

where K_β is a constant, is valid for sufficiently high values of Reynolds

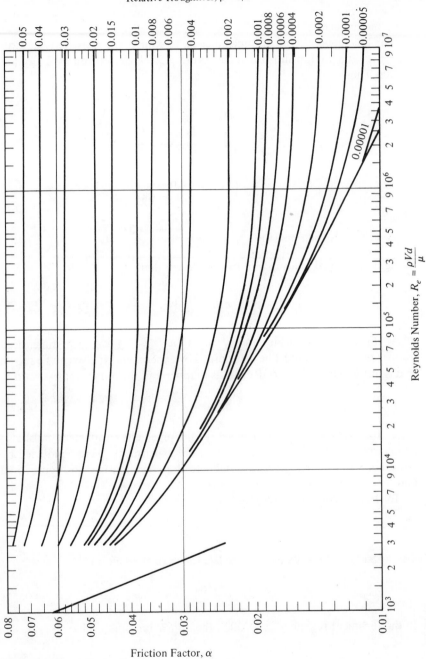

Relative Roughness, $\beta = \epsilon/d$

Reynolds Number, $R_e = \dfrac{\rho V d}{\mu}$

Friction Factor, α

(a)

116

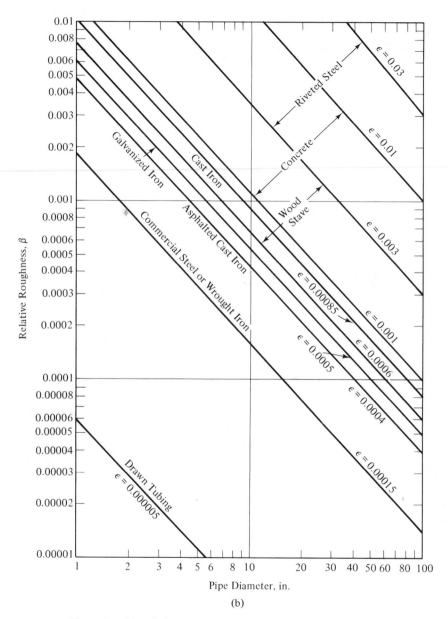

Figure 5.6. (a) Friction factor α as a function of the Reynolds number on pipe diameter. The laminar-turbulent transition can occur at a Reynolds number of 2300.[4] (b) relative roughness for commercial tubing and piping.[4] ϵ denotes the dimension of roughness. (L.F. Moody, "Friction Factors for Pipe Flow" *Trans. A.S.M.E.*, Vol. 66, No. 8, 1944, p. 671.)

number. The laminar case will be discussed in this section and the turbulent
case will be discussed in the next section.

The Reynold's number

$$R_e = \frac{\rho V d}{\mu} = \frac{\gamma V d}{g \mu} \tag{5.25}$$

μ is the viscosity and the volume flow rate

$$f_2 = AV = \frac{\pi d^2}{4} V. \tag{5.26}$$

Combine Equations (5.20), (5.23), (5.25), and (5.26), to show that for laminar
flow

$$e_1 = e_2 + R f_2 \tag{5.27a}$$

where

$$R = \frac{128 l \mu}{\pi d^4}. \tag{5.27b}$$

Transform Equations (5.19) and (5.27a), to the Laplace domain to find

$$\begin{bmatrix} E_1 \\ F_1 \end{bmatrix} = \begin{bmatrix} 1 & R \\ 0 & 1 \end{bmatrix} \begin{bmatrix} E_2 \\ F_2 \end{bmatrix}. \tag{5.28}$$

As will be shown, the resistance for turbulent flow does not satisfy this
simple linear relationship.

Of course, any real hydraulic pipe section will simultaneously display
resistive, inertance, and capacitive (compressible) properties which are distributed uniformly along the length of the pipe. The distributed parameter
models, discussed in Chapter 4, are more accurate descriptions than those
discussed in the present section.

The coefficient of viscosity, μ, which appears in Equation (5.27b), is a
strong function of temperature and is an important source of parameter
variation in hydraulic control systems. When petroleum oils are used, the
viscosity can vary by a factor of twenty over the normal range of operating
temperature. The viscosity of silicone-based fluids varies a good deal less and
these fluids are less flammable than oils; however, the oils provide the side
benefit of lubrication. See Figure 5.10.

An important effect not discussed in this text but of some importance to
the interested reader is that of "entrance length" effects.[4]

5.3 Turbulent Flow Resistance and Other Nonlinear Resistive Impedances; The Concept of Linearization

Combine Equations (5.20), (5.24), and (5.26) to show that, for turbulent flow,

$$e_1 - e_2 = K_T f_2^2, \tag{5.29}$$

where

$$K_T = \frac{8K_\beta \gamma l}{g\pi^2 d^5}. \tag{5.30}$$

Since Equation (5.29) relates a drop in effort to a flow, it defines a resistive impedance; however, for the first time in this text, the relationship is non-linear. The constant K_β is the asymptotic value of α for a given β (see Figure 5.6a).

Turbulent flow in a straight section of pipe is not the only source of nonlinear resistive impedances. Examples of such impedances are indicated in Figures 5.7a and 5.7b. For these and other types of discontinuities, the

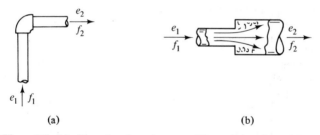

(a)   (b)

Figure 5.7. (a) Pipe bend resistance, (b) resistive impedance associated with an abrupt change in cross-sectional area.

common practice is to assume that

$$e_1 - e_2 = K_R \rho \frac{V^2}{2}, \tag{5.31}$$

where K_R is a dimensionless constant to be determined theoretically or by experiment. Values of K_R can be found in textbooks,[5] handbooks, or vendors' brochures (see Table 5.1). *Caution:* When references are consulted to determine K_R, one must be careful to determine what velocity is used in defining Equation (5.31). When the velocity V_2 is used as the reference velocity for the hydraulic component illustrated in Figure 5.7b, Shames[4] provides the theoretical result

$$K_R = \left\{1 - \frac{A_2}{A_1}\right\}^2. \tag{5.32}$$

Table 5.1 provides values of this loss coefficient for other types of components.

Equation (5.31) can be put into the form of Equation (5.29) by defining

$$K_T = \frac{8K_R\rho}{\pi^2 d^4} = \frac{8K_R\gamma}{g\pi^2 d^4}.$$ (5.33)

Table 5.1
LOSS COEFFICIENTS FOR HYDRAULIC COMPONENTS[5]
(From *Chemical Engineers' Handbook* 4th ed., McGraw-Hill Book Co., 1963, p. 33)

Type of fitting or valve	Loss coefficient, K_R
45° ell	
Standard	0.35
Long radius	0.20
90° ell	
Standard	0.75
Long radius	0.45
Square or miter	1.3
180° bend, close return	1.5
Tee, standard	
Along run, branch blanked off	0.4
Used as ell, entering run	1.3
Used as ell, entering branch	1.5
Coupling	0.04
Union	0.04
Gate valve	
Open	0.20
$\frac{3}{4}$ open	0.90
$\frac{1}{2}$ open	4.5
$\frac{1}{4}$ open	24.0
Diaphragm valve	
Open	2.3
$\frac{3}{4}$ open	2.6
$\frac{1}{2}$ open	4.3
$\frac{1}{4}$ open	21.0
Globe valve	
Open	6.4
$\frac{1}{2}$ open	9.5
Pipeline entrance	
Square edged	0.50
Well rounded	0.04
Sudden contraction	
$A_2/A_1 = 0.1$	0.37
$A_2/A_1 = 0.3$	0.32
$A_2/A_1 = 0.5$	0.22
$A_2/A_1 = 0.7$	0.10
$A_2/A_1 = 0.9$	0.02

Nonlinear resistive relations such as that of Equation (5.29) are quite common. For such nonlinear dynamics, the analyst cannot use the transfer matrix and Laplace transform techniques to produce *exact* analyses; however, in many cases, one can replace Equation (5.29) by an approximate linear relation which is valid in some restricted range of the e and f variables and thereby extend the usefulness of the linear analysis techniques discussed in previous chapters. This approximation procedure is called *linearization* or *extended linear analysis*. This linearization procedure is not valid for all nonlinear relations; in fact, the procedure is valid only for those nonlinear relations which may be expanded in a Taylor series (refer to Appendix B).

Begin the linearization of the resistive-type relation, Equation (5.29), by introducing the notion of a *nominal operating point* at which values of effort and flow are denoted e^0 and f^0. The nominal operating point is defined as the desired steady state values of the effort and flow variables. These values might be design specifications or the result of steady state analysis; in any case, it is assumed a priori that the feedback control system will maintain the efforts and flows very near the nominal operating point and thereby allow one to neglect higher-order terms in a Taylor series expansion about the nominal operating point.

Denote the pressure drop

$$e_1 - e_2 = e, \tag{5.34}$$

so that Equation (5.29) can be rewritten as

$$e = K_T f_2^2. \tag{5.35}$$

Denote the nominal operating point e^0 and f_2^0 and notice that

$$e^0 = K_T \{f_2^0\}^2. \tag{5.36}$$

Assume that the control system is so effective that one need only retain the first-order term from the Taylor series expansion of Equation (5.35), i.e.,

$$e \simeq e^0 + \frac{de}{df}\bigg|_{f^0} (f_2 - f_2^0). \tag{5.37}$$

It follows that

$$e - e^0 \simeq 2K_T f_2^0 (f_2 - f_2^0). \tag{5.38}$$

The convention is to use the nominal operating point as the zero level of the effort and flow variables so that Equation (5.38) becomes

$$e \simeq R_T f_2, \tag{5.39}$$

where

$$R_T = 2K_T f_2^0 = \frac{2e^0}{f_2^0}, \quad \textit{dynamic turbulent resistance.} \quad (5.40)$$

The second equality follows from Equation (5.36). The approximate equation (5.39) is linear and leads to the definition of a transfer matrix similar to that of Equation (5.28).

Notice the counterintuitive result that the dynamic turbulent resistance is twice the value of the static resistance e^0/f_2^0.

This extended linear analysis allows the analyst to apply transfer matrix and Laplace transform methods. It is important to underscore an implication of the above procedure: Since the efforts (pressures) at different points in a hydraulic network will have different nominal values, they will also have *different zero reference levels*. The same point can be made for the flow variables. The analyst must always remember that during an extended linear analysis, he is dealing with *deviations from the nominal operating point* and he will avoid much confusion.

5.4   Other Examples of Fluid Capacitive Impedance—Compliance and Accumulators; Fluid Properties

Fluid *compliance* refers to the tendency of so-called *incompressible* fluids to compress slightly under pressure. This effect is very important in high-speed position and/or velocity control systems. Idealize the model of the pipe section illustrated in Figure 5.8 by neglecting fluid inertia and pipe resistance.

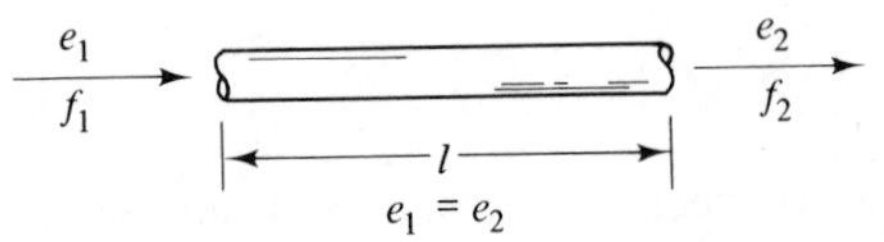

Figure 5.8. Fluid compliance.

Let B denote the bulk modulus, let A denote the cross-sectional area of the pipe section, and let τ denote the volume of the pipe section. The bulk modulus is defined by the relation

$$\int \{f_1 - f_2\} \, dt = \frac{\tau}{B} e_2 = \frac{Al}{B} e_2. \quad (5.41)$$

In the Laplace domain, this relation may be rearranged to read

$$F_1 = F_2 + CsE_2, \quad (5.42)$$

where

$$C = \frac{Al}{B}. \tag{5.43}$$

Equation (5.42) indicates that compliance is analogous to a shunted capacitive impedance.

Accumulators are inserted into hydraulic systems to provide fluid makeup and energy reserve during peak acceleration periods (one is reminded of a flywheel). For the accumulator illustrated in Figure 5.9, denote force on,

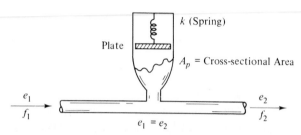

Figure 5.9. Accumulator.

displacement of, and area of the plate by e_p, x_p, and A_p. Notice that

$$e_p = kx_p.$$

If one transforms this equation to the Laplace domain and substitutes hydraulic variables for mechanical variables, he finds that

$$A_p E_2 = k\frac{[F_1 - F_2]}{A_p s}. \tag{5.44}$$

Since $e_1 = e_2$, it follows that

$$\begin{bmatrix} E_1 \\ F_1 \end{bmatrix} = \begin{bmatrix} 1 & 0 \\ Cs & 1 \end{bmatrix}\begin{bmatrix} E_2 \\ F_2 \end{bmatrix}, \tag{5.45}$$

where

$$C = \frac{A_p^2}{k}, \tag{5.46}$$

so that the accumulator presents a shunted capacitive impedance.

To conclude this introductory description of hydraulic impedances, a brief summary of the fluid properties γ, μ, and B will be provided since the values of these variables must be known in order to calculate the values of fluid capacitance (compliance), resistance, and inertance. The bulk modulus B is an important source of parameter variation in hydraulic systems because of *foaming* (air entrainment). Air entrainment is especially troublesome when oil is the working fluid, in which case, B may fall to one third of the value indicated in Table 5.2.

Table 5.2
APPROXIMATE VALUES OF WEIGHT, DENSITY, AND BULK MODULUS
FOR COMMON FLUIDS

Fluid	γ (*lb/cu ft*)	B (*lb/sq in.*)
Water	62.4	320,000
Mercury	846	3,900,000
Oil (SAE 30)	57.5	250,000
Carbon tetrachloride	128	200,000

The viscosity, μ, of commonly encountered fluids is indicated in Figure 5.10. Note the strong dependence on temperature. Note also that μ increases with increasing gas temperature but decreases with increasing liquid temperature.

A good estimate of weight density, γ, of a fluid can be obtained by multiplying the specific gravities indicated in Figure 5.10 by the weight density of water (62.4 lb/cu ft).

After the basic impedances are known, the dynamic systems analysis proceeds as before.

Example 5.1

For the system illustrated in Figure 5.11, neglect inertance and compliance and find the transfer function $H(s)/F_0(s)$.

Work forward to find

$$\begin{bmatrix} E_0 \\ F_0 \end{bmatrix} = \begin{bmatrix} 1 & 0 \\ C_1 s & 1 \end{bmatrix} \begin{bmatrix} E \\ F \end{bmatrix},$$

so that

$$F_0 = C_1 sE + F. \tag{i}$$

H is related to E so that F is extraneous and represents a loading effect which can be eliminated by working backward to find

$$\begin{bmatrix} E_f \\ 0 \end{bmatrix} = \begin{bmatrix} 1 & 0 \\ -C_2 s & 1 \end{bmatrix} \begin{bmatrix} 1 & -R_3 \\ 0 & 1 \end{bmatrix} \begin{bmatrix} 1 & 0 \\ \dfrac{-1}{R_2} & 1 \end{bmatrix} \begin{bmatrix} 1 & -R_1 \\ 0 & 1 \end{bmatrix} \begin{bmatrix} E \\ F \end{bmatrix},$$

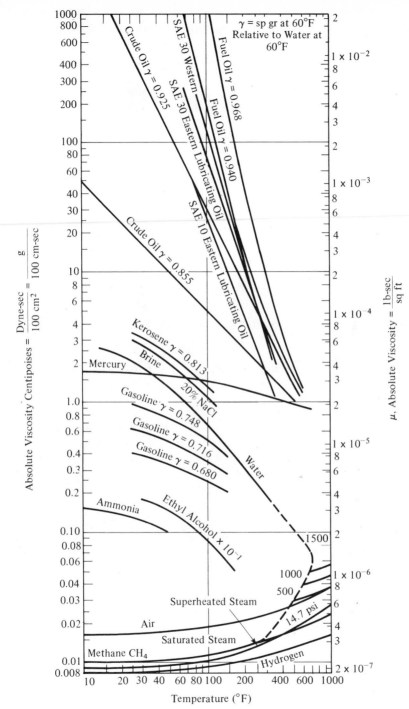

Figure 5.10. Viscosity of common fluids.[5] (From R.L. Daugherty and A.C. Ingersoll, *Fluid Mechanics*, McGraw-Hill Book Co., 1954.)

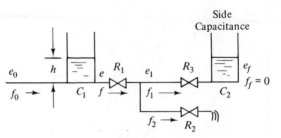

Figure 5.11. Hydraulic storage system.

so that '

$$
\begin{bmatrix} E_f \\ 0 \end{bmatrix} = \begin{bmatrix} \cdots & \cdots \\ -C_2 s - \dfrac{(1 + R_3 C_2 s)}{R_2} & R_1 C_2 s + \dfrac{(R_1 + R_2)}{R_2}(1 + R_3 C_2 s) \end{bmatrix} \begin{bmatrix} E \\ F \end{bmatrix}.
$$

Thus, it follows from the bottom equation that

$$
F = \frac{1 + (R_2 + R_3)C_2 s}{R_1 + R_2 + [R_3(R_1 + R_2) + R_1 R_2]C_2 s} E. \tag{ii}
$$

Combine Equations (i) and (ii) to find

$$
F_0 = \left[\frac{C_1 s \{R_1 + R_2 + [R_3(R_1 + R_2) + R_1 R_2]C_2 s\} + 1 + (R_2 + R_3)C_2 s}{R_1 + R_2 + [R_3(R_1 + R_2) + R_1 R_2]C_2 s} \right] E.
$$

Thus, since $\gamma H = E$, it follows that

$$
\frac{H(s)}{F_0(s)}
$$
$$
= \frac{R_1 + R_2 + [R_3(R_1 + R_2) + R_1 R_2]C_2 s}{\gamma\{1 + [(R_2 + R_3)C_2 + (R_1 + R_2)C_1]s + [R_3(R_1 + R_2) + R_1 R_2]C_2 C_1 s^2\}}.
$$

* * *

5.5 Low Mach Number Flow of Compressible Fluids; Pneumatic Circuits

A convention used by gas dynamics specialists is that if the velocities of a gas are some small fraction of the velocity of sound, the flow may be treated as nearly incompressible.[1] If this convention is adopted, the calculation of inertance and resistance proceeds exactly as before. However, the calculation of compliance is a different matter. One must begin by stating the type of compression process involved. For instance, consider *isothermal* compression in the component illustrated in Figure 5.12. Since $e_1 = e = e_2$, the tank is a shunt impedance.

It is assumed that the flow is so low that the pressure, temperature, and gas quantity are well described by that equilibrium condition called the *ideal gas law;* i.e.,

$$e_2 \tau = m \mathfrak{R}_0 T, \qquad (5.47)$$

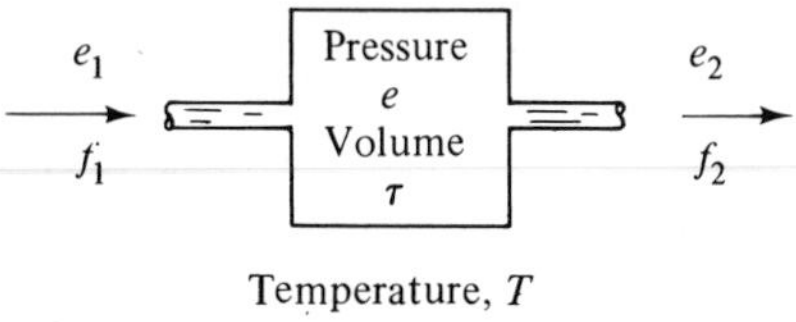

Figure 5.12. Pneumatic isothermal compliance.

where

$$\mathfrak{R}_0 = \text{gas constant,}$$
$$T = \text{absolute temperature (constant),}$$
$$m = \text{quantity of gas,}$$
$$\tau = \text{volume of the pneumatic component.}$$

Values of $\mathfrak{R}_0$ for some gases are provided in Table 5.3.

Table 5.3
PROPERTIES OF COMMON GASES[5]

Gas	$\mathfrak{R}_0$, Gas constant (ft-lb/lb_m-°R)	r, Ratio of specific heats
Air	53.35	1.40
Carbon dioxide	35.12	1.29
Helium	386.3	1.67
Hydrogen	766.5	1.41
Nitrogen	55.15	1.40
Oxygen	48.29	1.39
Water vapor	85.8	1.33

By definition

$$m = \rho \int_0^t [f_1(\xi) - f_2(\xi)] \, d\xi,$$

where ξ is the dummy time variable and ρ is a mass density. Thus, if ρ is taken to be approximately constant and equal to some reference density ρ_0, the above equation can be transformed to find

$$E_2 = \frac{\mathfrak{R}_0 T \rho_0}{\tau s} [F_1(s) - F_2(s)]$$

or

$$F_1 = CsE_2 + F_2, \tag{5.48}$$

where the *isothermal capacitance*

$$C = \frac{\tau}{\mathfrak{R}_0 T \rho_0}. \tag{5.49}$$

It follows from the ideal gas law that

$$\rho_0 = \frac{e_0}{\mathfrak{R}_0 T_0}, \tag{5.50}$$

where e_0 and T_0 are arbitrarily selected reference pressure and temperature. Equation (5.48) could be rewritten

$$C = \frac{\tau T_0}{e_0 T}. \tag{5.51}$$

One would derive different equations for compliance if adiabatic or some other type of compression is assumed.

Of course, any real pneumatic line will have distributed resistance compliance and also inertance properties. One may use the distributed parameter transfer matrices discussed in Chapter 4 or transfer matrices based on one of a large number of lumped models, two of which are illustrated in Figure 5.13.

$$e_1 \quad \underline{\quad} \boxed{C} \underline{\quad}^{R}\!\!\bowtie\!\!\underline{\quad} \quad e_2$$
$$f_1 \qquad\qquad\qquad\qquad f_2$$

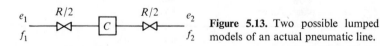

$$e_1 \quad \overset{R/2}{\bowtie} \boxed{C} \overset{R/2}{\bowtie} \quad e_2$$
$$f_1 \qquad\qquad\qquad\qquad f_2$$

Figure 5.13. Two possible lumped models of an actual pneumatic line.

5.6 A Numerical Procedure for Obtaining Steady Flow Conditions*

Consider the problem of calculating the steady state values of pressures and flows in a hydraulic network. The procedure will be illustrated by its application to the particular network sketched in Figure 5.14. The first step

---

* This section contains reference material and may be omitted during the first reading of the text.

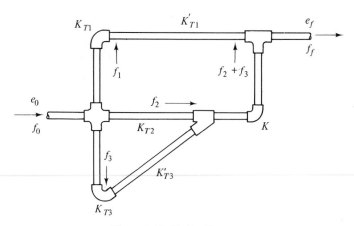

Figure 5.14. Hydraulic network.

is to ignore all inertances and compliances since these impedances have no influence on steady state conditions (a slightly more complicated statement is required for electrical networks which can contain series capacitors). The next step is to identify all of the *independent* flows *within* the network. For the above example, these flows are $f_1, f_2,$ and f_3. The next step is to write a set of equations which reflect the fact that the pressures at flow junctions have unique values: If there are n independent flows, there will be $n-1$ independent algebraic equations resulting from this step. Begin with the assumption that all flows are turbulent so that Equation (5.29) applies; for the above example

$$\{K_{T1} + K'_{T1}\}f_1^2 = K_{T2}f_2^2 + K\{f_2 + f_3\}^2 \tag{5.52}$$

and

$$K_{T2}f_2^2 = \{K_{T3} + K'_{T3}\}f_3^2. \tag{5.53}$$

Since there are three unknowns, a third independent equation must be written.

The third equation will involve the input to the network, which may be one of two quantities: the flow $f_0 = f_f$ or the pressure drop

$$e = e_0 - e_f. \tag{5.54}$$

If the flow is the input, the third equation is the conservation equation·

$$f_0 = f_1 + f_2 + f_3. \tag{5.55}$$

If the pressure drop is the input, the third equation follows from Equation (5.29), namely,

$$e = \{K_{T1} + K'_{T1}\}f_1^2 \tag{5.56}$$

(e could also have been written in terms of the other flows).

Some interesting aspects of the problem can be illustrated by considering one of the above cases in some detail. Consider the case when pressure drop is the input; define

$$x_k = \frac{f_k}{\sqrt{e}}, \tag{5.57}$$

and show that Equations (5.52), (5.53), and (5.56) can be rewritten

$$K_{T2}x_2^2 + K\{x_2 + x_3\}^2 = 1,$$
$$K_{T2}x_2^2 - \{K_{T3} + K'_{T3}\}x_3^2 = 0, \tag{5.58}$$
$$\{K_{T1} + K'_{T1}\}x_1^2 = 1.$$

The advantage of introducing the variable x_k is that the above system of equations can be solved without choosing specific values for the pressure drop, e. The same type of advantage is also often obtained for the case of flow input by defining

$$y_k = \frac{f_k}{f_0}. \tag{5.59}$$

Some analytical simplifications in the nonlinear equations [e.g., Equation (5.58)] can usually be obtained. For instance, for the above example, x_1 can be obtained immediately from the last of Equations (5.58). Another example of the simplifications that can often be made is provided by the flow input case; notice that one of the flows can be written in terms of f_0 and the other internal flows by manipulating Equation (5.55). This flow can then be eliminated from Equations (5.52) and (5.53) in order to reduce the number of unknowns.

After the simplifications have been performed, the analyst must then solve two or more simultaneous, nonlinear (quadratic) equations. A very efficient numerical routine, called generalized Newton iteration, is described in Appendix B. This routine should prove very useful in the solution of steady state hydraulic network problems. An initial guess of the volume flow rates can be obtained by calculating the rates associated with linear laminar flow.

Finally, after the analyst has determined the volume flow rates, he must calculate all relevant Reynolds numbers. If some numbers fall below 2300, he must reformulate his equations by using the laminar resistive relations for the appropriate pipe sections.

5.7 State Variable Analysis of Nonlinear Dynamic Systems*

The concept of generalized (nonlinear) dynamic systems axioms was introduced in Equations (3.55)–(3.57) and the technique of the state variable analysis of linear dynamic systems was introduced in Section 4.8. In the present section, the state variable analysis of nonlinear dynamic systems will be discussed.

Notice that Equation (5.29) is an example of the generalized resistive axiom

$$e = e_1 - e_2 = \phi_R(f). \qquad (5.60)$$

Another example of a basic nonlinear relation is the storage tank illustrated in Figure 5.15. Denote the cross-sectional area at height (head) h by $A(h)$;

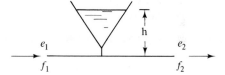

Figure 5.15. Hydraulic storage tank for which the cross-sectional area varies with h.

since

$$\frac{dq}{dh} = A(h),$$

it follows that

$$q = \int_0^h A(\xi) \, d\xi$$

or

$$q = \int_0^{\gamma e_2} A(\xi) \, d\xi. \qquad (5.61)$$

Clearly,

$$q = \phi_c(e_2), \qquad (5.62)$$

where ϕ_c denotes a nonlinear compliance-like functional relationship. Components described by nonlinear axioms are quite common in all types of physical dynamic systems; for instance, the relation between force and velocity in the presence of coulomb friction (see Figure 4.12) is another particular example of the general resistive relation of Equation (5.60). Still another example of this type of relation is the dashpot whose characteristics

---

* The material presented in this section may be omitted during the first reading of the text.

are illustrated in Figure 5.16. This type of dashpot is purposely designed for use in vehicle suspensions.

The generalized inductance (inertia) relationship has the form

$$\lambda = \phi_I(f). \qquad (5.63)$$

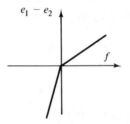

Figure 5.16. Dashpot with nonlinear operating characteristics.

To develop state variable equations for dynamic systems of components, some of which are described by relations of the form of Equation (5.60) or (5.62) or (5.63), the recommended procedure is

1. For each compliance, write a state variable equation of the form

$$\dot{q}_k = f_{1k} - f_{2k}, \qquad (5.64)$$

2. For each inertia, write a state variable equation of the form

$$\dot{\lambda}_k = e_{1k} - e_{2k}. \qquad (5.65)$$

[Equations (5.64) and (5.65) follow from the definition of the tetrahedron of state.]

3. Use relations (5.60), (5.62), and (5.63) to write a set of nonlinear algebraic equations which relate the variables on the right-hand sides of Equations (5.64) and (5.65) to the state variables (λ_k and q_k) or to inputs.

The technique will now be illustrated by example.

Example 5.2

For the hydraulic system of Example 5.1 (Figure 5.11), determine the state variable equations when all resistances are nonlinear (i.e., all flows are turbulent). There are two energy storages so that the state variables are the two volumes q_1 and q_2 for which

$$\dot{q}_1 = f_0 - f, \qquad (5.66)$$
$$\dot{q}_2 = f_1. \qquad (5.67)$$

The extraneous variables f and f_1 must be eliminated from these equations.

The nonlinear resistive relations are

$$e_1 = \phi_{R_2}(f_2)$$

and

$$e - e_1 = \phi_{R_1}(f),$$
$$e_1 - e_f = \phi_{R_3}(f_1). \tag{5.68}$$

Since

$$e = \frac{1}{C_1} q_1,$$

$$e_f = \frac{1}{C_1} q_2, \tag{5.69}$$

$$f_2 = f - f_1,$$

it follows from Equation (5.68) that

$$\phi_{R_1}(f) + \phi_{R_2}(f - f_1) = \frac{1}{C_1} q_1 \tag{5.70}$$

and

$$\phi_{R_2}(f - f_1) - \phi_{R_3}(f_1) = \frac{1}{C_2} q_2. \tag{5.71}$$

These equations are nonlinear relations between the extraneous variables and the state variables which could be solved by generalized Newton iteration (see Appendix B). Equations (5.66) and (5.67), together with Equations (5.70) and (5.71), form a complete set of state variable equations.

* * *

Example 5.3

Repeat Example 5.2, but assume that the tanks are nonlinear capacitors [see Figure 5.15 and Equation (5.62)]. For this case, the first two equations of Equations (5.69) must be replaced by the nonlinear functional relations

$$e = \phi_{C_1}^{-1}(q_1),$$
$$e_f = \phi_{C_2}^{-1}(q_2), \tag{5.72}$$

so that the final relations between extraneous and state variables are

$$\phi_{R_1}(f) + \phi_{R_2}(f - f_1) = \phi_{C_1}^{-1}(q_1),$$
$$\phi_{R_2}(f - f_1) - \phi_{R_3}(f_1) = \phi_{C_2}^{-1}(q_2). \tag{5.73}$$

* * *

It is clear from the above examples that the final result of the state variable analysis will be a set of simultaneous, first-order differential equations of the general form

$$\dot{\mathbf{x}} = \boldsymbol{\phi}(\mathbf{x}, \mathbf{u}), \tag{5.74}$$

where $\boldsymbol{\phi}(\cdot)$ is some nonlinear vector function, $\mathbf{u}$ is a vector of input flows and/or efforts, and the state vector

$$\mathbf{x} = \begin{bmatrix} \mathbf{q} \\ \cdots \\ \boldsymbol{\lambda} \end{bmatrix}. \tag{5.75}$$

Equation (5.74) is the starting point for one technique of extended linear analysis. First, define the *nominal operating point*, $\mathbf{x}°$, as the *equilibrium point*, i.e.,

$$\dot{\mathbf{x}}° = \mathbf{0} = \boldsymbol{\phi}(\mathbf{x}°, \mathbf{u}(\infty)), \tag{5.76}$$

where $\mathbf{u}(\infty)$ is the vector of final values of the input variables. In many cases, Equation (5.76) can be solved by generalized Newton iteration for $\mathbf{x}°$; the second step is to assume, a priori, that a control system will be designed so that $\mathbf{x}(t)$ will always be very near the nominal operating point $\mathbf{x}°$. This assumption allows the analyst to expand the right-hand side of Equation (5.74) in a Taylor series about $\mathbf{x}°$ and retain only first-order terms in $\mathbf{x} - \mathbf{x}°$ to find

$$\boldsymbol{\phi}(\mathbf{x}, \mathbf{u}) \simeq \boldsymbol{\phi}\{\mathbf{x}°, \mathbf{u}(\infty)\} + \mathcal{g}_x(\mathbf{x} - \mathbf{x}°) + \mathcal{g}_u(\mathbf{u} - \mathbf{u}(\infty)),$$

where the Jacobian matrices

$$(\mathcal{g}_x)_{ik} = \frac{\partial \phi_i}{\partial x_k}(\mathbf{x}°, \mathbf{u}(\infty)) \tag{5.77}$$

and

$$(\mathcal{g}_u)_{ik} = \frac{\partial \phi_i}{\partial u_k}(\mathbf{x}°, \mathbf{u}(\infty)). \tag{5.78}$$

See Appendix B, Equation (B.56), for details. Combine the above approximation with Equations (5.74) and (5.76) to find

$$\dot{\mathbf{x}} \simeq \mathcal{g}_x(\mathbf{x} - \mathbf{x}°) + \mathcal{g}_u(\mathbf{u} - \mathbf{u}(\infty)). \tag{5.79}$$

If $\mathbf{x}°$ is redefined as the zero reference level for the state variables and $\mathbf{u}(\infty)$ as the zero reference levels for the input variables, then the above equation reduces to

$$\dot{\mathbf{x}} \simeq \mathcal{g}_x\mathbf{x} + \mathcal{g}_u\mathbf{u}. \tag{5.80}$$

The analysis of this linearized state variable equation was discussed in Section 2.10.

References

1. KIRSHNER, J.M., ed., *Fluid Amplifiers*, McGraw-Hill, New York, 1966.

2. LETHAM, D. L., "Fluidic System Design," *Machine Design*, a series of articles beginning Feb. 2, 1966.

3. PAYNTER, H. M., *Analysis and Design of Engineering Systems*, M.I.T. Press, Cambridge, Mass., 1961.

4. SHAMES, I. H., *Mechanics of Fluids*, McGraw-Hill, New York, 1962.

5. WHITAKER, S., *Introduction to Fluid Mechanics*, Prentice-Hall, Englewood Cliffs, N.J., 1968.

Problems

5.1. Derive Equation (5.14b) from Equation (5.2).

5.2. Explain, in your own words, why inertance increases with decreasing cross-sectional area.

5.3. Show that the Reynolds number

$$R_e = \frac{4\gamma f}{g\pi \, d\mu}. \tag{5.81}$$

5.4. Denote the laminar resistance of a fluid at temperature θ by R_θ. Plot the ratio $R_\theta/R_{100°}$ for (a) S.A.E. 30 western oil, (b) water, and (c) air and for the temperature range

$$100°\text{F} \leq \theta \leq 200°\text{F}.$$

5.5. Denote the inertance of a fluid at temperature θ by L_θ. Plot the ratio $L_\theta/L_{100°}$ for the same fluids and the same temperature range used in the calculations for Problem 5.4 [for air the density can be calculated using Equation (5.50) and Table 5.3].

5.6. For the pipe section illustrated in Figure 5.4, assume that both inertance and laminar resistive impedance exist and that $e_2(t) \equiv 0$. Determine the transfer function $F_2(s)/E_1(s)$. Denote the gain of this transfer function by K_θ and the time constant T_θ. Plot $K_\theta/K_{100°}$ and $T_\theta/T_{100°}$ for the three fluids and for the temperature range used in the calculations for Problems 5.4 and 5.5.

5.7. A colleague of the author designed the manifold and sensor location, indicated in Figure 5.17, in order to measure the gas pressure, e_0. When the system was constructed, a high-frequency oscillatory pressure was measured which was believed not to be truly representative of the pressure, e_0. The author suggested that the dimension d be increased. When this was done the oscillations disappeared. Explain the illness and the cure.

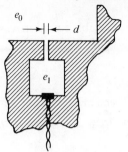

Pressure Sensor

Figure 5.17. Gas pressure measuring device.

5.8. The conduit carrying a fluid stream will have an elasticity associated with it. In many cases (e.g., water hammer) conduit elasticity plays an important role. Model an elastic conduit or pipe as an accumulator and relate the equivalent compliance constant, C, to the dimensions and physical properties of the pipe.

5.9. Refer to Figure 4.5 and Equation (4.11). Determine an equation for the constant b in terms of return pipe dimensions and oil properties.

5.10. Determine the transfer function $H_1(s)/F(s)$ for both of the normal operating conditions shown in Figure 5.18. Find $h_1(t)$ when $F(s) = 1/s$ for both cases. Both capacitors are $1.0 \text{ ft}^5/\text{lb}$.

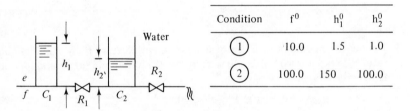

Condition	f^0	h_1^0	h_2^0
①	10.0	1.5	1.0
②	100.0	150	100.0

Figure 5.18. Turbulent flow.

5.11. Find the transfer function

$$G(s) \triangleq \frac{H_2(s)}{F_i(s)}$$

for the *side capacitance* system illustrated in Figure 5.19.

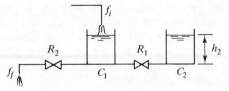

Figure 5.19. System with side capacitance.

5.12. Find the transfer function

$$G(s) \triangleq \frac{H_2(s)}{F_i(s)}$$

for the *cascade* system illustrated in Figure 5.20.

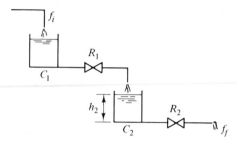

Figure 5.20. Cascade connection.

5.13. Find the transfer function

$$G(s) \triangleq \frac{H_2(s)}{F_i(s)}$$

for the *coupled* system illustrated in Figure 5.21.

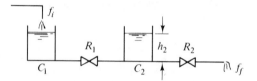

Figure 5.21. Bilaterally coupled connection.

5.14. For the ideal case, the power states on either side of a dashpot are related by Equation (4.16). In high-speed position control systems the following parameters become significant:

1. The housing inertia.
2. The compressibility of the oil.
3. The elastic properties of the output shaft.
4. Leakage of oil.

Determine the transfer matrix relating the power state P_1 to the power state P_2 for this case. See Figure 5.22.

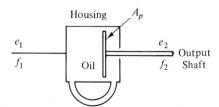

Figure 5.22. Real dashpot.

5.15. Derive the partial differential equations of Problem 2.9.

5.16. Determine the linearized capacitance for the case of adiabatic expansion defined by $pe^{-r} = $ constant, where r is the ratio of specific heats.

5.17. Consider the two physical systems shown in Figures 5.23 and 5.24 and the three inputs shown in Figures 5.25a, 5.25b, and 5.25c. Using only the initial and final value theorems, *sketch* the response of both systems to all three inputs.

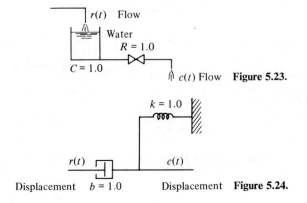

Figure 5.23.

Figure 5.24.

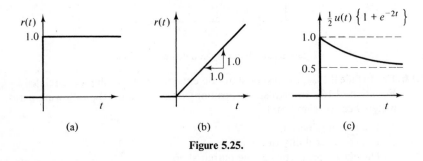

Figure 5.25.

5.18. Let $C_1 = 1.0$ ft^5/lb, $C_2 = 1.0$ ft^5/lb, $R_1 = 1.0$ lb-sec/ft^5, and $R_2 = 0.5$ lb-sec/ft^5 and suppose that

$$f_1(t) = \tfrac{1}{2}u(t).$$

Assuming I.Q. conditions, what is $h_2(t)$ for the systems of the previous three problems? Plot all three responses on the same graph.

5.19. Sufficient information is given in Figure 5.26 for you to determine the transfer functions for four systems. *Sketch* the *step* response for the systems. Use the initial and final value theorems.

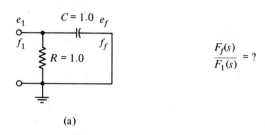

$$\frac{F_f(s)}{F_1(s)} = ?$$

(a)

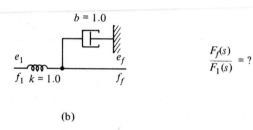

$$\frac{F_f(s)}{F_1(s)} = ?$$

(b)

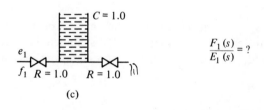

$$\frac{F_1(s)}{E_1(s)} = ?$$

(c)

$$\frac{C(s)}{R(s)} = \frac{1-s}{1+s} \qquad c(t) = ?$$

(d)

Figure 5.26.

Dynamic Thermal Systems and Systems with Time Delays*   6

This chapter is an introduction to the analysis of dynamic thermal systems. The transfer function analysis of distributed parameter systems is further explored, and the state space representations of such systems are introduced.

6.1 Transfer State Variables

For heat transfer networks the following arbitrary assignment is made for the effort and flow variables:

$$
\begin{aligned}
e &= \text{temperature} \quad &\text{(effort)}, \\
f &= \text{flow rate of heat} \quad &\text{(flow)},
\end{aligned}
\tag{6.1a}
$$

although some analysts would use entropy rate as the flow variable.

Notice that the vector

$$
\mathbf{P} = \begin{bmatrix} e \\ f \end{bmatrix}
\tag{6.1b}
$$

cannot be called a *power state* since the product $e \cdot f$ does not represent power in a heat transfer network. Indeed, $f(t)$ itself is a power. When discus-

*This chapter may be omitted during the first reading of the text.

sing heat transfer networks in this text, **P** will be referred to as the *transfer state*.

6.2 The Impedances of Solid Heat Transfer Networks

It is often sufficiently accurate to make lumped approximations of heat transfer networks in order to derive dynamic models. When the system being analyzed does not contain a forced convection component, one can construct an analogous electric network and proceed in the analysis as outlined in the previous chapters.

The basic elements of such networks are derived from certain idealizations. The first idealization approximately describes the conduction of heat in metals, namely, the *solid with infinite conductivity and finite capacity*. Use the notation indicated in Figure 6.1 and show that

$$e_1 = e_2,$$

$$\int_0^t (f_1 - f_2)\, d\zeta = C_p e_2,$$

$$C_p = \text{heat capacity.}$$

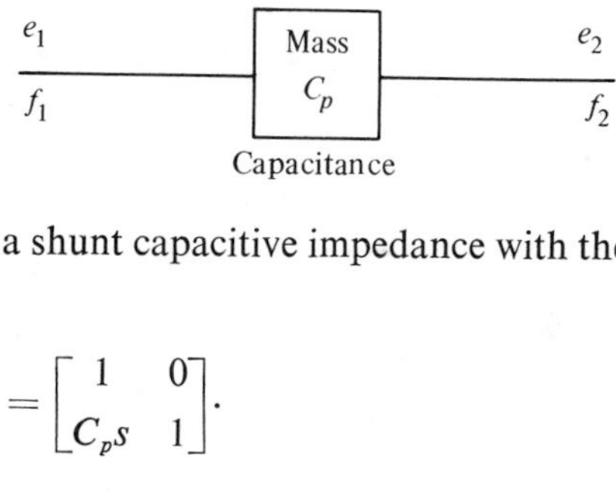

Figure 6.1. Idealized solid (infinite conductivity).

Thus, the above is analogous to a shunt capacitive impedance with the transfer matrix

$$\mathbf{J} = \begin{bmatrix} 1 & 0 \\ C_p s & 1 \end{bmatrix}. \tag{6.2a}$$

C_p is the specific heat capacity times the mass of the solid, i.e.,

$$\begin{aligned} C_p &= c_p M \\ &= c_p \rho V. \end{aligned} \tag{6.2b}$$

ρ is mass density and V is the volume of the solid.

The next idealization is that of the *solid with zero heat capacity*. If k denotes thermal conductivity, then for the solid illustrated in Figure 6.2, the relation between the e and f variables is

$$\frac{kA}{L}(e_2 - e_1) = f_2$$

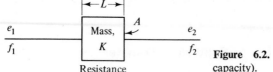

Figure 6.2. Idealized solid (zero capacity).

and

$$f_1 = f_2.$$

A is the cross-sectional area and L is the length of the conductive path. Define the *thermal resistance:*

$$R = \frac{L}{kA}. \tag{6.3}$$

Then the above idealized solid is analogous to a series impedance with the transfer matrix

$$\mathcal{T} = \begin{bmatrix} 1 & R \\ 0 & 1 \end{bmatrix}. \tag{6.4}$$

Of course, real masses have finite heat capacitance and thermal conductivity so that a more realistic lumped approximation for a real solid is as shown in Figure 6.3. Thermal properties of some common materials are provided in Table 6.1.

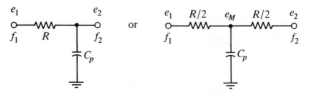

Figure 6.3. Two possible models for a real solid (e_M = solid temperature).

Many solids will interface with a fluid. This interface may often be modeled with a *free convection resistance*. Consider the mass-fluid interface illustrated in Figure 6.4. There is no storage of heat at this interface so that

$$f_1 = f_2.$$

If U denotes the *convective overall heat transfer* coefficient,[2] then a heat balance leads to

$$UA(e_1 - e_2) = f_2.$$

Table 6.1

THERMAL PROPERTIES[2]

Material	ρ (lb/cu ft)	c_p (Btu/lb-°F)	k (Btu/hr-ft-°F)
Aluminum	169	0.214	118
Iron	493	0.108	42
Steel	487	0.113	31
(1% carbon)			
Copper	559	0.0915	223
Asbestos	36	0.195	0.111
(212°F)			
Glass wool	12.5	0.16	0.023
Concrete	119–144	0.21	0.47–0.81

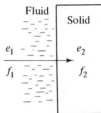

Figure 6.4. Fluid-solid interface.

Define the free convection resistance:

$$R_c = \frac{1}{UA}. \qquad (6.5a)$$

The above equations are used to show that the transfer matrix

$$\mathcal{T} = \begin{bmatrix} 1 & R_c \\ 0 & 1 \end{bmatrix} \qquad (6.5b)$$

may be used to describe the interface.

There are no heat transfer elements which are analogous to a series capacitance or to an inductance in series or shunt.

6.3 Analysis of Dynamic Heat Transfer Systems

To develop dynamic models for heat transfer systems, one may use the same concepts that were used in previous chapters. This technique is best illustrated by example. Notice that, in the following example, the analogous electric network is sketched in order to aid the analyst. As is shown in the example, the ambient temperature is taken as the *ground* temperature (therefore as the *zero reference temperature*).

Example 6.1

For the system shown in Figure 6.5, let $\dot{q}(t)$ be the rate of energy dissipated by the heater, and assume that the fluid temperature is the output. Find the transfer function

$$G(s) = \frac{E_2(s)}{\dot{Q}(s)}.$$

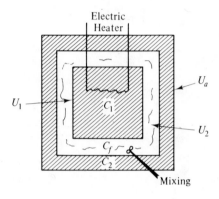

Figure 6.5. Dynamic thermal system network for Example 6.1.

Define the resistances

$$R_0 = \frac{L_1}{kA_0},$$

$$R_1 = \frac{1}{U_1A_1},$$

$$R_2 = \frac{1}{U_2A_2},$$

$$R_a = \frac{1}{U_aA_a},$$

$$R_M = \frac{L_M}{kA_M}.$$

Taking the ambient temperature as ground, the thermal system is analogous to the ladder sketched in Figure 6.6. Work forward to find

$$\begin{bmatrix} E \\ \dot{Q} \end{bmatrix} = \begin{bmatrix} 1 & R_0 \\ 0 & 1 \end{bmatrix}\begin{bmatrix} 1 & 0 \\ C_1s & 1 \end{bmatrix}\begin{bmatrix} 1 & R_1 \\ 0 & 1 \end{bmatrix}\begin{bmatrix} 1 & 0 \\ C_fs & 1 \end{bmatrix}\begin{bmatrix} E_2 \\ F_2 \end{bmatrix}$$

$$= \begin{bmatrix} (1 + R_0C_1s)(1 + R_1C_fs) + R_0C_fs & R_0 + R_1(1 + R_0C_1s) \\ C_1s(1 + R_1C_fs) + C_fs & 1 + R_1C_1s \end{bmatrix}\begin{bmatrix} E_2 \\ F_2 \end{bmatrix}.$$

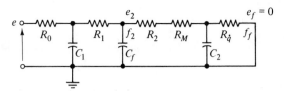

Figure 6.6. Electric analogy of the network shown in Figure 6.5.

Thus

$$\dot{Q}(s) = [C_1 s(1 + R_1 C_f s) + C_f s]E_2 + [1 + R_1 C_1 s]F_2.$$

To eliminate the extraneous transfer state variable, F_2, derive an input impedance equation. Work backward to find

$$\begin{bmatrix} 0 \\ F_f \end{bmatrix} = \begin{bmatrix} 1 & -R_a \\ 0 & 1 \end{bmatrix}\begin{bmatrix} 1 & 0 \\ -C_2 s & 1 \end{bmatrix}\begin{bmatrix} 1 & -R_M \\ 0 & 1 \end{bmatrix}\begin{bmatrix} 1 & -R_2 \\ 0 & 1 \end{bmatrix}\begin{bmatrix} E_2 \\ F_2 \end{bmatrix}$$

or

$$\begin{bmatrix} 0 \\ F_f \end{bmatrix} = \begin{bmatrix} 1 + R_a C_2 s & -R_a - (R_M + R_2)(1 + R_a C_2 s) \\ -C_2 s & 1 + (R_M + R_2)C_2 s \end{bmatrix}\begin{bmatrix} E_2 \\ F_2 \end{bmatrix}.$$

Thus, the desired input impedance relation is

$$F_2 = \frac{1 + R_a C_2 s}{R_a + (R_M + R_2)(1 + R_a C_2 s)}E_2.$$

Thus,

$$\frac{\dot{Q}(s)}{E_2(s)} = C_1 s(1 + R_1 C_f s) + C_f s + \frac{(1 + R_1 C_1 s)(1 + R_a C_2 s)}{R_a + (R_M + R_2)(1 + R_a C_2 s)}$$

or

$$G(s) = \frac{E_2(s)}{\dot{Q}(s)}$$

$$= \frac{R_a + (R_M + R_2)(1 + R_a C_2 s)}{(1 + R_1 C_1 s)(1 + R_a C_2 s) + [R_a + (R_M + R_2)}{(1 + R_a C_2 s)][C_f s + C_1 s(1 + R_1 C_f s)]}.$$

* * *

6.4   Heat Transfer in the Presence of a
Carrier (*Thermal*) Fluid

In Chapter 5, the concept of a hydraulic fluid was introduced as an idealization wherein changes in mechanical energies (kinetic and flow work) were
far more important than changes in internal thermal energy, i. In this section,
the extreme opposite case is introduced wherein changes in internal thermal
energies outweigh any other energy transformations except σ/ρ. Fluids for
which such an idealization applies are called *carrier* or *thermal* fluids. There
are many important practical examples where neither the hydraulic nor the
carrier approximations are sufficiently accurate for dynamic modeling. These
situations are characterized by strong mechanical-thermal interactions. Flows
of gases in turbines and in sonic nozzles are specific examples not well
modeled by either the hydraulic or carrier approximations.

For the carrier approximation, Equation (5.4b) becomes

$$\frac{dN}{dt}\bigg|_{\text{control volume}} = \frac{dQ}{dt} - \oiint_{\substack{\text{control} \\ \text{surface}}} \{\rho i + p\}\vec{\mathbf{V}}\cdot d\vec{\mathbf{A}}, \tag{6.6}$$

where N is total thermal energy stored in the control volume, p denotes
pressure, and Q is the heat energy entering the control surface. This equation
should be compared to Equation (5.10).

Consider any portion of the control surface for which p and ρ are constant so that

$$\iint \{i\rho + p\}\vec{\mathbf{V}}\cdot d\vec{\mathbf{A}} = \{i\rho + p\}w, \tag{6.7}$$

where the volume flow rate is defined by

$$w = \left| \iint \vec{\mathbf{V}}\cdot d\vec{\mathbf{A}} \right|. \tag{6.8}$$

A different notation is introduced for volume flow rate and pressure in order
to emphasize the fact that, unlike the case of hydraulic fluids, these variables
are not power state variables.

Another useful approximation[3] is that the specific *enthalpy* is

$$i + \frac{p}{\rho} = c_p(e - e_R), \tag{6.9}$$

where

$$c_p = \text{specific heat at constant pressure,}$$
$$e = \text{fluid temperature,} \tag{6.10}$$
$$e_R = \text{reference state temperature.}$$

A similar approximation is that the total internal thermal energy is

$$N = Mc_v(e - e_R), \tag{6.11}$$

where

$$c_v = \text{specific heat at constant volume,*}$$
$$M = \text{total mass of fluid in the control volume.} \tag{6.12}$$

Consider the system illustrated in Figure 6.7 wherein the thermal energy input to a system is that supplied by a carrier. To analyze this system, one

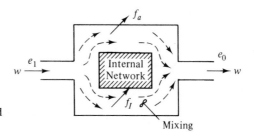

Figure 6.7. Network with forced convection to a carrier fluid.

proceeds as follows: assume that fluid enters at temperature e_i and flow rate w and leaves at temperature e_0 and flow rate w. f_a is the rate of heat loss to surroundings and f_I is the rate of heat supplied to a thermal network embedded in fluid but into which no fluid flows. Adopt the notation

$$e_f = \text{temperature of fluid in mixing tank,}$$

and make the assumption that all of the above approximations apply and that the mixing in the tank is complete, so that

$$e_0 = e_f. \tag{6.13}$$

Also take e_R to be $0°$. Under these assumptions, Equation (6.6) becomes

$$C\frac{de_f}{dt} = -f_I - f_a + pwc_p(e_i - e_0), \tag{6.14}$$

*If the fluid is incompressible, then $c_v = c_p$.

where the *capacitance* is

$$C = Mc_v. \tag{6.15}$$

Thus,

$$\rho w c_p(e_i - e_0) = C\frac{de_0}{dt} + f_a + f_I. \tag{6.16}$$

For an extended linear analysis, this equation is linearized assuming that only small changes occur in w, e_i, e_0, f_a, and f_I. Use superscript 0 for nominal conditions and show that

$$\rho w^0 c_p(e_i^0 - e_0^0) = C\frac{de_0^0}{dt} + f_a^0 + f_I^0.$$

Choose these nominal conditions so that

$$\frac{de_0^0}{dt} = 0, \tag{6.17a}$$

i.e., so that

$$f_a^0 + f_I^0 = \rho w^0 c_p(e_i^0 - e_0^0). \tag{6.17b}$$

For small changes, denoted by δ, Equation (6.16) becomes

$$\rho(w^0 + \delta w)c_p(e_i^0 + \delta e_i - e_0^0 - \delta e_0)$$
$$= C\frac{de_0^0}{dt} + C\frac{d\,\delta e_0}{dt} + f_a^0 + \delta f_a + f_I^0 + \delta f_I. \tag{6.18}$$

Neglect products of variations and use Equations (6.17b) and (6.18) to show that

$$\rho c_p(e_i^0 - e_0^0)\,\delta w + \rho c_p w^0(\delta e_i - \delta e_0) = C\frac{d\,\delta e_0}{dt} + \delta f_a + \delta f_I. \tag{6.19}$$

Define the constants

$$R_w = \frac{1}{\rho c_p w^0}, \tag{6.20}$$

$$T = CR_w, \tag{6.21}$$

and

$$K = \frac{e_1^0 - e_0^0}{w^0}. \tag{6.22}$$

Take the Laplace transform of Equation (6.19) and substitute Equations

(6.20)–(6.22) to obtain

$$(1 + Ts) \delta E_0(s) = \delta E_i(s) + K \delta W - R_w \delta F_a - R_w \delta F_I. \tag{6.23}$$

Finally, drop the δ notation to find the *mixing tank equation*,

$$\boxed{(1 + Ts)E_0(s) = E_i(s) + KW(s) - R_w F_a(s) - R_w F_I(s).} \tag{6.24}$$

F_a and/or F_I can often be eliminated in favor of the temperature E_0 by using techniques outlined in Section 6.1. Many lumped parameter analyses can be performed starting with Equation (6.24).

Example 6.2

Consider the carrier fluid-pipe system illustrated in Figure 6.8. Water is assumed to enter at a constant flow rate (i.e., $\delta w \equiv 0$). Find

$$G(s) \triangleq \frac{E_0(s)}{E_i(s)}. \tag{i}$$

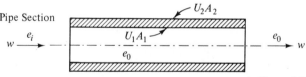

Pipe Section

$U_2 A_2$

$U_1 A_1$

$w \xrightarrow{\quad e_i \quad}$ $\xrightarrow{\quad e_0 \quad} w$

e_0

k, Thermal Conductivity

Ambient (Ground) Temperature, e_a

Figure 6.8. Lumped parameter model of the flow of a carrier fluid in a pipe.

Define the constants

$$R_1 = \frac{1}{U_1 A_1}, \qquad R_2 = \frac{1}{U_2 A_2},$$

$$R = \frac{L}{AK}, \qquad C_w = \text{heat capacity of wall}.$$

L denotes the length of the pipe and A is an *effective* heat transfer area. The analogous electrical network for the solid portion of the system is shown in Figure 6.9. It can be shown that[2]

$$A = -2\pi L^2 \ln\left\{\frac{r_1}{r_2}\right\}. \tag{6.25}$$

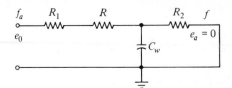

Figure 6.9. Analogy for a portion of the dynamic thermal system of Figure 6.8.

r_1 is the inside radius and r_2 is the outside radius of the pipe. Work backward to find

$$\begin{bmatrix} 0 \\ F \end{bmatrix} = \begin{bmatrix} 1 & -R_2 \\ 0 & 1 \end{bmatrix}\begin{bmatrix} 1 & 0 \\ -C_w s & 1 \end{bmatrix}\begin{bmatrix} 1 & -R \\ 0 & 1 \end{bmatrix}\begin{bmatrix} 1 & -R_1 \\ 0 & 1 \end{bmatrix}\begin{bmatrix} E_0 \\ F_a \end{bmatrix}$$

$$= \begin{bmatrix} 1 + R_2 C_w s & -R_2 - (R + R_1)(1 + R_2 C_w s) \\ -C_w s & 1 + (R + R_1)C_w s \end{bmatrix}\begin{bmatrix} E_0 \\ F_a \end{bmatrix}.$$

Thus,

$$F_a = \frac{1 + R_2 C_w s}{R_2 + (R + R_1)(1 + R_2 C_w s)} E_0. \tag{ii}$$

Since $F_I = 0$ and $\delta w = 0$, it follows from the Equation (6.24) that

$$(1 + Ts)E_0(s) = E_i(s) - \frac{R_w + (1 + R_2 C_w s)}{R_2 + (R + R_1)(1 + R_2 C_w s)} E_0(s),$$

whence

$$G(s) \triangleq \frac{E_0(s)}{E_i(s)} = \frac{R_2 + (R + R_1)(1 + R_2 C_w s)}{R_w(1 + R_2 C_w s) + (1 + Ts)[R_2 + (R + R_1)(1 + R_2 C_w s)]}. \tag{iii}$$

$$* \quad * \quad *$$

Example 6.3

Consider the dynamic thermal system illustrated in Figure 6.10. The basic equation is

$$(1 + Ts)E_0 = E_i + KW - R_w F_a - R_w F_I. \tag{i}$$

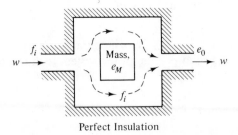

Perfect Insulation

Figure 6.10. Internal solid dynamic thermal system.

Assume that e_i^0 is constant so that $e_i \equiv 0$ and that f_a^0 is constant so that $f_a \equiv 0$.

Find the transfer functions

$$\frac{E_0(s)}{W(s)}$$

and

$$\frac{E_m(s)}{W(s)}.$$

For the internal solid network, the electric analogy is shown in Figure 6.11.

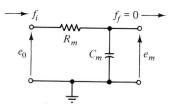

Figure 6.11. Analogy of the internal solid in Figure 6.10.

(R_m due to convection and mass conduction)

Work backward to find

$$
\begin{bmatrix} E_m \\ 0 \end{bmatrix} = \begin{bmatrix} 1 & 0 \\ -C_m s & 1 \end{bmatrix} \begin{bmatrix} 1 & -R_m \\ 0 & 1 \end{bmatrix} \begin{bmatrix} E_0 \\ F_I \end{bmatrix}
$$

$$
= \begin{bmatrix} 1 & -R_m \\ -C_m s & 1 + R_m C_m s \end{bmatrix} \begin{bmatrix} E_0 \\ F_I \end{bmatrix}.
$$

Thus,

$$E_m = E_0 - R_m F_I$$

and

$$F_I = \frac{C_m s}{1 + R_m C_m s} E_0.$$

Thus,

$$E_m = \frac{1}{1 + R_m C_m s} E_0. \tag{ii}$$

From the basic Equation (i), it is then easily shown that

$$\left[\frac{R_w C_m s}{1 + R_m C_m s} + (1 + Ts) \right] E_0 = KW.$$

Finally,

$$G_0 = \frac{E_0(s)}{W(s)} = \frac{K(1 + R_m C_m s)}{1 + (T + R_m C_m + R_w C_m)s + R_m C_m T s^2}.$$

It then follows from Equation (ii) that

$$G_m = \frac{E_m(s)}{W(s)} = \frac{K}{1 + (T + R_m C_m + R_w C_m)s + R_m C_m T s^2}$$

for both cases; the gain $= K$.

If $w(t) = u(t)$, it follows from the initial value theorem that

$$e_0(0^+) = 0, \qquad \dot{e}_0(0^+) = K,$$
$$e_m(0^+) = 0, \qquad \dot{e}_m(0^+) = 0.$$

*   *   *

6.5 Distributed Parameter Models of Dynamic Thermal Systems[1]

To this point, only lumped parameter models have been discussed. Distributed parameter models of dynamic thermal systems are also easily developed and will now be introduced. As indicated in Section 4.7, distributed models provide the more accurate descriptions of dynamic phenomena but are associated with difficult mathematical manipulations.

Begin the discussion with an analysis of the *one-dimensional dispersive transmission line* for which the e and f variables are related in the manner indicated in Figure 6.12. Here

$$\mathfrak{R} = \text{resistance per unit length,}$$

$$\mathfrak{C} = \text{capacitance per unit length.}$$

The transmission line can be used to describe the transfer of electrical signals, in certain cases. For present purposes, it is useful to take the trans-

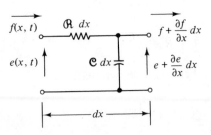

Figure 6.12. Differential element of a dispersive transmission line.

mission line to be the distributed parameter counterpart of the lumped parameter thermal model illustrated in Figure 6.3.

Use transfer matrices and Laplace transformation to show that

$$
\begin{bmatrix} E(x, s) \\ F(x, s) \end{bmatrix} = \begin{bmatrix} 1 & \Re\, dx \\ 0 & 1 \end{bmatrix} \begin{bmatrix} 1 & 0 \\ \mathfrak{C}\, dxs & 1 \end{bmatrix} \begin{bmatrix} E + \dfrac{\partial E}{\partial x} dx \\ F + \dfrac{\partial F}{\partial x} dx \end{bmatrix}
$$

$$
= \begin{bmatrix} 1 + \Re\mathfrak{C}(dx)^2 s & \Re\, dx \\ \mathfrak{C}\, dxs & 1 \end{bmatrix} \begin{bmatrix} E + \dfrac{\partial E}{\partial x} dx \\ F + \dfrac{\partial F}{\partial x} dx \end{bmatrix}.
$$

Neglect terms of $(dx)^2$ to find

$$\frac{\partial E}{\partial x} = -\Re F, \tag{6.26}$$

$$\frac{\partial F}{\partial x} = -\mathfrak{C}sE. \tag{6.27}$$

[Compare with Equations (4.55) and (4.56).] In the time domain, these equations become

$$\frac{\partial e}{\partial x} = -\Re f, \tag{6.28}$$

$$\frac{\partial f}{\partial x} = -\mathfrak{C}\frac{\partial e}{\partial t}. \tag{6.29}$$

These equations are easily combined to show that

$$\frac{\partial^2 e}{\partial x^2} = \Re\mathfrak{C}\frac{\partial e}{\partial t}, \tag{6.30}$$

which is known as the one-dimensional *diffusion equation*. It follows from Equations (6.26) and (6.27) and the solution of Problem 2.9 that the basic transfer matrix equation is

$$
\begin{bmatrix} E(0, s) \\ F(0, s) \end{bmatrix} = \begin{bmatrix} \cosh\{\sqrt{\Re\mathfrak{C}s}\,x\} & \sqrt{\dfrac{\Re}{\mathfrak{C}s}}\sinh\{\sqrt{\Re\mathfrak{C}s}\,x\} \\ \sqrt{\dfrac{\mathfrak{C}s}{\Re}}\sinh\{\sqrt{\Re\mathfrak{C}s}\,x\} & \cosh\{\sqrt{\Re\mathfrak{C}s}\,x\} \end{bmatrix} \begin{bmatrix} E(x, s) \\ F(x, s) \end{bmatrix}.
$$

$$\tag{6.31}$$

Example 6.4

Heat conduction in a solid: Determine the transfer function

$$\frac{E(x, s)}{F(0, s)} \triangleq G(s)$$

for the solid subjected to the input and boundary conditions illustrated in Figure 6.13. Work forward [use Equation 6.31)] and show that

$$\begin{bmatrix} E(0, s) \\ F(0, s) \end{bmatrix} = \begin{bmatrix} \cdots & \cdots \\ \sqrt{\dfrac{\mathcal{C}s}{\mathcal{R}}} \sinh\{\sqrt{\mathcal{R}\mathcal{C}s}x\} & \cosh\{\sqrt{\mathcal{R}\mathcal{C}s}x\} \end{bmatrix} \begin{bmatrix} E(x, s) \\ F(x, s) \end{bmatrix};$$

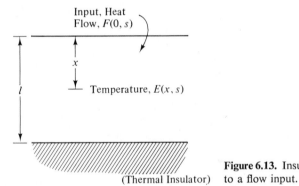

Figure 6.13. Insulated solid subjected to a flow input.

thus,

$$F(0, s) = \sqrt{\frac{\mathcal{C}s}{\mathcal{R}}} \sinh\{\sqrt{\mathcal{R}\mathcal{C}s}x\}E(x, s) + \cosh\{\sqrt{\mathcal{R}\mathcal{C}s}x\}F(x, s). \qquad \text{(i)}$$

To eliminate the extraneous power state variable, $F(x, s)$, work backward to find

$$\begin{bmatrix} E_f(l, s) \\ 0 \end{bmatrix} = \begin{bmatrix} \cdots & \cdots \\ -\sqrt{\dfrac{\mathcal{C}s}{\mathcal{R}}} \sinh\{\sqrt{\mathcal{R}\mathcal{C}s}(l - x)\} & \cosh\{\sqrt{\mathcal{R}\mathcal{C}s}(l - x)\} \end{bmatrix} \begin{bmatrix} E(x, s) \\ F(x, s) \end{bmatrix};$$

thus,

$$F(x, s) = \sqrt{\frac{\mathcal{C}s}{\mathcal{R}}} \frac{\sinh\{\sqrt{\mathcal{R}\mathcal{C}s}(l - x)\}}{\cosh\{\sqrt{\mathcal{R}\mathcal{C}s}(l - x)\}} E(x, s). \qquad \text{(ii)}$$

Combining Equations (i) and (ii) leads to the following result:

$$G(s) = \frac{E(x, s)}{F(0, s)} = \frac{1}{\sqrt{\frac{\mathcal{C}s}{\mathcal{R}}} \sinh\{\sqrt{\mathcal{R}\mathcal{C}s}x\} + \tanh\{\sqrt{\mathcal{R}\mathcal{C}s}(l - x)\}\cosh\{\sqrt{\mathcal{R}\mathcal{C}s}x\}}.$$

$$(6.32)$$

* * *

The generalized procedure for derivation of the basic transfer matrix equation for any distributed parameter component is described below:

Step 1. Hypothesize a micro-lumped model with series element parameters quantified as impedance per unit length and shunt element parameters quantified as conductance per unit length.

Step 2. Multiply the micro-lumped transfer matrices together, and neglect terms of $(dx)^2$ to obtain equations of the form

$$\frac{\partial E}{\partial x} = -Z_1(s)F, \tag{6.33a}$$

$$\frac{\partial F}{\partial x} = -\frac{1}{Z_2(s)}E. \tag{6.33b}$$

Z_1 should have the units of impedance per unit length, and $1/Z_2$ should have the units of conductance per unit length.

Step 3. It follows from the results of Problem 2.9 that

$$\begin{bmatrix} E(0, s) \\ F(0, s) \end{bmatrix} = \begin{bmatrix} \cosh\left\{\sqrt{\frac{Z_1}{Z_2}}x\right\} & \sqrt{Z_1 Z_2}\sinh\left\{\sqrt{\frac{Z_1}{Z_2}}x\right\} \\ \frac{1}{\sqrt{Z_1 Z_2}}\sinh\left\{\sqrt{\frac{Z_1}{Z_2}}x\right\} & \cosh\left\{\sqrt{\frac{Z_1}{Z_2}}x\right\} \end{bmatrix} \begin{bmatrix} E(x, s) \\ F(x, s) \end{bmatrix}.$$

$$(6.34)$$

The occurrence of transcendental functions of s in the basic transfer matrix can lead to considerable mathematical complexity.

6.6 Distributed Parameter Models of Dynamic Thermal and Diffusion Systems Interacting with a Carrier Fluid; The *Percolation* Process [5,6]

The analyst who wishes to derive distributed parameter models of the interaction of carrier fluids with solid networks can start with Equation (6.6) (provided the carrier approximation is valid). The procedure will be illustrated with the derivation of some one-dimensional models of this general type. Takahashi has supplied a more comprehensive set of examples.[5]

Consider a section of *differential length* of the pipe illustrated in Figure 6.8. This section is illustrated in Figure 6.14. It will be assumed that variations occur only in the axial (i.e., the x) direction so that a one-dimensional model will be derived. It is assumed that the approximations summarized in Equations (6.7), (6.9), and (6.11) are valid.

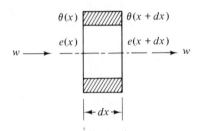

$\theta(x)$ $\theta(x + dx)$

$e(x)$ $e(x + dx)$

$w \longrightarrow$ $\longrightarrow w$

$|\!\!\leftarrow dx \rightarrow\!\!|$

Figure 6.14. Differential section of the pipe section illustrated in Figure 6.8. θ denotes solid temperature and e denotes fluid temperature.

Take the control surface to be just inside the inner surface of the pipe; then

$$\oiint_{\substack{\text{control}\\\text{surface}}} \{\rho i + p\}\vec{\mathbf{V}} \cdot d\mathbf{A} = \rho c_p[-e(x) + e(x + dx)]w. \qquad (6.35)$$

Expand $e(x + dx)$ in a Taylor series and show that

$$\oiint_{\substack{\text{control}\\\text{surface}}} \{\rho i + p\}\vec{\mathbf{V}} \cdot d\mathbf{A} = \rho c_p w \frac{\partial e}{\partial x}\, dx. \qquad (6.36)$$

Also show that

$$\left.\frac{dN}{dt}\right|_{\text{control volume}} = \mathcal{C}\, dx \frac{\partial e}{\partial t}, \qquad (6.37)$$

where

$$\mathcal{C} = \text{fluid capacitance per unit length}$$
$$\text{[related to } C \text{ of Equation (6.15)]}. \qquad (6.38)$$

The term dQ/dt in Equation (6.6) is interpreted as the flow of heat from the pipe to the fluid. Use the notation

$$\frac{1}{\mathcal{R}_C} = \text{convective conductance per unit length at the inner}$$
$$\text{surface of the pipe [refer to Equation (6.5)]}. \qquad (6.39)$$

Then, for the fluid control volume

$$\frac{dQ}{dt} = \frac{dx}{\Re_C}[\theta - e]. \tag{6.40}$$

Combine Equations (6.36), (6.37), and (6.40) to show that Equation (6.6) reduces to

$$e\frac{\partial e}{\partial t} = \frac{1}{\Re_C}[\theta - e] - \rho c_p w \frac{\partial e}{\partial x},$$

which may be rewritten as

$$\frac{\partial e}{\partial t} + \frac{\rho c_p w}{e}\frac{\partial e}{\partial x} = \frac{1}{\Re_C e}[\theta - e]. \tag{6.41}$$

This equation must be coupled with an equation for the rate of change of $\theta(t)$ in order to complete the model. It is left as an exercise for the reader to show that

$$\frac{\partial \theta}{\partial t} = \frac{1}{\Re_C e_W}[e - \theta] - \frac{1}{\Re e_W}\theta, \tag{6.42}$$

where

$$e_W = \text{capacitance of the pipe per unit length}, \tag{6.43}$$

$$\frac{1}{\Re} = \text{convective conductance per unit length}$$
$$\text{at the outer surface of the pipe}, \tag{6.44}$$

and it is assumed that

1. The conductivity of the pipe is so high that no variations in θ occur in the radial direction and that heat conduction in the radial direction is insignificant. (6.45)

2. The ambient temperature is taken as the reference temperature and is constant in the x direction. (6.46)

Denote the *effective velocity*

$$v = \frac{\rho c_p w}{e}. \tag{6.47}$$

If the fluid is incompressible, $c_v = c_p$ and

$$v = \frac{\rho w}{\mathfrak{M}},$$

where

$$\mathfrak{M} = \text{Fluid mass per unit length}$$

$$= \rho A,$$

$$A = \text{Cross-sectional area of the pipe.}$$

Thus, for incompressible fluids

$$v = \frac{w}{A}. \tag{6.48}$$

In any case, Equation (6.41) can be rewritten

$$\frac{\partial e}{\partial t} + v\frac{\partial e}{\partial x} = \frac{1}{\mathfrak{R}_c\mathfrak{C}}[\theta - e]. \tag{6.49}$$

Example 6.5

Rederive the transfer function

$$G(s) = \frac{E_0(s)}{E_i(s)}$$

for the pipe section illustrated in Figure 6.8 using the one-dimensional distributed parameter model.

Introduce the dimensionless variables

$$\tau = \frac{vt}{L}, \tag{6.50}$$

$$\eta = \frac{x}{L} \tag{6.51}$$

(L is the total length of the pipe) and the dimensionless constants

$$a = \frac{L}{v\mathfrak{R}_c\mathfrak{C}}, \tag{6.52}$$

$$b = \frac{L}{v\mathfrak{R}_c\mathfrak{C}_w}, \tag{6.53}$$

$$d = \frac{L}{v\mathfrak{R}\mathfrak{C}_w}, \tag{6.54}$$

and show that Equations (6.42) and (6.49) become

$$\frac{\partial\theta}{\partial\tau} = b[e - \theta] - d\theta \tag{6.55}$$

and

$$\frac{\partial e}{\partial \tau} + \frac{\partial e}{\partial \eta} = a[\theta - e].$$
(6.56)

Take the Laplace transform of these equations (in the dimensionless time, τ), assume initial quiescence, and show that

$$s\Theta = b[E - \Theta] - d\Theta$$

and

$$\frac{dE}{d\eta} = a[\Theta - E] - sE.$$

Combine these equations to show that

$$\Theta = \frac{b}{s + b + d}E$$
(6.57)

and

$$\frac{dE}{d\eta} = -\left[\frac{s^2 + (a + b + d)s + ad}{s + b + d}\right]E.$$
(6.58)

Notice that Equation (6.57) defines a transfer function relation between pipe temperature and fluid temperature. It follows from Equation (6.58) that

$$E(\eta, s) = e^{-A(s)\eta}E(0, s),$$
(6.59)

where $A(s)$ is the coefficient of E on the right-hand side of Equation (6.58). Note that

$$E_0(s) = E(1, s)$$

since $\eta = 1$ when $x = L$. It follows that the desired transfer function

$$G(s) = e^{-A(s)},$$
(6.60)

where

$$A(s) = \frac{s^2 + (a + b + d)s + ad}{s + b + d}.$$
(6.61)

Compare Equation (6.60) with Equation (iii) of Example 6.2. The comparison is not strictly valid since fluid capacitance was neglected in the former example.

$$* \quad * \quad *$$

The exchange of energy of a carrier fluid with its conduit is referred to as a *percolation* process and occurs in many practical situations.

6.7 State Variable Models for the Transfer of Heat in Solids; Distributed *Modes*

If the analyst has decided that a lumped parameter model of heat transfer dynamics is sufficiently accurate and he has constructed the analogous electric circuit, a state variable model can be determined in the manner indicated in Section 4.8. Since inertance components do not appear in dynamic heat transfer systems, the task of constructing state variable heat transfer models is made somewhat easier.

To illustrate the procedure, consider the conduction of heat in a one-dimensional rod. The analogous network for a two-lump model of the rod is illustrated in Figure 6.15.

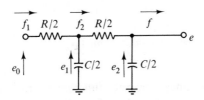

Figure 6.15. Two-lump model of a conducting rod.

The capacitive relations are

$$\frac{C}{2}\frac{de_1}{dt} = f_1 - f_2,$$

$$\frac{C}{2}\frac{de_2}{dt} = f_2 - f.$$

$$(6.62)$$

To eliminate the extraneous f variables note that

$$f_1 = \frac{e_0 - e_1}{R/2},$$

$$f_2 = \frac{e_1 - e_2}{R/2},$$

$$(6.63)$$

so that the state variable equations become

$$\frac{d}{dt}\begin{bmatrix} e_1 \\ e_2 \end{bmatrix} = \begin{bmatrix} \dfrac{-8}{RC} & \dfrac{4}{RC} \\ \dfrac{4}{RC} & \dfrac{-4}{RC} \end{bmatrix}\begin{bmatrix} e_1 \\ e_2 \end{bmatrix} + \begin{bmatrix} \dfrac{4}{RC} & 0 \\ 0 & -1 \end{bmatrix}\begin{bmatrix} e_0 \\ f \end{bmatrix} \qquad (6.64)$$

If n lumps had been used instead of two, Equations (6.62) would have been

$$\frac{C}{n}\frac{de_k}{dt} = f_k - f_{k+1}, \quad k = 1, 2, \ldots, n.$$

Equations (6.63) would become

$$f_k = \frac{e_{k-1} - e_k}{R/n}, \quad k = 1, 2, \ldots, n.$$

And the state variable equations would become

$$\frac{d}{dt}\begin{bmatrix} e_1 \\ e_2 \\ e_3 \\ \cdot \\ \cdot \\ \cdot \\ e_n \end{bmatrix} = \begin{bmatrix} \dfrac{-2n^2}{RC} & \dfrac{n^2}{RC} & 0 & \cdots & 0 \\ \dfrac{n^2}{RC} & \dfrac{-2n^2}{RC} & \dfrac{n^2}{RC} & \cdots & 0 \\ 0 & \dfrac{n^2}{RC} & \dfrac{-2n^2}{RC} & \cdots & 0 \\ \cdot & \cdot & \cdot & & \cdot \\ \cdot & \cdot & \cdot & & \cdot \\ \cdot & \cdot & \cdot & & \cdot \\ 0 & 0 & 0 & \cdots & \dfrac{-n^2}{RC} \end{bmatrix}\begin{bmatrix} e_1 \\ e_2 \\ e_3 \\ \cdot \\ \cdot \\ \cdot \\ e_n \end{bmatrix} + \begin{bmatrix} \dfrac{n^2}{RC} & 0 \\ 0 & 0 \\ \cdot & \cdot \\ \cdot & \cdot \\ \cdot & \cdot \\ 0 & -1 \end{bmatrix}\begin{bmatrix} e_0 \\ f \end{bmatrix}.$$

$$(6.65)$$

One would expect that as n is increased, the above representation would be made more accurate.

An alternative state variable representation can be developed by starting with a modification of Equation (6.30).[4] The great advantage of the alternative representation is that distributed effects are modeled. A complete discussion of the trade-offs involved in the use of the representations must be delayed until Chapter 14.

The description begins with an analysis of Equation (6.30). It is assumed that the temperature of the left end is held at zero, i.e.,

$$e(0, t) = 0, \tag{6.66}$$

and that the flow at the right end is held at zero (i.e., the right end is insulated), i.e., the temperature gradient

$$\frac{\partial e}{\partial x}(L, t) = 0. \tag{6.67}$$

The solution obtained, for this unforced case, is referred to as the *homogeneous* solution and will be used later when the forced case is discussed.

Attempt a solution of the form

$$e(x, t) = \psi(x)\phi(t). \tag{6.68}$$

For obvious reasons, this method is known as the *method of separation of variables*. Substitute Equation (6.68) into Equation (6.30) and find

$$\frac{1}{\psi(x)} \frac{d^2\psi(x)}{dx^2} = \frac{\Re\mathfrak{C}}{\phi(t)} \frac{d\phi(t)}{dt}. \tag{6.69}$$

The interesting point is that the left-hand side of this equation is a function only of x and the right-hand side is a function only of t. The equality can be preserved only if both sides are equal to the same constant, say $-\omega^2$ (the reason that $+\omega^2$ cannot be used will be given later). It follows that

$$\frac{d^2\psi}{dx^2} + \omega^2\psi = 0, \tag{6.70}$$

and that

$$\frac{d\phi}{dt} + \frac{\omega^2}{\Re\mathfrak{C}}\phi = 0. \tag{6.71}$$

Clearly,

$$\psi = A \cos(\omega x) + B \sin(\omega x), \tag{6.72}$$

$$\phi = Ce^{-\omega^2 t/\Re\mathfrak{C}}, \tag{6.73}$$

where A, B, C are arbitrary constants.

To satisfy Equation (6.66), choose

$$A = 0,$$

so that

$$e(x, t) = De^{-\omega^2 t/\Re\mathfrak{C}} \sin(\omega x), \tag{6.74}$$

where D is an arbitrary constant. The problem that remains is to satisfy Equation (6.67). This can be accomplished by selecting the proper value for ω [the reader should verify that this task could not be accomplished if both sides of Equation (6.69) were set equal to $+\omega^2$]. By applying Equation (6.67) to Equation (6.74) it is easily shown that the proper choice of ω is

$$\omega_k = \frac{k\pi}{L2},$$

$$k = \text{Odd integer.}$$

Another, more convenient, way to write the same equation is

$$\omega_k = \frac{(2k - 1)\pi}{2L},$$ (6.75)

$$k = \text{Any integer.}$$ (6.76)

Thus, there are a countably infinite number of values of ω which will do the job. Equation (6.30) is linear so that superposition applies. Thus, the most general solution is the sum of the solutions corresponding to the different values of ω_k, that is,

$$e(x, t) = \sum_{k=1}^{\infty} D_k e^{-\omega^2_k t / \Re \mathfrak{C}} \sin(\omega_k x),$$ (6.77)

which is the desired homogeneous solution. For the unforced case, the D_k could be chosen to satisfy some initial conditions (temperature distribution) for the bar.

The more interesting case is the forced case. In this illustrative example, it will be assumed that a temperature, denoted $e_0(t)$, is imposed on the left end of the bar and that a heat flow, denoted $f_1(t)$, is imposed on the right end of the bar. The analyst has two choices: Modify conditions (6.66) and (6.67) or keep these conditions and assume that the inputs are generated just inside the ends of the bar. The latter procedure is adopted here because the homogeneous solution derived above can be exploited. It should be noted that errors near the ends of the rod will occur, but as will become apparent, the extent of the erroneous temperature distribution can be made negligibly small. The modifications that must be made in Equations (6.28) and (6.29) are

$$\frac{\partial e}{\partial x} = -\Re f + e_0(t)\,\delta(x)^+,$$ (6.78)

$$\frac{\partial f}{\partial x} = -\mathfrak{C}\frac{\partial e}{\partial t} - f_1(t)\,\delta(x - L)^-,$$ (6.79)

where $\delta(\ )$ denotes a delta function in the variable x. It is left as an exercise to verify that these delta functions must appear as they do in these equations. The final minus sign in Equation (6.79) denotes the fact that f_1 is taken positive when energy is extracted from the rod. The superscript $+$ in Equation (6.78) denotes the fact that the delta appears at a value of x slightly greater than zero. Similarly, the superscript $-$ in Equation (6.79) denotes the fact that the delta occurs slightly before $x = L$.

When Equations (6.78) and (6.79) are combined in a "formal" (nonrigorous) fashion, the result is

$$\Re C \frac{\partial e}{\partial t} = \frac{\partial^2 e}{\partial x^2} - e_0(t)\frac{d\delta(x)^+}{dx} - \Re f_1(t)\,\delta(x - L)^-. \tag{6.80}$$

From what has occurred to this point, there is no reason to believe that any meaning, either physical or mathematical, can be associated with the derivative of a delta function. A completely rigorous treatment of such functions is well beyond the scope of the present discussion, and such terms will be handled in a formal manner. The reader is referred to the literature for a more complete discussion.[7]

Equation (6.80) is the forced version of Equation (6.30). Taking a clue from Equation (6.77) a solution of the form

$$e(x, t) = \sum_{k=1}^{\infty} z_k(t)\,\sin\,(\omega_k x) \tag{6.81}$$

is attempted. Notice that conditions (6.66) and (6.67) are satisfied by this guessed solution. The $z_k(t)$ are to be determined.

Substitute Equation (6.81) into (6.80) and find

$$\Re C \sum_{k=1}^{\infty} \dot{z}_k(t)\,\sin\,(\omega_k x)$$

$$= -\sum_{k=1}^{\infty} z_k(t)\omega_k^2\,\sin\,(\omega_k x) - e_0(t)\frac{d\delta(x)^+}{dx} - \Re f_1(t)\,\delta(x-L)^-. \tag{6.82}$$

A remarkable property of the sine functions, called the *orthogonality* property, will now be exploited. Use elementary calculus or a table of integrals to show that

$$\int_0^L \sin\left[\frac{(2k-1)\pi x}{2L}\right]\sin\left[\frac{(2l-1)\pi x}{2L}\right]dx = \int_0^L \sin\,(\omega_k x)\sin\,(\omega_l x)\,dx$$

$$= \begin{cases} 0, & l \neq k, \\ \dfrac{L}{2}, & l = k, \end{cases} \tag{6.83}$$

where *l and k are any two integers*.

Multiply Equation (6.82) by $\sin\,(\omega_l x)$ and integrate over x between 0 and

$+L$. It follows from Equation (6.83) that

$$\frac{L\Re C \dot{z}_l(t)}{2} = -\frac{\omega_l^2 L z_l(t)}{2} - e_0(t) \int_0^L \sin(\omega_l x)\frac{d\,\delta(x)^+}{dx}\,dx$$

$$-\Re f_1(t) \int_0^L \sin(\omega_l x)\,\delta(x - L)^-\,dx. \tag{6.84}$$

By the definition of the delta function

$$\int_0^L \sin(\omega_l x)\,\delta(x - L)^-\,dx = \sin(\omega_l L)$$

$$= \sin\frac{(2l - 1)\pi}{2}$$

$$= (-1)^{l-1}. \tag{6.85}$$

Proceed in a formal way and integrate the first integral in Equation (6.84) by parts to show that

$$\int_0^L \sin(\omega_l x)\frac{d\,\delta(x)^+}{dx}\,dx = \delta(x)^+ \sin(\omega_l x)\bigg|_0^L - \int_0^L \delta(x)^+\omega_l \cos(\omega_l x)\,dx.$$

It follows from Equation (2.54) that the first term on the right-hand side of this equation is zero so that

$$\int_0^L \sin(\omega_l x)\frac{d\,\delta(x)^+}{dx}\,dx = -\int_0^L \delta(x)^+\omega_l \cos(\omega_l x)\,dx = -\omega_l \cos(0) = -\omega_l. \tag{6.86}$$

Combine Equations (6.84)–(6.86) to show that

$$\dot{z}_l = \frac{-\omega_l^2}{\Re C}z_l + \frac{2\omega_l}{\Re C L}e_0(t) + \frac{2(-1)^l}{C L}f_1(t), \quad l = 1, 2, 3, \ldots, \infty. \tag{6.87}$$

As with the representation of Equation (6.65), an infinite number of Equations (6.87) must be retained in order to achieve perfect accuracy. In any practical situation, the analyst will retain a finite number, say n, of these equations. For example, if the first n of these equations are retained, the matrix (state variable) equation is of the form

$$
\frac{d}{dt}
\begin{bmatrix}
z_1 \\
z_2 \\
z_3 \\
\cdot \\
\cdot \\
\cdot \\
z_n
\end{bmatrix}
=
\begin{bmatrix}
\dfrac{-\omega_1^2}{\Re C} & 0 & 0 & \cdots & 0 \\
0 & \dfrac{-\omega_2^2}{\Re C} & 0 & \cdots & 0 \\
0 & 0 & \dfrac{-\omega_3^2}{\Re C} & \cdots & 0 \\
\cdot & \cdot & & & \\
\cdot & \cdot & & & \\
\cdot & \cdot & & & \\
0 & 0 & & \cdots & \dfrac{-\omega_n^2}{\Re C}
\end{bmatrix}
\begin{bmatrix}
z_1 \\
z_2 \\
z_3 \\
\cdot \\
\cdot \\
\cdot \\
z_n
\end{bmatrix}
$$

$$
+ \frac{2}{CL}
\begin{bmatrix}
\dfrac{\omega_1}{\Re} & -1 \\
\dfrac{\omega_2}{\Re} & 1 \\
\dfrac{\omega_3}{\Re} & -1 \\
\cdot & \cdot \\
\cdot & \cdot \\
\cdot & \cdot \\
\dfrac{\omega_n}{\Re} & (-1)^n
\end{bmatrix}
\begin{bmatrix}
e_0 \\
f_1
\end{bmatrix}.
\tag{6.88}
$$

Equation (6.88) should be compared with Equation (6.65). The z_i are called *distributed modes*. Notice that the *modal equations* (6.87) are decoupled and can be solved individually and *analytically* using the techniques introduced in Chapter 2. The temperature at any point, x, can be obtained from Equation (6.81). As discussed above, Equation (6.81) is in error for e at $x = 0$ and for f at $x = L$.

A general method for finding the proper initial conditions for the modal variables is fairly easy to obtain and is left to the Problems at the end of the chapter.

6.8 The *Dead Time* Element

Consider a section of pipe which is perfectly insulated and for which the side walls have negligible heat capacitance. The section is illustrated in Figure 6.16. Assume that longitudinal mixing is negligible so that the temperature

$$
e_0(t) = e_i(t - \tau),
\tag{6.89}
$$

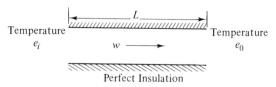

Figure 6.16. Dead time element.

where the *holdup* or *dead* time

$$\tau = \frac{L}{v} = \frac{LA}{w}. \tag{6.90}$$

It follows from the shifting theorem that

$$E_0(s) = e^{-\tau s} E_i(s),$$

so that the transfer function

$$\frac{E_0(s)}{E_i(s)} = e^{-\tau s} \tag{6.91}$$

This exponential function of s is the characteristic transfer function of the so-called dead time process. Obviously, the i transfer state is unilaterally coupled to the 0 transfer state.

Dead time can often be a bothersome element for analysis and design and is rarely synthesized intentionally. The design of a system with a dead time element is often unavoidable. For instance, it may not be possible to place a sensor at the precise location of an important output variable. For example, suppose that for the pipe section in Figure 6.16, a thermocouple can be placed at position 0 but not at position i. Clearly, this physical consideration leads to the introduction of a dead time element into the dynamic system.

Another example of dead time occurring as the result of sensor placement is provided in Problem 2.10.

There is at least one important case where the designer would in a sense introduce the dead time element intentionally. On occasion, distributed elements such as electrical cables and pneumatic transmission lines have negligible resistance and may be modeled as indicated in Figure 6.17. It is easily shown that equations similar to Equations (6.33a) and (6.33b) can be derived to indicate that

$$Z_1(s) = \mathcal{L}s \tag{6.92}$$

and

$$Z_2(s) = \frac{1}{\mathcal{C}s}. \tag{6.93}$$

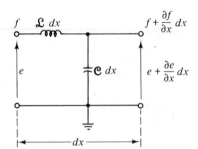

Figure 6.17. Differential element of the *lossless* transmission line.

If the transmission line is of length L, it follows from Equation (6.34) that

$$
\begin{bmatrix} E(0, s) \\ F(0, s) \end{bmatrix} = \begin{bmatrix} \cosh\{\sqrt{\mathcal{L}\mathcal{C}}sL\} & \sqrt{\dfrac{\mathcal{L}}{\mathcal{C}}}\sinh\{\sqrt{\mathcal{L}\mathcal{C}}sL\} \\ \sqrt{\dfrac{\mathcal{L}}{\mathcal{C}}}\sinh\{\sqrt{\mathcal{L}\mathcal{C}}sL\} & \cosh\{\sqrt{\mathcal{L}\mathcal{C}}sL\} \end{bmatrix}\begin{bmatrix} E(L, s) \\ F(L, s) \end{bmatrix}
$$

$$(6.94)$$

Suppose that the measurement system is as illustrated in Figure 6.18. Use Equation (6.94) and standard transfer matrix procedures to show that

$$
\frac{E(L, s)}{F(0, s)} = \frac{1}{\cosh\{\sqrt{\mathcal{L}\mathcal{C}}sL\} + (1/R_c)(\mathcal{L}/\mathcal{C})^{1/2}\sinh\{\sqrt{\mathcal{L}\mathcal{C}}sL\}}. \qquad (6.95)
$$

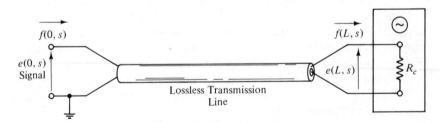

Figure 6.18. Transmission to an instrument via a lossless line.

Suppose now that the designer may *tune* the system by choosing

$$
R_c = \sqrt{\frac{\mathcal{L}}{\mathcal{C}}}. \qquad (6.96)
$$

For this case, Equation (6.95) becomes

$$
\frac{E(L, s)}{E(0, s)} = e^{-\tau s}, \qquad (6.97)
$$

where

$$\tau = \sqrt{\mathfrak{L}\mathfrak{C}}L. \tag{6.98}$$

In other words, with the system tuned, the only distortion in the signal is a simple delay.

References

1. BROWN, F., "A Unified Approach to the Analysis of Uniform One Dimensional Distributed Systems," *A.S.M.E. Paper 66-WA/AUT-20*, 1966.

2. ECKERT, E., and DRAKE, R., JR., *Heat and Mass Transfer*, McGraw-Hill, New York, 1959.

3. HOLMAN, J., *Thermodynamics*, McGraw-Hill, New York, 1969.

4. KARNOPP, D., and ROSENBERG, R., *Analysis and Simulation of Multiport Systems*, M.I.T. Press, Cambridge, Mass., 1968.

5. TAKAHASHI, Y., "Transfer Function Analysis of Heat Exchange Processes," in *Automatic and Manual Control*, Butterworth's, London, 1952.

6. TAKAHASHI, Y., RABINS, M. J., and AUSLANDER, D. M., *Control*, Addison-Wesley, Reading, Mass., 1970.

7. ZADEH, L., and DESOER, C., *Linear System Theory: The State Space Approach*, McGraw-Hill, New York, 1963.

Problems

6.1. Determine an equation for resistance per unit length, $\mathfrak{R}$, and compliance per unit length, $\mathfrak{C}$, in terms of the fundamental parameters ρ, c_p, and k.

6.2. Derive Equation (6.42).

6.3. Rederive Equation (6.77) for the condition

$$e(L, t) = 0$$

instead of that indicated in Equation (6.67).

6.4. Integrate Equation (6.78) with respect to x over a short section near the left end of the rod and indicate why this equation might be correct. Proceed in the same way with Equation (6.79) near the right end of the rod.

6.5. Rederive Equation (6.87) when $e(L, t)$ rather than $f(L, t)$ is an input. Use the result of Problem 6.3.

6.6. Use Equations (6.81) and (6.83) to derive initial conditions for the $z_k(t)$ when $e(x, 0)$ is known.

6.7. Rederive Equation (6.32) but use one of the lumped models indicated in Figure 6.3. Calculate gains and time constants for all the materials indicated in Table 6.1.

6.8. Assume that the conductor shown in Figure 6.19 has zero heat capacity and derive the transfer function

$$\frac{E_0(s)}{E_i(s)},$$

where $e_0(t)$ is the outlet temperature of the cold fluid and $e_i(t)$ is the inlet temperature of the hot fluid.

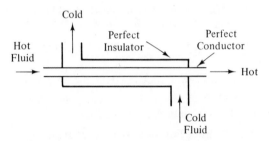

Figure 6.19. *Counterflow* heat exchanger.

6.9. Repeat Problem 6.8 for the parallel flow exchanger shown in Figure 6.20.

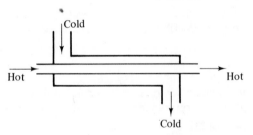

Figure 6.20. *Parallel flow* heat exchanger.

6.10. Determine the transfer function

$$\frac{T_F(s)}{A(s)}.$$

See Figure 6.21.

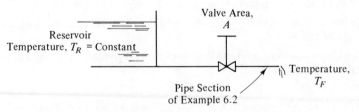

Figure 6.21. Combined hydraulic fluid-carrier fluid problem.

Prime Movers, Sensors, and Feedback Compensators

7

This chapter is the final chapter on dynamic systems analysis. While quite similar in form to Chapters 3–6, this chapter contains the connecting link between dynamic systems analysis and succeeding chapters on simulation and automatic feedback control.

The first part of the chapter contains a discussion of *transducers*, which are devices which convert power in one form (electrical or mechanical or hydraulic or pneumatic, etc.) into another form. The prime movers and sensors discussed in Chapter 1 are examples of such devices and will be the subject of some discussion.

The second part of the chapter provides an introduction to block diagram algebra and rudimentary automatic control theory.

7.1 Electromechanical Transducers[1, 3, 6, 10]

Consider the current-carrying coil in a magnetic field which is illustrated in Figure 7.1. Denote the electrical (voltage-current) power state by e and f and the mechanical (torque-angular velocity) power state by e_m and f_m. Use the relations indicated in Figure 7.1 to show that

$$e_m = nabB \cos (\theta)f. \qquad (7.1)$$

Define

$$g(t) = nabB(t) \cos \{\theta(t)\}, \qquad (7.2)$$

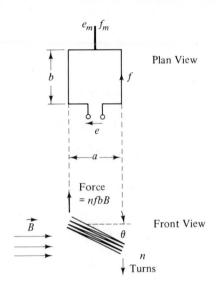

Plan View

Front View

Figure 7.1. Current-carrying coil in a uniform magnetic field, $\vec{B}$. The force per unit length is the vector cross product of current into the $\vec{B}$ field.[5]

so that Equation (7.1) may be rewritten

$$e_m = g(t)f \qquad (7.3)$$

if the field, $\vec{B}$, is uniform. No net work will be done on the coil by the magnetic field[5] so that

$$ef = e_m f_m, \qquad (7.4)$$

whence

$$ef = g(t)ff_m$$

or

$$f_m = \frac{1}{g(t)}e, \qquad (7.5)$$

so that

$$\begin{bmatrix} e_m \\ f_m \end{bmatrix} = \begin{bmatrix} 0 & g(t) \\ \dfrac{1}{g(t)} & 0 \end{bmatrix} \begin{bmatrix} e \\ f \end{bmatrix}. \qquad (7.6)$$

As was indicated in Chapter 3,

$$\begin{bmatrix} e \\ f \end{bmatrix} = \begin{bmatrix} 0 & g(t) \\ \dfrac{1}{g(t)} & 0 \end{bmatrix} \begin{bmatrix} e_m \\ f_m \end{bmatrix}. \qquad (7.7)$$

Equation (7.5) could also have been deduced from Faraday's law (voltage is equal to the rate of change of magnetic flux in the circuit).

Notice that the magnitude of the magnetic field, $B(t)$, influences the flow of power between the mechanical and electrical power states without a direct exchange of energy between either of these power states and the source of the magnetic field. One is reminded of the discussion of the symmetric rotating body in Section 4.6.

If one of the mechanical power state variables is the input variable, the device illustrated in Figure 7.1 is called a *generator*, and if the magnetic field or one of the electrical power state variables is the input variable, the device is called a d-c *motor*.

The electric coil circuit illustrated in Figure 7.1 is called the *armature*. The field, B, may be created by a permanent magnet or, more often, by another circuit which will be called the field circuit.

The *armature-controlled* d-c motor is a device wherein the field, B, is constant and the input is an electrical signal in the armature circuit. The standard symbol for this device is illustrated in Figure 7.2. These devices can

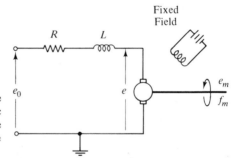

Figure 7.2. Standard symbol for the fixed field, armature-controlled d-c motor. The resistance and inductance are associated with the armature winding.

be designed with several coils so that θ is, effectively, a constant. Then, as is illustrated in Equation (7.2), g is also a constant. For this case Equation (7.7) can be transformed to the Laplace domain to find

$$\begin{bmatrix} E \\ F \end{bmatrix} = \begin{bmatrix} 0 & g \\ \dfrac{1}{g} & 0 \end{bmatrix} \begin{bmatrix} E_m \\ F_m \end{bmatrix}. \qquad (7.8)$$

Notice the characteristic gyrator matrix. Since the armature coil has resistance and inductance, the input to the circuit is $e_0(t)$ and not $e(t)$. Thus, the mechanical power state is related to the input power state by

$$\begin{bmatrix} E_0 \\ F \end{bmatrix} = \begin{bmatrix} 1 & R + Ls \\ 0 & 1 \end{bmatrix} \begin{bmatrix} 0 & g \\ \dfrac{1}{g} & 0 \end{bmatrix} \begin{bmatrix} E_m \\ F_m \end{bmatrix}. \qquad (7.9)$$

It follows from Equation (7.8) that the mechanical impedances will be reflected back through the motor into the electrical circuit. Thus the input signal must be of sufficiently high power to drive the mechanical system. For this reason a power amplifier, whose output is e_0, is usually put in cascade with an armature-controlled d-c motor.

An alternative device is one for which the armature current, f, is kept constant and control is effected by manipulating the value of the magnitude of the magnetic field, $B(t)$. This device is called a *field-controlled* d-c motor, and the standard symbol is illustrated in Figure 7.3. Since the field, B, is proportional to field current, f_B, it follows from Equation (7.1) that

$$e_m = K_F f_B, \tag{7.10a}$$

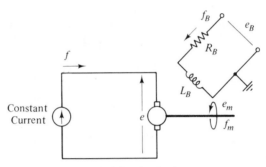

Figure 7.3. Standard symbol for the fixed armature current, field-controlled d-c motor.

where K_F is a proportionality constant. However, since the field does not exchange energy with either the armature circuit or the mechanical system, the voltage across the field inductance does not depend in any way on the current, f, or on the angular velocity, f_m. Thus, Equation (7.10a) is the only relation between a field variable and a variable in some other part of the motor. A useful way of thinking about this type of connection is that information is transferred from the field to the mechanical network but not energy. One says that a *unilateral coupling* or an *active bond*[10] exists between the field circuit and mechanical network. When the connection between two parts of a dynamic system is described by a transfer matrix, energy can flow and impedances can be reflected. This type of connection is called a *bilateral coupling* or a *passive bond*.[10] Notice that in a d-c motor the armature circuit and mechanical networks are bilaterally coupled regardless of the type of control, armature or field.

The standard symbol for a field-controlled *generator* is illustrated in Figure 7.4. The mechanical shaft is driven at constant speed; since the *B* field is proportional to the field circuit current, f_B, it follows from Equations (7.1) and (7.5) that

$$e = K_G f_B. \tag{7.10b}$$

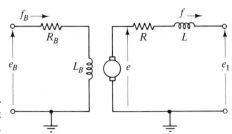

Figure 7.4. Field-controlled generator. The mechanical network (not shown) is driven at constant speed.

This relation describes an active bond between the field and armature circuits. If the current, e, is used to generate a voltage in the armature circuit of a second-stage generator, the device is called an *amplidyne*.[4] Since relatively small power is required to drive the field circuit, field-controlled generators are power amplifiers.

A *rate generator* or *tachometer* is a constant field generator which is used to measure the angular velocity of a mechanical network. The standard symbol is indicated in Figure 7.5. Equation (7.6) applies to the description of

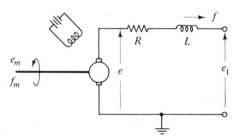

Figure 7.5. Tachometer.

a tachometer so that a passive bond (bilateral coupling) exists between the mechanical network and the armature circuit. Since the armature coil has resistance, the relationship between input and output power states is

$$\begin{bmatrix} E_m \\ F_m \end{bmatrix} = \begin{bmatrix} 0 & g \\ \dfrac{1}{g} & 0 \end{bmatrix} \begin{bmatrix} 1 & R + Ls \\ 0 & 1 \end{bmatrix} \begin{bmatrix} E_1 \\ F \end{bmatrix} \qquad (7.11)$$

The tachometer is used quite often as a sensor in feedback control systems.

Alternating-current motors and tachometers are widely used and operate on a somewhat different principle.[5] The reader is referred to the literature for a description of the operation of these devices.[1] Fine descriptions of other electromechanical sensor-transducer devices, such as the moving plate capacitor,[3] are also to be found in the literature.

Example 7.1

A mechanical system is instrumented with a tachometer in the manner shown in Figure 7.6. Neglect armature resistance and inductance (R_0 is the input impedance of the oscilloscope) and determine the transfer function,

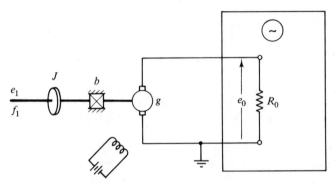

Figure 7.6. Instrumented mechanical system.

$F_1(s)/E_1(s)$. Work backward to find

$$
\begin{bmatrix} E_0 \\ 0 \end{bmatrix} = \begin{bmatrix} 1 & 0 \\ -\dfrac{1}{R_0} & 1 \end{bmatrix} \begin{bmatrix} 0 & g \\ \dfrac{1}{g} & 0 \end{bmatrix} \begin{bmatrix} 1 & -(b + Js) \\ 0 & 1 \end{bmatrix} \begin{bmatrix} E_1 \\ F_1 \end{bmatrix}
$$

$$
= \begin{bmatrix} \cdots & \cdots \\ \dfrac{1}{g} & \dfrac{-(b + Js)}{g} - \dfrac{g}{R_0} \end{bmatrix} \begin{bmatrix} E_1 \\ F_1 \end{bmatrix}
$$

Thus,

$$
\frac{F_1}{E_1} = \frac{K}{Ts + 1}, \tag{7.12}
$$

where

$$
K = \frac{R_0}{g^2 + R_0 B}, \qquad T = \frac{J R_0}{g^2 + R_0 B}. \tag{7.13}
$$

Notice that the instrumentation has modified both the static and the dynamic characteristics of the mechanical system!

*   *   *

Example 7.2

Denote the angular displacement of the rotational inertia, illustrated in Figure 7.7, by $\theta_2(t)$ and find the transfer function, $\Theta_2(s)/E_1(s)$. Work forward

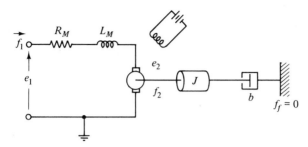

Figure 7.7. Mechanical network driven by an armature-controlled
d-c motor.

to find

$$
\begin{bmatrix} E_1 \\ F_1 \end{bmatrix} = \begin{bmatrix} 1 & R_M \\ 0 & 1 \end{bmatrix} \begin{bmatrix} 1 & L_M s \\ 0 & 1 \end{bmatrix} \begin{bmatrix} 0 & g \\ \dfrac{1}{g} & 0 \end{bmatrix} \begin{bmatrix} E_2 \\ F_2 \end{bmatrix}
$$

$$
= \begin{bmatrix} \dfrac{1}{g}(R_M + L_M s) & g \\ \cdots & \cdots \end{bmatrix} \begin{bmatrix} E_2 \\ F_2 \end{bmatrix}.
$$

Thus,

$$
E_1 = \frac{1}{g}(R_M + L_M s)E_2 + gF_2. \tag{i}
$$

Work backward to find

$$
\begin{bmatrix} E_f \\ 0 \end{bmatrix} = \begin{bmatrix} 1 & 0 \\ \dfrac{-1}{b} & 1 \end{bmatrix} \begin{bmatrix} 1 & -Js \\ 0 & 1 \end{bmatrix} \begin{bmatrix} E_2 \\ F_2 \end{bmatrix} = \begin{bmatrix} 1 & -Js \\ \dfrac{-1}{b} & 1 + \dfrac{Js}{b} \end{bmatrix} \begin{bmatrix} E_2 \\ F_2 \end{bmatrix}
$$

and find

$$
E_2 = (b + Js)F_2.
$$

From Equation (i)

$$
E_1 = \left[\frac{1}{g}(R_M + L_M s)(b + Js) + g \right] F_2,
$$

so that

$$
\frac{F_2}{E_1} = \frac{g}{L_M J s^2 + (bL_M + JR_M)s + g^2 + R_M b}.
$$

Since

$$
F_2 = s\Theta_2,
$$

it follows that

$$
\frac{\Theta_2(s)}{E_1(s)} = \frac{g}{s[L_M J s^2 + (bL_M + JR_M)s + g^2 + R_M b]}.
$$

* * *

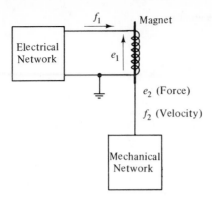

Figure 7.8. Solenoid.

Another important electromechanical device is the solenoid illustrated in Figure 7.8. The current creates a magnetic field in the coil so that for a limited range of the variables

$$e_2 = kf_1, \qquad (7.14)$$

where k is a proportionality constant. Concurrently, the motion of a magnet generates a back emf so that for a limited range

$$e_1 = k_0 f_2, \qquad (7.15)$$

where k_0 is a proportionality constant. Conservation of energy can be used to show that $k_0 = k$. (Why?) Therefore,

$$\begin{bmatrix} E_1 \\ F_1 \end{bmatrix} = \begin{bmatrix} 0 & k \\ \dfrac{1}{k} & 0 \end{bmatrix} \begin{bmatrix} E_2 \\ F_2 \end{bmatrix}. \qquad (7.16)$$

Note the characteristic gyrator form.

The range of applicability of Equations (7.14) and (7.15) is so small that the solenoid is usually used as an on-off (*bang-bang*) device. Also, the coil will have resistance and inductance which must be accounted for in a more realistic model.

7.2 Hydromechanical Transducers[7, 8, 9, 11]

The discussion in this section is devoted to devices at the interfaces of mechanical and hydraulic networks. It will be assumed that the hydraulic approximation applies (see Chapter 5).

The simplest hydromechanical device is the piston cylinder arrangement illustrated in Figure 7.9. Denote the piston area by A_p and neglect (1) mass of piston, (2) leakage flow, (3) fluid inertia, (4) fluid resistance, and (5) fluid compliance, and use the definition of pressure to show that

$$e_1 = A_p e_2. \qquad (7.17)$$

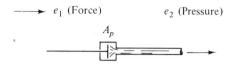

Figure 7.9. Piston and cylinder. $\longrightarrow f_1$ (Velocity) f_2 (Flow)

Since the fluid is incompressible, it follows that

$$f_1 = \frac{f_2}{A_p}. \tag{7.18}$$

Thus, the Laplace domain matrix equation relating the mechanical power state to the fluid power state is

$$\begin{bmatrix} E_1 \\ F_1 \end{bmatrix} = \begin{bmatrix} A_p & 0 \\ 0 & \dfrac{1}{A_p} \end{bmatrix} \begin{bmatrix} E_2 \\ F_2 \end{bmatrix}. \tag{7.19}$$

Note the characteristic transformer matrix. It is easily shown that

$$\begin{bmatrix} E_2 \\ F_2 \end{bmatrix} = \begin{bmatrix} \dfrac{1}{A_p} & 0 \\ 0 & A_p \end{bmatrix} \begin{bmatrix} E_1 \\ F_1 \end{bmatrix}. \tag{7.20}$$

A very important type of hydromechanical transducer is the variable displacement pump illustrated in Figure 7.10. As the shaft rotates the rotating pistons are forced in and out of the cylinders by the stationary yoke. A complex manifold (not shown) directs the pressurized oil into a hydraulic network. Very high pressures ($\simeq 5000$ psi) can be developed with this device, which is often used in applications requiring very high power. Denote

$$r_y = \text{radius of yoke,}$$
$$\alpha = \text{yoke angle,}$$
$$N = \text{number of pistons,}$$
$$A = \text{area of pistons.}$$

In one rotation of the shaft, a volume of oil equal to $2NAr_y \sin \alpha$ will be forced into the hydraulic network. Thus,

$$f_2 = \frac{2NAr_y \sin \alpha}{2\pi} f_1 \tag{7.21}$$

or

$$f_2 = nf_1, \tag{7.22}$$

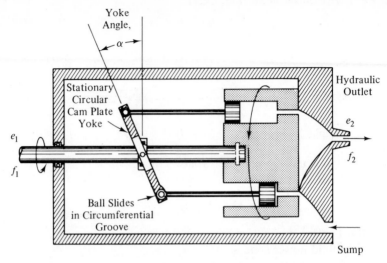

Figure 7.10. Variable displacement piston pump. The pistons and cylinder block rotate with the mechanical input shaft, while the yoke remains stationary. The *sump* is a low-pressure oil supply.

where

$$n = \frac{NAr_y \sin \alpha}{\pi}. \tag{7.23}$$

If the yoke angle were kept constant and if the pump were 100% efficient, then

$$\begin{bmatrix} e_1 \\ f_1 \end{bmatrix} = \begin{bmatrix} n & 0 \\ 0 & \dfrac{1}{n} \end{bmatrix} \begin{bmatrix} e_2 \\ f_2 \end{bmatrix}. \tag{7.24}$$

Note the characteristic transformer matrix. However, the usual configuration is to maintain f_1 constant (usually driven by an a-c motor which operates efficiently at constant speed) and to use a variable yoke angle as the input. For this case, it follows from Equation (7.21) that

$$f_2 = K_p \sin \alpha \approx K_p \alpha, \tag{7.25}$$

where

$$K_p = \frac{NAr_y f_1}{2\pi}. \tag{7.26}$$

Equation (7.25) defines an active bond (unilateral coupling) between the yoke displacement and the hydraulic power state.

Many other types of variable displacement pumps can be constructed.[1] A *hydraulic motor* is simply, conceptually at least, a pump turned around. That is, a motor is a mechanical output device driven by a hydraulic input. In actual practice, the yoke angle is constant so that the mechanical and hydraulic subsystems are connected by a bilateral coupling described by

$$
\begin{bmatrix} e_2 \\ f_2 \end{bmatrix} = \begin{bmatrix} \dfrac{1}{n} & 0 \\ 0 & n \end{bmatrix} \begin{bmatrix} e_1 \\ f_1 \end{bmatrix},
\tag{7.27}
$$

where n is defined by Equation (7.23). A hydraulic pump followed by a hydraulic motor is called a *hydraulic transmission*.

A quite different type of hydromechanical transducer is the *centrifugal pump*. In this device the mechanical input shaft rotates vanes in a fluid sump and the fluid is pressurized and forced through circumferential output ports by centrifugal accelerations. Clearly, the output pressure is a monotone increasing function of shaft velocity, since centrifugal acceleration is such a function of angular velocity. For a limited range of the variables, the output pressure is

$$
e_2 \simeq g f_1,
$$

where g is a proportionality constant and f_1 is the angular velocity of the shaft. Proceed in the usual manner to show that for the centrifugal pump

$$
\begin{bmatrix} e_1 \\ f_1 \end{bmatrix} = \begin{bmatrix} 0 & g \\ \dfrac{1}{g} & 0 \end{bmatrix} \begin{bmatrix} e_2 \\ f_2 \end{bmatrix}.
\tag{7.28}
$$

Compare this equation with Equation (7.24).

The valve (see the illustration in Figure 7.11) is an often-encountered hydromechanical device. The mechanical displacement, x, results in regulation in the flow, $f_2 = f = f_1$. Empirically, one finds

$$
f = KA\sqrt{e},
\tag{7.29}
$$

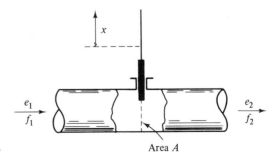

Figure 7.11. Hydraulic valve.　　　　　　　Area A

where K is a constant, A is the flow area (which is a function of x), and

$$e = e_1 - e_2. \tag{7.30}$$

The relation (7.29) can be linearized in order to use extended linear analysis. Retaining only linear terms from a Taylor series expansion results in

$$f - f^0 \simeq K\sqrt{e^0}\{A - A^0\} + \frac{KA^0}{2\sqrt{e^0}}\{e - e^0\}, \tag{7.31}$$

where the superscript 0 indicates the nominal values of the variables which also satisfy Equation (7.29), i.e.,

$$f^0 = KA^0\sqrt{e^0}. \tag{7.32}$$

Define

$$R_v = \frac{2\sqrt{e^0}}{KA^0} = \frac{2e^0}{f^0} \tag{7.33a}$$

and

$$C_v = \frac{2e^0}{A^0}. \tag{7.33b}$$

Redefine the nominal values of the variables as the zero reference values and show that Equation (7.31) becomes

$$e_1 - e_2 \simeq R_v f - C_v A. \tag{7.34}$$

This relation defines unilateral coupling between valve displacement and the hydraulic network.

Example 7.3

A device which regulates the position, velocity, or acceleration of an inertia is called a *servomechanism*. An open loop, hydraulic servomechanism is illustrated in Figure 7.12. Determine the transfer function, $X(s)/A(s)$. It follows that

$$E_1 - E_2 = R_v F - C_v A.$$

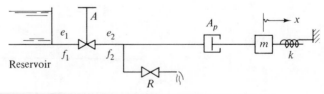

Figure 7.12. Hydromechanical servomechanism.

Since e_1 is a constant *head* and since its value is to be interpreted as a deviation from its nominal value, $E_1 = 0$; thus,

$$C_v A = R_v F_2 + E_2. \tag{i}$$

Work forward to relate E_2 and F_2 to X:

$$\begin{bmatrix} E_2 \\ F_2 \end{bmatrix} = \begin{bmatrix} 1 & 0 \\ \dfrac{1}{R} & 1 \end{bmatrix} \begin{bmatrix} \dfrac{1}{A_p} & 0 \\ 0 & A_p \end{bmatrix} \begin{bmatrix} 1 & ms \\ 0 & 1 \end{bmatrix} \begin{bmatrix} 1 & \dfrac{k}{s} \\ 0 & 1 \end{bmatrix} \begin{bmatrix} 0 \\ sX \end{bmatrix}$$

$$= \begin{bmatrix} \dfrac{1}{A_p} & 0 \\ \dfrac{1}{RA_p} & A_p \end{bmatrix} \begin{bmatrix} 1 & \dfrac{k}{s} + ms \\ 0 & 1 \end{bmatrix} \begin{bmatrix} 0 \\ sX \end{bmatrix}$$

$$= \begin{bmatrix} \cdots & \dfrac{(k + ms^2)}{A_p s} \\ \cdots & A_p + \dfrac{(k + ms^2)}{RA_p s} \end{bmatrix} \begin{bmatrix} 0 \\ sX \end{bmatrix}$$

Combine this relation with Equation (i) to find

$$C_v A = \left\{ \left[\frac{k + ms^2}{A_p s} \right] + R_v A_p + \frac{R_v(k + ms^2)}{RA_p s} \right\} sX.$$

Thus, the desired transfer function

$$\frac{X(s)}{A(s)} = \frac{C_v R A_p}{(R_v + R)ms^2 + RR_v A_p^2 s + (R_v + R)k}.$$

$$\ast \quad \ast \quad \ast$$

Example 7.4

An electrohydromechanical servomechanism is illustrated in Figure 7.13. Determine the transfer function, $X(s)/E(s)$. Assume that the valve stem moves in and out of the fluid stream without resistance. Working forward to the valve stem,

$$\begin{bmatrix} E \\ F \end{bmatrix} = \begin{bmatrix} 1 & R_0 + Ls \\ 0 & 1 \end{bmatrix} \begin{bmatrix} 0 & g \\ \dfrac{1}{g} & 0 \end{bmatrix} \begin{bmatrix} 1 & \dfrac{k_0}{s} \\ 0 & 1 \end{bmatrix} \begin{bmatrix} 0 \\ F_v \end{bmatrix}$$

$$= \begin{bmatrix} \cdots & \dfrac{k_0(R_0 + Ls)}{sg} + g \\ \cdots & \cdots \end{bmatrix} \begin{bmatrix} 0 \\ F_v \end{bmatrix}.$$

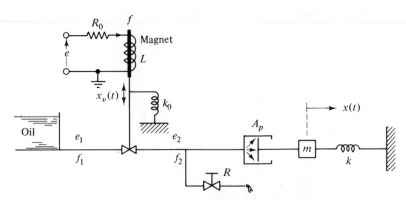

Needle Valve (Variable Resistance)

Figure 7.13. Position servo.

Thus,

$$E = \frac{[k_0(R_0 + Ls) + g^2s]}{sg} F_v = \frac{[k_0R_0 + (k_0L + g^2)s]}{g} X_v.$$

The valve area $A = k_v X_v$, where k_v is a proportionality constant; thus,

$$\frac{A(s)}{E(s)} = \frac{g/k_v}{k_0R_0 + (k_0L + g^2)s}. \tag{i}$$

Now consider the valve-mechanical system. Work forward to find

$$\begin{bmatrix} E_2 \\ F_2 \end{bmatrix} = \begin{bmatrix} 1 & 0 \\ \dfrac{1}{R} & 1 \end{bmatrix} \begin{bmatrix} \dfrac{1}{A_p} & 0 \\ 0 & A_p \end{bmatrix} \begin{bmatrix} 1 & Ms + \dfrac{k}{s} \\ 0 & 1 \end{bmatrix} \begin{bmatrix} 0 \\ F_m \end{bmatrix}$$

$$= \begin{bmatrix} \cdots & \dfrac{Ms^2 + k}{A_p s} \\ \cdots & A_p + \dfrac{Ms^2 + k}{RA_p s} \end{bmatrix} \begin{bmatrix} 0 \\ sX \end{bmatrix}.$$

Thus,

$$E_2 = \frac{Ms^2 + k}{A_p} X, \qquad F_2 = \frac{Ms^2 + RA_p^2s + k}{RA_p} X. \tag{ii}$$

From Equation (7.34), it follows that

$$E_1 - E_2 = R_v F_2 - C_v A. \tag{iii}$$

Again, $E_1 = 0$ so that combining Equations (i)–(iii) leads to the conclusion that

$$\frac{C_v g E(s)}{k_v\{k_0 R_0 + (kL + g^2)s\}} = \left\{\frac{M[1 + (R_v/R)]s^2 + R_v A_p^2 s + k[1 + (R_v/R)]}{A_p}\right\} X(s).$$

Finally,

$$\frac{X(s)}{E(s)} = \frac{A_p C_v g/k_v}{[kR + (kL + g^2)s]\{M[1 + (R_v/R)]s^2 + R_v A_p^2 s + k[1 + (R_v/R)]\}}.$$

*   *   *

A commonly encountered type of hydraulic-powered position control device (servomechanism) is illustrated in Figure 7.14. The reader is encouraged to identify components with the functional blocks in Figure 1.1. The spool valve is generally considered to move without resistance (i.e., $e_1 = 0$), although a small *Bernoulli force*[8], caused by variations in fluid velocity, is exerted on it. Also, a coulomb-like friction force, called *stiction*[8], can be exerted on the spool. Stiction is often overcome by causing the spool to move in a superimposed, small-amplitude oscillation called *dither*[8].

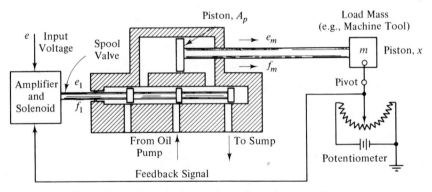

Figure 7.14. Electro-hydro-mechanical position control system.

Equation (7.29) applies to the spool valve but the linearization (7.34) no longer applies because (1) the nominal conditions $A^0 = 0 = f^0$ are usually selected, and (2) there are actually two valves (one to pump, the other to sump) which operate simultaneously.

Morse indicates the appropriate starting point for an extended linear analysis.[8] His derivation is reproduced here. It is assumed that the areas of the pump valve and the sump valve are proportional to spool displacement, x_1. It follows from Equation (7.29) that the flow from pump to piston is

$$f_p = K_p x_1 \sqrt{e_p}, \tag{7.35}$$

where e_p is the pressure drop from pump to piston and K_p is a constant of proportionality. Similarly, the flow from piston to sump is

$$f_s = K_s x_1 \sqrt{e_s},\tag{7.36}$$

where e_s is the appropriate pressure drop. The following assumptions are made:

1. All compliances are neglected so that

$$f_p = f_s,\tag{7.37}$$

2. All resistances are neglected so that

$$e_p + e_s = e_0 - e_L,\tag{7.38}$$

where e_0 is the supply pressure and e_L is the pressure on the pump side of the piston.

Using these assumptions, Equations (7.35) and (7.36) can be manipulated to show that

$$f_L = k_L x_1 \sqrt{e_0 - e_L},\tag{7.39}$$

where

$$k_L = \sqrt{\frac{K_p^2 K_s^2}{K_p^2 + K_s^2}},\tag{7.40}$$

and f_L is the flow in the cylinder. It follows that

$$A_p f_m = k_L x_1 \sqrt{e_0 - \frac{e_m}{A_p}},$$

so that

$$f_m = \frac{k_L}{A_p^{3/2}} x_1 \sqrt{A_p e_0 - e_m}.\tag{7.41}$$

This relation can be linearized to obtain a relation between spool displacement, x_1, and the mechanical power state (f_m and e_m). It is usually assumed that the supply pressure e_0 is a constant. Equation (7.41) indicates an active bond between spool and piston. Morse[8] indicates that for most valves, the constant K, defined by Equation (7.29), is given by

$$K = 0.63 \sqrt{\frac{2}{\rho}},\tag{7.42}$$

where ρ denotes the fluid density.

The reader is referred to the literature for a discussion of other hydraulic devices such as the *jet pipe* servo.[8,9,11]

7.3 Pneumatic to Mechanical Transducers

Pneumatic devices are applied a great deal because of their simplicity and because compressed air supplies (15–30 psi) are often readily available. The power output of pneumatic prime movers is usually intermediate between electrical motors and hydraulic prime movers.

A passive transducer is the bellows device illustrated in Figure 7.15. The *bellows constant*, k_B, is defined by

$$k_B x_2 = A_B e_1 - e_2. \tag{7.43}$$

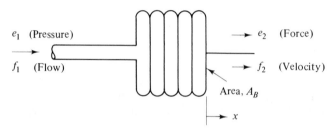

Figure 7.15. Bellows.

If one neglects compressibility, it follows that

$$f_1 = A_B f_2, \tag{7.44}$$

and since $X_2(s) = F_2(s)/s$, it follows that, in the Laplace domain,

$$\begin{bmatrix} E_1 \\ F_1 \end{bmatrix} = \begin{bmatrix} \dfrac{1}{A_B} & \dfrac{k_B}{A_B s} \\ 0 & A_B \end{bmatrix} \begin{bmatrix} E_2 \\ F_2 \end{bmatrix}. \tag{7.45}$$

Notice that the above transfer matrix equals the matrix product

$$\begin{bmatrix} \dfrac{1}{A_B} & 0 \\ 0 & A_B \end{bmatrix} \begin{bmatrix} 1 & \dfrac{k_B}{s} \\ 0 & 1 \end{bmatrix}, \tag{7.46}$$

which is the matrix product associated with a transformer followed by a series capacitance. Clearly, for a bellows, the pneumatic and mechanical power states are bilaterally coupled. The bellows volume often must be modeled as a compliance.

One of the most useful pneumatic devices is the *flapper-nozzle* illustrated in Figure 7.16.

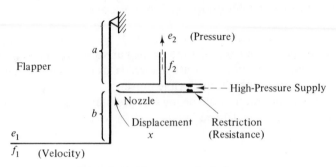

Figure 7.16. Flapper-nozzle device.

Let x be the displacement at the orifice of the flapper. The approximation that is usually made is that

$$e_2 = k_F x, \qquad (7.47)$$

where k_F is usually a very large number; this relation is valid for a very limited range of displacement, x. Notice that for small displacements

$$x = \frac{a}{a+b} x_1. \qquad (7.48)$$

Hydraulic flapper-nozzles are also used. The operating characteristic of such a device is illustrated in Figure 7.17. Typical values for hydraulic flapper-nozzles are $L = 0.001$ in. and $e_1 = 3000$ psi. This figure illustrates the range of applicability of Equation (7.47).

Equation (7.47) defines a unilateral coupling (active bond) between the flapper and the pneumatic (or hydraulic) power state. Thus, the movement of the flapper does not require a great deal of power. This fact makes the flapper-

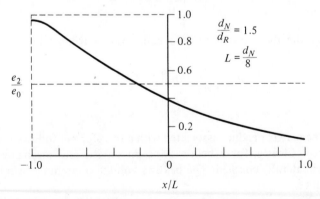

Figure 7.17. Operating characteristics of a hydraulic flapper-nozzle.[8] e_0 is supply pressure, L is the nominal distance between flapper and nozzle, d_N is nozzle diameter, and d_R is restriction diameter.

nozzle an ideal power amplifier for many types of measurements. The displacement, x_1, can be generated by many different sources. A few examples are

1. The expansion or contraction of a metal in response to temperature change.
2. The expansion or contraction of certain plastic materials in response to humidity changes.
3. The deflection of a platform or conveyor belt in response to changes in weight of a supported load.

Thus the flapper-nozzle serves both as a prime mover and a sensor of temperature, humidity, or weight or any other variable that can be indicated by a deflection.

7.4 Examples of Other Types of Transducers

Many other transduction devices are available.[1,2,7,8,10,11] A very few of them are discussed in this section.

The *lead screw* converts mechanical rotation energy to mechanical translational energy and is illustrated in Figures 7.18a and 7.18b. Consider the lead screw illustrated in Figure 7.18b and neglect (1) inertias and (2) frictional forces. The angle, γ, is defined such that

$$e_T \cos \gamma = e_2.$$

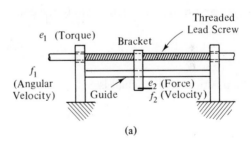

(a)

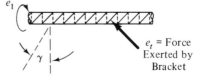

Figure 7.18. (a) Lead screw assembly, (b) lead screw.

(b)

But

$$re_T \sin \gamma = e_1,$$

where r is the radius of the lead screw; thus,

$$e_1 = r \tan (\gamma) e_2. \tag{7.49}$$

The tangential velocity of the lead screw is rf_1 so that

$$f_2 = r \tan (\gamma) f_1. \tag{7.50}$$

Finally,

$$\begin{bmatrix} e_1 \\ f_1 \end{bmatrix} = \begin{bmatrix} n & 0 \\ 0 & \dfrac{1}{n} \end{bmatrix} \begin{bmatrix} e_2 \\ f_2 \end{bmatrix}, \tag{7.51a}$$

where

$$n = r \tan \gamma. \tag{7.51b}$$

Note the characteristic transformer-like matrix and that the mechanical power states are bilaterally coupled.

The lead screw is a transducer which converts rotary power to translational power. Another similar type of transducer, which is used as a sensor of angular velocity rather than as a power transmitter, is the *flyball governor* (see Chapter 1). The governor is illustrated in Figures 7.19a and 7.19b. Let $\vec{\mathbf{u}}_x$, $\vec{\mathbf{u}}_y$, and $\vec{\mathbf{u}}_z$ be unit vectors along the coordinate axes. Notice that

$$\frac{d}{dt}\vec{\mathbf{u}}_x = f_1\vec{\mathbf{u}}_y, \qquad \frac{d}{dt}\vec{\mathbf{u}}_y = -f_1\vec{\mathbf{u}}_x; \qquad \frac{d}{dt}\vec{\mathbf{u}}_z = \vec{\mathbf{0}}. \tag{7.52}$$

The position of the flyball is

$$\vec{\mathbf{R}} = l \sin \theta \vec{\mathbf{u}}_z - (r + l \cos \theta)\vec{\mathbf{u}}_y. \tag{7.53}$$

Thus, the momentum of the flyball is

$$m\dot{\vec{\mathbf{R}}} = ml \cos (\theta)\dot{\theta}\vec{\mathbf{u}}_z + mf_1(r + l \cos \theta)\vec{\mathbf{u}}_x + ml \sin (\theta)\dot{\theta}\vec{\mathbf{u}}_y, \tag{7.54}$$

and the time rate of change of flyball momentum is

$$\frac{d}{dt}\{m\dot{\vec{\mathbf{R}}}\} = m\{l \cos (\theta)\ddot{\theta} - l \sin (\theta)\dot{\theta}^2\}\vec{\mathbf{u}}_z$$
$$- m\{f_1l \sin (\theta)\dot{\theta} - \dot{f}_1(r + l \cos \theta) + f_1l \sin (\theta)\dot{\theta}\}\vec{\mathbf{u}}_x$$
$$+ m\{f_1^2(r + l \cos \theta) + l \sin (\theta)\ddot{\theta} + l \cos (\theta)\dot{\theta}^2\}\vec{\mathbf{u}}_y. \tag{7.55}$$

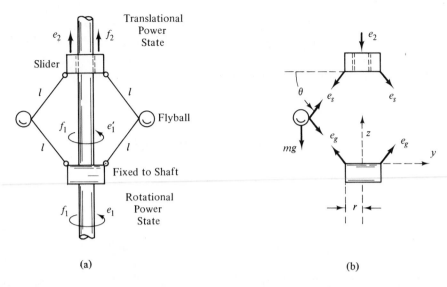

(a)   (b)

Figure 7.19. (a) Flyball transducer. The slider is usually spring-loaded (see Figure 1.4.), (b) free body diagram. The x-axis points out of the diagram, and the x-y-z coordinate system rotates with the shaft.

The force acting on the flyball is

$$\vec{e} = \{e_s \sin\theta - e_g \sin\theta - mg\}\vec{u}_z$$
$$+ \left\{\frac{e_1 - e_1'}{r + l\cos\theta}\right\}\vec{u}_x + \{e_s \cos\theta + e_g \cos\theta\}\vec{u}_y. \quad (7.56)$$

Newton's second axiom is used to equate the right-hand sides of Equations (7.55) and (7.56) so that

$$(e_s - e_g)\sin\theta = ml\cos(\theta)\ddot{\theta} - ml\sin(\theta)\dot{\theta}^2 + mg;$$
$$(e_s + e_g)\cos\theta = mf_1^2(r + l\cos\theta) + ml\sin(\theta)\ddot{\theta} + ml\cos(\theta)\dot{\theta}^2, \quad (7.57)$$
$$\frac{e_1 - e_1'}{(r + l\cos\theta)} = mf_1(r + l\cos\theta) - 2mf_1 l\sin(\theta)\dot{\theta}.$$

It follows from the first two of the above equations that

$$2\sin(\theta)\cos(\theta)e_s = ml\ddot{\theta} + mg\cos\theta + m\sin\theta(r + l\cos\theta)f_1^2. \quad (7.58)$$

Assume the slider is massless and show that

$$2\sin(\theta)e_s = -e_2,$$

so that

$$-e_2 = mg + \frac{ml\ddot{\theta}}{\cos \theta} + \frac{m \sin \theta(r + l \cos \theta)}{\cos \theta} f_1^2. \tag{7.59}$$

The z coordinate of the slider is $2l \sin \theta$ so that

$$f_2 = +2l \cos (\theta)\dot{\theta}$$

and

$$\dot{f}_2 = -2l \sin (\theta)\dot{\theta}^2 + 2l \cos (\theta)\ddot{\theta}.$$

Combine these equations with Equations (7.57) and (7.59) to show that

$$e_1 = e_1' + m(r + l \cos \theta)^2 \dot{f}_1 - m \tan \theta(r + l \cos \theta)f_1 f_2 \tag{7.60}$$

and

$$e_2 = -mg - \frac{m \tan \theta}{4l \cos^3 \theta} f_2^2 - m \tan \theta(r + l \cos \theta)f_1^2$$

$$- \frac{m}{2 \cos^2 \theta} \dot{f}_2, \tag{7.61}$$

which are the full nonlinear relationships between rotational and translational power state variables. There are several alternative ways to linearize these equations; for instance, if it is assumed that the terms containing the translational velocity, f_2, and the translational acceleration, $\dot{f}_2$, are negligibly small in Equations (7.60) and (7.61), then these equations become

$$e_1 \simeq e_1' + m(r + l \cos \theta)^2 \dot{f}_1 \tag{7.62}$$

and

$$e_2 \simeq -mg - m \tan \theta(r + l \cos \theta)f_1^2. \tag{7.63}$$

Equation (7.63) defines a unilateral coupling between the rotational and translational power states. This approximate description of the coupling is a result of eliminating the small bilateral terms in Equations (7.60) and (7.61).

The translational power state (e_2, f_2) is usually connected to a series spring and then to a spool valve or some other prime mover's throttle. Viscous friction between the slider and the shaft may be modeled as a series mechanical resistance.

Fluidic devices are pneumatic devices with no moving mechanical parts.[7,9] A fluidic transducer is illustrated in Figure 7.20. The device is rigidly attached to a mechanical component whose angular velocity, f_1, is to be sensed. As indicated in Figure 7.20, an angular velocity fluctuation will cause a fluctuation in the output port pressure. The mechanical power state

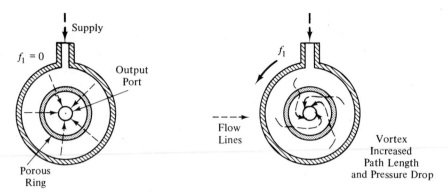

Figure 7.20. Fluidic vortex *rate gyro*.[7]

is unilaterally coupled to the pneumatic power state at the output port. For a limited range, the output pressure is

$$e_2 = -Kf_1, \tag{7.64}$$

where K is a proportionality constant.

A unilateral coupling is a highly desirable type of connection for a sensor because impedances are not reflected and loading effects do not influence the behavior of the system being instrumented. Thermocouples and potentiometers are examples of devices which, in certain ideal situations, do form active bonds (unilateral couplings) and are used quite often as sensors. Another example of such a device is the thermal to pneumatic transducer illustrated in Figure 7.21. If the mass of air in the volume τ can be kept constant and if the thermal resistance of the walls is neglected, the ideal gas law can be used to show that

$$e_2 = \frac{R}{\tau}e_1. \tag{7.65}$$

A typical application of this sensor is indicated in Figure 7.21. The temper-

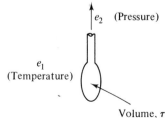

e_2 (Pressure)

e_1
(Temperature)

Volume, τ

Figure 7.21. Thermal to pneumatic transducer.

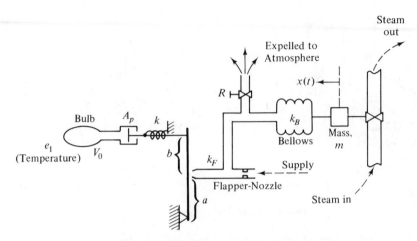

Figure 7.22. Temperature control system.

ature, e_1, is influenced by the rate of flow of steam. Figure 7.22 illustrates the feedback portion of an automatically controlled temperature system.

7.5   Block Diagram Notation and Algebra

Block diagrams are precise graphical representations of mathematical relations. These diagrams are convenient devices for summarizing and communicating technical information. An alternative notation, called a *signal flow graph*, is discussed in other texts.[9,11] Block diagrams may indicate the relations between time domain variables or the relations between Laplace domain variables. Obviously, if nonlinear relations are to be described, time domain block diagrams must be used.

Laplace domain block diagrams will be introduced first. The mathematical relation

$$X(s) = U(s) \pm B(s) \tag{7.66}$$

is represented by the symbol indicated in Figure 7.23. The arrows indicate inputs and outputs of the summing junction. The junction may simply represent a mathematical relation or it may represent a physical device called a *comparator* (see Figure 1.1). Electronic comparators are discussed in

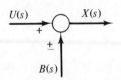

Figure 7.23. *Summing junction.*

Chapter 8. It is required that all signals be unilaterally coupled in order for the symbol in Figure 7.23 to have an unambiguous meaning. The transfer function relation

$$X(s) = G(s)U(s) \qquad (7.67)$$

is represented by the symbol indicated in Figure 7.24. Of course, $G(s)$ could be a simple constant. U or X could be connected to other blocks or summing junctions. In either case, the connection must be unilateral (i.e., an active bond) in order for the block diagram to have a precise, unambiguous meaning.

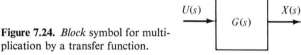

Figure 7.24. *Block* symbol for multiplication by a transfer function.

Finally, the identity relation

$$X = X_0 = X_1 \qquad (7.68)$$

is symbolized in the manner indicated in Figure 7.25. The symbol is suggestive of the identity symbol used in circuit diagrams; however, unlike circuit diagrams, all variables indicated about a pickoff point must be unilaterally coupled to other devices symbolized in a block diagram.

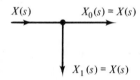

Figure 7.25. Identity symbol or *pickoff point.*

In the present chapter, it is sufficient to develop a few examples of block diagrams by concentrating on the unilateral couplings (active bonds) which occur quite naturally in dynamic systems.

Example 7.5

Construct a block diagram of the temperature control feedback path indicated in Figure 7.22. Denote the flapper displacement, $x_f(t)$, and notice that active bonds occur at

1. The temperature-sensing bulb,
2. The flapper-nozzle, and
3. The steam valve.

Thus, a valid block diagram is that indicated in Figure 7.26. The transfer functions, $G_m(s)$ and $G_p(s)$, can be determined using the transfer matrix techniques discussed in previous chapters. Notice that one of the power

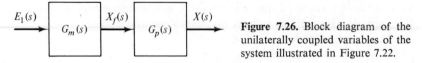

Figure 7.26. Block diagram of the unilaterally coupled variables of the system illustrated in Figure 7.22.

state variables at an active bond is zero. This fact is used to determine input impedances and transfer functions.

$$* \quad * \quad *$$

Example 7.6

Construct a block diagram of the servomechanism illustrated in Figure 7.14. Insufficient detail about the "amplifier and solenoid" device is included in Figure 7.14 to allow the analyst to construct a block diagram. To proceed, it is surmised that an electronic comparator subtracts the potentiometer voltage, denoted $e_p(t)$, from the reference voltage to produce an *error* voltage, denoted $v_e(t)$. It is further surmised that the error voltage regulates the output voltage of a power amplifier, which in turn regulates the position of a spring-loaded solenoid rigidly connected to the spool valve. It is assumed that active bonds exist at the potentiometer, comparator, power amplifier, and spool. The block diagram is illustrated in Figure 7.27. Compare this diagram with Figure 1.1. The transfer functions G_c, G_p, and H can be determined using the usual techniques. H and G_c are probably constants for this case.

$$* \quad * \quad *$$

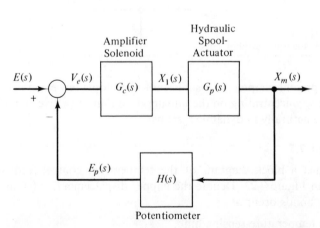

Figure 7.27. Block diagram of the servomechanism illustrated in Figure 7.14.

As noted above, block diagrams of time domain variables can also be constructed. The summing junction and pickoff point have the same meaning in the block diagrams of time domain variables as they have in block diagrams of Laplace domain variables. The block symbol, however, is given a more general interpretation in the case of time domain diagrams. A block indicates the *operation upon* an input by a physical component to produce an output. Examples of operation are the nonlinear algebraic operation

$$x = \phi(u) \tag{7.69a}$$

and the calculus type of operation

$$x(t) = \int_0^t u(\xi)\, d\xi. \tag{7.69b}$$

Two common ways of representing Equation (7.69a) graphically are illustrated in Figure 7.28a. Notice that in one of the methods, a sketch of the graph of x vs. u is provided. The common method for representing Equation (7.69b) is to use the integral sign in the manner indicated in Figure 7.28b.

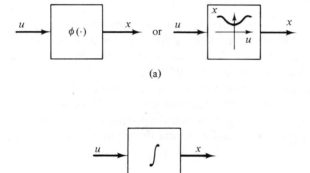

(a)

(b)

Figure 7.28. (a) Nonlinear algebraic operation, (b) operation of integration.

"Block diagram algebra" refers to the process of replacing a group of block diagram symbols by a single equivalent block. For instance, the block diagram structure illustrated in Figure 7.27 is a very common structure, and it is useful to know an equation for the transfer function for an equivalent single block. To derive this equation, notice that

$$X_m(s) = G_c(s)G_p(s)V_e(s)$$

and

$$V_e(s) = E(s) - H(s)X_m(s).$$

Combine these equations to show that

$$X_m(s) = \frac{G_c(s)G_p(s)}{1 + G_c(s)G_p(s)H(s)} E(s),$$

so that the equivalent single transfer function is

$$G(s) = \frac{G_c(s)G_p(s)}{1 + G_c(s)G_p(s)H(s)} \tag{7.70}$$

and the block diagram of Figure 7.27 may be replaced by the simplified diagram of Figure 7.29.

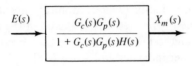

Figure 7.29. Block diagram equivalent of Figure 7.27.

7.6 Feedback Improves Linearity

It must be emphasized that the relationships between output and input for many of the above prime movers are valid for extremely limited ranges of input. For instance, the flapper-nozzle may typically be operated within the input displacement range of ± 0.03 in. Feedback may be employed to improve the linearity of the device, i.e., make it operable for a greater range of input displacements. For the flapper-nozzle, this feedback is implemented as shown in Figure 7.30. Let x_1 denote the displacement at point 1, x_B denote the bellows displacement, and x denote the input displacement; then from superposition

$$x = -\frac{a}{a+b}x_B + \frac{b}{a+b}x_1.$$

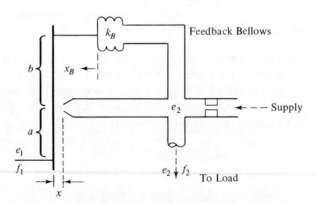

Figure 7.30. Flapper-nozzle with mechanical feedback.

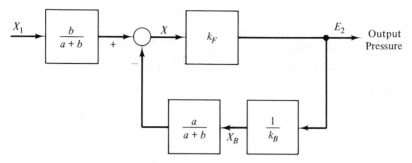

Figure 7.31. Block diagram for the system illustrated in Figure 7.30.

The block diagram for this system is illustrated in Figure 7.31. Thus, using Equation (7.70),

$$\frac{X}{X_1} = \left(\frac{b}{a+b}\right)\left\{\frac{1}{1 + (k_F/k_B)[a/(a+b)]}\right\}$$

or

$$\frac{X}{X_1} = \frac{b}{a + b + k_F(a/k_B)}. \tag{7.71}$$

Since k_F is usually a very large number, it follows that x approaches 0. This is highly desirable because it keeps x within the linear operating range of the device for a large range of displacements x_1.

To see what is lost in order to gain increased linearity, determine the transfer function E_2/X_1. Use Equation (7.70) to show that

$$\frac{E_2}{X_1} = \left(\frac{b}{a+b}\right)\frac{k_F}{\{1 + (k_F/k_B)[a/(a+b)]\}}$$

or

$$\frac{E_2}{X_1} = \frac{b}{[(a+b)/k_F] + (a/k_B)}. \tag{7.72}$$

Thus, as $k_F \to \infty$, $E_2/X_1 \to k_B b/a$. This gain is much smaller than k_F; thus, in using feedback, one trades gain for linearity.

7.7 Use of Feedback to Synthesize P.I.D. Controllers (Compensators)

The typical feedback control system is illustrated in Figure 7.32. Compare this diagram with Figure 1.1. A commonly employed controller (also called a *compensator*) is the P.I.D. (proportional, integral, derivative) controller,[2] which is designed to have the transfer function of the form

$$\frac{M(s)}{E(s)} = k_c\left(1 + T_D s + \frac{1}{T_I s}\right). \tag{7.73}$$

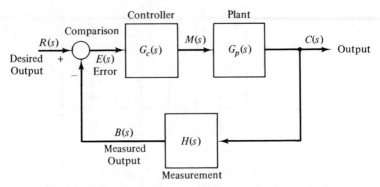

Figure 7.32. Block diagram of the typical feedback control system.

The constants are given the names

$$k_c = proportional\ sensitivity,$$

$$T_D = rate\ time,$$

$$\frac{1}{T_I} = reset. \tag{7.74}$$

Notice that the output of the controller, $M(s)$, is proportional to $E(s)$, the transform of the derivative of $e(t)$ and the transform of the integral of $e(t)$. The adjustable parameters, k_c, T_D, and T_c are set in such a way as to improve the performance of the overall system.

These controllers can be synthesized by employing feedback with a flapper nozzle–passive network arrangement. As an example, consider the P-D controller shown in Figure 7.33. Denote the flow into the needle valve by F, model the valve as a resistance and the bellows as a compliance, and show

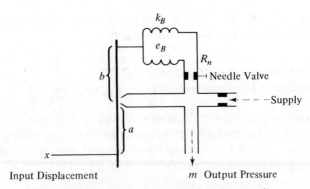

Figure 7.33. Synthesis of a pneumatic P-D controller.

that

$$\begin{bmatrix} M \\ F \end{bmatrix} = \begin{bmatrix} 1 & R_n \\ 0 & 1 \end{bmatrix}\begin{bmatrix} 1 & 0 \\ C_Bs & 1 \end{bmatrix}\begin{bmatrix} E_B \\ 0 \end{bmatrix} = \begin{bmatrix} 1 + R_nC_Bs & R_n \\ \cdots & \cdots \end{bmatrix}\begin{bmatrix} E_B \\ 0 \end{bmatrix},$$

so that

$$\frac{E_B(s)}{M(s)} = \frac{1}{1 + R_nC_Bs} = \frac{1}{1 + T_{nB}s}. \tag{7.75}$$

Modify the block diagram of Figure 7.31 as indicated in Figure 7.34.

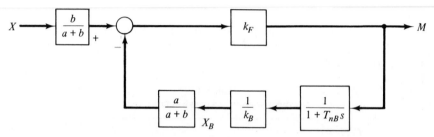

Figure 7.34. Modified block diagram of the flapper-nozzle system shown in Figure 7.31.

It follows from Equation (7.70) that

$$\frac{M(s)}{X(s)} = \frac{b}{a+b}\left(\frac{k_F}{1 + \{ak_F/[(a+b)k_B(1+T_{nB}s)]\}}\right).$$

If it is assumed that k_F is very large, then

$$\frac{M(s)}{X(s)} = \frac{bk_B}{a}(1 + T_{nB}s), \tag{7.76}$$

which is the transfer of a P-D controller, where

$$\text{Proportional sensitivity:} \quad k_c = \frac{bk_B}{a},$$
$$\text{Rate time:} \quad T_D = R_nC_B. \tag{7.77}$$

k_B and C_B are constants, but the b/a ratio and the valve resistance, R_n, can be varied so as to provide desirable control action for a large number of different types of *plants* or *controlled objects*.

The derivation of theoretical values for controller constants is the object of succeeding chapters on automatic control. At this point, it suffices to describe a quick experimental procedure which proves to be adequate for a surprisingly large number of feedback systems. In this experimental procedure the control system is constructed and the following closed loop experiment is performed;

1. Set T_D and $1/T_I$ to zero,
2. Slowly increase the gain until the output is a sine-wave-like function of time (i.e., constant amplitude cycling).
3. Record this setting of gain (denoted k_u and called the *ultimate sensitivity*) and the frequency of oscillation (denoted ω_u).

Define

$$P_u = \frac{2\pi}{\omega_u}.$$
(7.78)

Certain references contain compendiums of recommended controller settings based on the measured values of P_u and ultimate sensitivity, k_u.[2] For example, consult Table 7.1.

Table 7.1
RECOMMENDED CONTROLLER SETTINGS[2]

Control action	k_c	T_D	$1/T_I$
P	$0.50k_u$	—	—
P-D	$0.60k_u$	$P_u/8.0$	—
P-I	$0.45k_u$	—	$1.2/P_u$
P-I-D	$0.60k_u$	$P_u/8.0$	$2.0/P_u$

References

1. AHRENDT, W., and SAVANT, C., *Servo-Mechanism Practice*, McGraw-Hill, New York, 1960.

2. CALDWELL, W., COON, G., and ZOSS, L., *Frequency Response for Process Control*, McGraw-Hill, New York, 1959.

3. CRANDALL, S., KARNOPP, D., et al., *Dynamics of Physical Systems*, McGraw-Hill, New York, 1967.

4. D'AZZO, J., and HOUPIS, C., *Feedback Control System Analysis and Synthesis*, McGraw-Hill, New York, 1966.

5. FEYNMAN, R., LEIGHTON, R., and SANCS, M., *The Feynman Lectures on Physics*, Vol. II: *Mainly Electromagnetism and Matter*, Addison-Wesley, Reading, Mass., 1964.

6. KARNOPP, D., and ROSENBERG, R., *Analysis and Simulation of Multiport Systems*, M.I.T. Press, Cambridge, Mass., 1968.

7. KIRSHNER, J., ed., *Fluid Amplifiers*, McGraw-Hill, New York, 1966.

8. MORSE, A., *Electrohydraulic Servomechanisms*, McGraw-Hill, New York, 1963.

9. OGATA, K., *Modern Control Engineering*, Prentice-Hall, Englewood Cliffs, N.J., 1970.

10. PAYNTER, H., M., *Analysis and Design of Engineering Systems*, M.I.T. Press, Cambridge, Mass., 1961.

11. TAKAHASHI, Y., RABINS, M. J., and AUSLANDER, D. M., *Control*, Addison-Wesley, Reading, Mass., 1970.

Problems

7.1. Derive Equation (7.5) using Faraday's axiom.

7.2. Derive an equation for the power that must be supplied by the constant current source of the field-controlled motor (see Figure 7.3). What power is supplied when $e_B = 0$?

7.3. In what way is the amplidyne (see the discussion of generators) analogous to the fluid cascade system illustrated in Figure 5.20.

7.4. Consider the velocity servo system of Figure 7.35. Demonstrate that the block diagram illustrated in Figure 7.36 applies. [Assume that the amplifier provides an active bond of $e(t)$ to $m(t)$.] Determine the transfer function

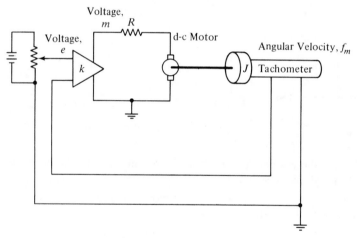

Figure 7.35. Velocity servo.

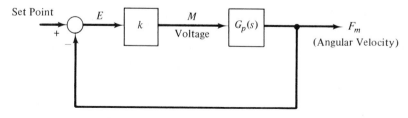

Figure 7.36. Block diagram of a velocity servo.

$G_p(s)$. Neglect any inductance or resistance or any other parameter not shown explicitly in Figure 7.35.

7.5. Consider the hydraulic transmission illustrated in Figure 7.37. Assume that the hydraulic line has compliance and resistance and that leakage occurs in the motor. Relate X_1 to E_m and F_m.

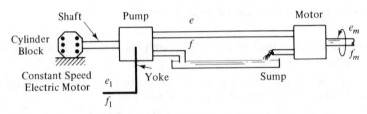

Figure 7.37. Hydraulic transmission.

7.6. Find the transfer function $H_1(s)/A(s)$ for the system illustrated in Figure 7.38.

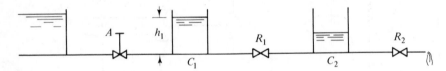

Figure 7.38. Head control.

7.7. Repeat the example of Problem 7.6, but interchange the positions of the valve and the resistance R_1.

7.8. Construct a block diagram of Ktesibios' device, illustrated in Figure 1.2; determine all transfer functions.

7.9. Why does e_2 point upward in Figure 7.19a and downward in Figure 7.19b?

7.10. Assume that Equations (7.62) and (7.63) apply, that the slider is spring-loaded, and that viscous friction exists between the slider and shaft. Relate the Laplace domain variables $F_1(s)$ to $E_2(s)$ and $F_2(s)$ (output of slider).

7.11. Typically, the device illustrated in Figure 7.39 is approximately linear for the range $-0.02'' < x_1 < 0.02''$. Devise a feedback arrangement that would extend the range of linearity. What is traded for this increased linearity? Assume that C_v is a very large number.

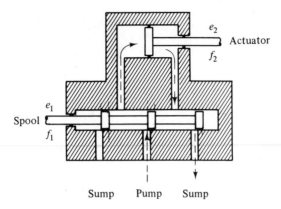

Figure 7.39. Spool-actuator combination.

7.12. Show that the system illustrated in Figure 7.40 is a synthesis of a P.I.D. controller. Find equations for proportional sensitivity, rate time, and reset time in terms of system constants.

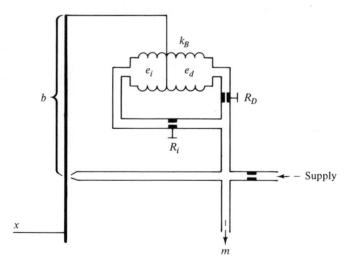

Figure 7.40. Pneumatic synthesis of a P.I.D. controller.

7.13. Take a cue from Figure 7.40 and sketch a feedback arrangement for the spool-actuator combination (Figure 7.39) that would provide P.I.D. control action.

7.14. The following data were taken for a water-mixing valve:

Spool position, $x(cm)$	Outlet temperature, $\theta\ (T - 100\,°F)$
0	0
1.0	+1
2.0	+3.5
3.0	+8.5
4.0	+14.5
5.0	+23
−1.0	−1
−3.0	−8.5
−5.0	−23

Imagine that the valve is placed in the closed loop system shown in Fig. 7.41. Plot θ vs. r for the closed loop on the same graph with θ vs. x in the open loop. What benefit has feedback provided?

Valve Position

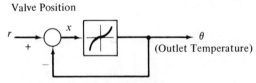

Figure 7.41. Fluid stream temperature control.

Simulation and Synthesis of Dynamic Systems: State Space Analysis*

8

The main purpose of this chapter is to introduce the techniques of synthesis and of computer simulation of dynamic systems. Particular emphasis is placed on analog computer simulations, although the techniques described may be equally well applied to digital computer simulation.

In the discussion of the development of computer simulations, it will prove very convenient to introduce the state space formalism for the description of dynamic systems. It should be pointed out that the state space formalism forms the basis for modern control theory. Thus, while learning to program computers, the beginning student will be introduced to some advanced topics.

The term *synthesis* refers to the construction of a system which is described by a predetermined and given transfer function. Synthesis is important to control system design because the transfer function of a compensator is often known before thought is given to its construction. The synthesis procedure discussed in this chapter is restricted to synthesis by electronic components.

8.1 The Basic Electric Analog Computer Component; the Operational Amplifier

The operational amplifier (Figure 8.1) is an active device which possesses the following highly desirable characteristics:

* The material in this chapter is not required reading for the chapters on automatic control and may be omitted during the first reading of the text.

1. The input and output are unilaterally coupled, i.e.,

$$e_0 = +Ke_g \quad (K \simeq 10^8). \tag{8.1}$$

2. High input impedance (10^8-ohm resistance). $\hspace{2cm}$ (8.2)
3. Low output impedance (10^{-4}-ohm resistance). $\hspace{1.5cm}$ (8.3)
4. Saturation current: 25 ma at 100 v (a rule of thumb is to
keep outputs below 100 v). $\hspace{5cm}$ (8.4a)
5. Noise level: 10^{-2} v. $\hspace{6cm}$ (8.4b)

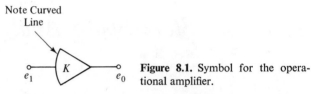

Note Curved
Line

Figure 8.1. Symbol for the operational amplifier.

It follows from Equations (8.1) and (8.4a) that an operational amplifier operates in a linear fashion for only a very small range of input voltage. This range is extended by using feedback in the manner indicated in Figure 8.2. Notice that

$$\frac{e_0}{K} = e_g. \tag{8.5}$$

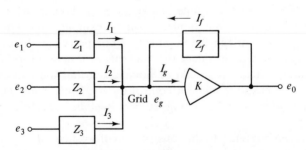

Figure 8.2. Amplifier circuit.

Since E_0 must be kept below 100 v to avoid saturation and since K is large, it follows that

$$\boxed{e_g \simeq 0.} \tag{8.6}$$

It will be assumed that the input impedance is sufficiently large so that one may assume that

$$\boxed{i_g \simeq 0.} \tag{8.7}$$

From Equation (8.7) and from a current balance at the grid, it follows that

$$I_1 + I_2 + I_3 = -I_f. \tag{8.8}$$

Thus, from the basic definition of impedance

$$\frac{E_1 - E_g}{Z_1} + \frac{E_2 - E_g}{Z_2} + \frac{E_3 - E_g}{Z_3} = -\frac{(E_0 - E_g)}{Z_f}. \tag{8.9}$$

From Equations (8.6) and (8.9) it follows that

$$\boxed{\textit{The basic equation:} \quad E_0 = -\frac{Z_f}{Z_1}E - \frac{Z_f}{Z_2}E_2 - \frac{Z_f}{Z_3}E_3.} \tag{8.10}$$

8.2 The Basic Building Blocks of Synthesis and of Computer Simulation

The *summer* (Figure 8.3) is the basic amplifier circuit wherein the impedances are all resistive. Thus, from Equation (8.10)

$$E_0 = -\frac{R_f}{R_1}E_1 - \frac{R_f}{R_2}E_2 - \frac{R_f}{R_3}E_3. \tag{8.11}$$

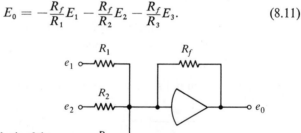

Figure 8.3. *Summer.* Synthesis of the comparator.

Thus, the output voltage is a linear combination of the input voltages (hence the name summer). Notice that the gain for each input may be varied by varying the resistor values. Also note the minus signs in Equation (8.11). The resistances commonly used are in the megohm range. For a typical computer, available fixed resistors are 10.0, 1.0, and 0.1 megohms; thus, gains of 0.1, 1.0, and 10.0 may be used. Clearly, the summer is the synthesis of the comparator.

Because summers are used so often, a standard symbol for them is employed. This symbol is shown in Figure 8.4.

The *integrator* (Figure 8.5) has a capacitor as a feedback impedance. From the basic equation (8.10),

$$E_0 = -\frac{1}{R_1Cs}E_1 - \frac{1}{R_2Cs}E_2 - \frac{1}{R_3Cs}E_3. \tag{8.12}$$

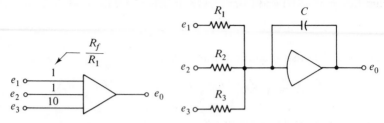

Figure 8.4. Standard symbol for the   **Figure 8.5.** Integrator circuit.
summer.

Thus, the output is the *integral* of a linear combination of the inputs. A typical range of capacitors, C, is 1.0, 0.1, and 0.01 μf.

Table 8.1 shows the range of possible gains [i.e., values of $1/RC$ in Equation (8.12)] for typical values of parameters.

Table 8.1
TIME CONSTANTS FOR TYPICAL VALUES
OF R AND C PARAMETERS (seconds)

R \ C	10^{-8}	10^{-7}	10^{-6}	(*farads*)
10^{+5}	1000	100	10.0	
10^{+6}	100	10.0	1.0	
(ohms)				

The standard symbol for the integrator is shown in Figure 8.6.

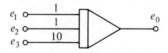

Figure 8.6. Standard symbol for the integrator.

As one might guess, the differentiator is synthesized by an operational amplifier circuit for which a capacitor is used as an input impedance (see Figures 8.7 and 8.8) so that

$$E_0 = -RCsE_1. \tag{8.13}$$

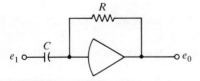

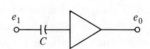

Figure 8.7. Differentiator circuit symbol.

Figure 8.8. Standard symbol for the differentiator.

The differentiator is, surprisingly, rarely used. To see why this configuration is avoided, suppose that

$$e_1 = A\cos(\omega t) \tag{8.14}$$

Then, from Equation (8.13)

$$e_0 = RC\,\omega A\sin(\omega t)$$

Thus, the amplitude is multiplied by the factor

$$RC\omega. \tag{8.15}$$

Thus, high-frequency noise would be greatly amplified. The unfavorable noise amplification properties of differentiators render them useless as building blocks in the simulation of dynamic systems. It is left as an exercise to show that integrators tend to *attenuate* high-frequency noise signals.

The final building block is the potentiometer. A typical operational amplifier circuit which contains a potentiometer is shown in Figure 8.9. Using Equation (8.6) and transfer matrix techniques, it is not difficult to show that

$$E_2 = \frac{x}{1 + (R/R_1)x(1 - x)}E_1, \tag{8.16}$$

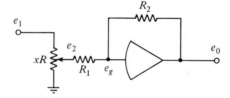

Figure 8.9. Computer circuit containing a potentiometer.

where R is the total resistance of the potentiometer and x is the percentage of R between the wiper and ground. Usually the ratio R/R_1 is kept small so that

$$E_2 \simeq xE_1, \quad x \leq 1.0. \tag{8.17}$$

From this point it will be assumed that Equation (8.17) is sufficiently accurate. The standard symbol for a potentiometer is illustrated in Figure 8.10.

The manner in which the above building blocks can be used to develop a computer simulation is best illustrated by example.

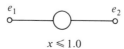

Figure 8.10. Standard symbol for the potentiometer.

Example 8.1

Suppose that it is desired to synthesize a dynamic system described by the transfer function

$$\frac{C(s)}{R(s)} = \frac{1}{s^2 + s + 1},\tag{i}$$

which may be rewritten

$$s^2C + sC + C = R.\tag{ii}$$

The following procedure is used to avoid differentiations [see the discussion above Equation (8.15)].

1. Divide by the coefficient of the highest power of s and solve for C:

$$C = \frac{1}{s^2}R - \frac{1}{s}C - \frac{1}{s^2}C.$$

2. Collect terms in like powers of s:

$$C = \frac{1}{s^2}[R - C] - \frac{1}{s}C.$$

3. *Nest* integrations; i.e., rearrange parentheses so that

$$C = \frac{1}{s}\left[\frac{1}{s}[R - C] - C\right].$$

4. Define new variables by working out through the nesting:

$$\text{State variables:} \quad \begin{cases} X_1 = \dfrac{1}{s}[R - C], \\[2mm] X_2 = \dfrac{1}{s}[X_1 - C] = C. \end{cases}\tag{iii}$$

The synthesis follows immediately from Equation (iii) and is shown in Figure 8.11. The synthesis of the dynamic system has been reduced to two integrations. Notice that differentiations are avoided. The summer in the feedback loop is added to account for the negative signs in Equations (8.11) and (8.12).

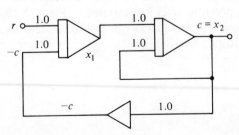

Figure 8.11. Synthesis for Example 8.1.

Notice that the circuit satisfies the

Rule of Thumb. All loops should have an odd number of amplifiers if the system being synthesized is to be inherently stable.

*   *   *

Example 8.2

In Example 8.1, a potentiometer was not required. For this example consider the system which is described by the transfer function $C(s)/R(s) = \alpha k/(s + \alpha)$, where α and k are constants. In operational form

$$\frac{1}{\alpha}sC + C = kR.$$

Divide by the highest power of s and solve for C to find

$$C = \frac{\alpha}{s}[kR - C]. \tag{i}$$

Using the basic building blocks, Equation (i) leads to the circuit of Figure 8.12. This circuit will provide a simulation of the first-order system provided that

$$\alpha, \alpha k \leq 1.0. \tag{ii}$$

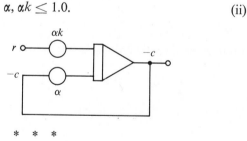

Figure 8.12. Computer circuit for Example 8.2.

*   *   *

The development of computer circuits for the simulation or synthesis of dynamic systems can be put on a more direct basis. This basis is presented in the next three sections. The question of initial conditions will also be resolved in these sections.

8.3 Phase Space Representation of Systems Described by Transfer Functions Without Numerator Dynamics

Recall that a dynamic system is said to have numerator dynamics if derivatives of the forcing function appear on the right-hand side of the differential equation. In this section, only systems described by transfer functions without numerator dynamics are discussed. Discussion of the more general case is delayed until a later section.

Consider the transfer function relationship

$$C(s) = \frac{k}{a_n s^n + a_{n-1} s^{n-1} + \cdots + a_0} R(s), \qquad (8.18)$$

where the a_i and k are constants. Notice that Equation (8.18) can be rewritten

$$a_n s^n C + a_{n-1} s^{n-1} C + \cdots + a_0 C = kR. \qquad (8.19)$$

The n *phase space* variables are defined by

$$X_1(s) = C(s) \qquad (8.20a)$$

and

$$X_2 = sX_1,$$
$$X_3 = sX_2,$$
$$X_4 = sX_3,$$
$$\cdot$$
$$\cdot$$
$$\cdot \qquad (8.20b)$$
$$X_{n-1} = sX_{n-2},$$
$$X_n = sX_{n-1}.$$

From the above definitions, it follows that

$$sX_n = s^2 X_{n-1} = s^3 X_{n-2} = \cdots = s^n X_1 = s^n C.$$

Use Equation (8.19) to show that

$$sX_n = -\frac{a_{n-1}}{a_n} X_n - \frac{a_{n-2}}{a_n} X_{n-1} - \cdots - \frac{a_0}{a_n} X_1 + \frac{k}{a_n} R. \qquad (8.21)$$

In summary, the n phase space variables satisfy the simultaneous equations

$$sX_1 = X_2,$$
$$sX_2 = X_3,$$
$$\cdot$$
$$\cdot$$
$$\cdot \qquad (8.22a)$$
$$sX_{n-1} = X_n,$$
$$sX_n = -\frac{a_0}{a_n} X_1 - \frac{a_1}{a_n} X_2 \cdots - \frac{a_{n-1}}{a_n} X_n + \frac{k}{a_n} R.$$

These equations can be rewritten in matrix form as follows:

$$
s \begin{bmatrix} X_1 \\ X_2 \\ \cdot \\ \cdot \\ \cdot \\ X_{n-1} \\ X_n \end{bmatrix} = \begin{bmatrix} 0 & 1 & 0 & \cdots & 0 \\ 0 & 0 & 1 & \cdots & 0 \\ \cdot & \cdot & \cdot & & \cdot \\ \cdot & \cdot & \cdot & \cdots & \cdot \\ \cdot & \cdot & \cdot & & \cdot \\ 0 & 0 & 0 & \cdots & 1 \\ \dfrac{-a_0}{a_n} & \dfrac{-a_1}{a_n} & & \cdots & \dfrac{-a_{n-1}}{a_n} \end{bmatrix} \begin{bmatrix} X_1 \\ X_2 \\ \cdot \\ \cdot \\ \cdot \\ X_{n-1} \\ X_n \end{bmatrix} + \begin{bmatrix} 0 \\ 0 \\ \cdot \\ \cdot \\ 0 \\ \dfrac{k}{a_n} \end{bmatrix} R. \quad (8.22b)
$$

Notice that the phase space Equations (8.22a) can also be written in the integral form

$$
X_1 = \frac{1}{s} X_2,
$$

$$
X_2 = \frac{1}{s} X_3,
$$

$$
X_n = \frac{1}{s}\left[\frac{-a_0}{a_n} X_1 \cdots \frac{-a_{n-1}}{a_n} X_n + \frac{k}{a_n} R \right]. \quad (8.23)
$$

Equations in this form lead directly to a synthesis of Equation (8.18) or to the

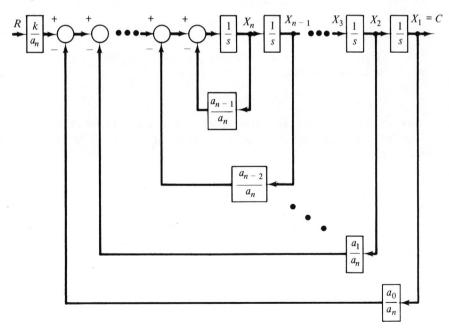

Figure 8.13. Block diagram of the synthesis of a dynamic system using the phase space representation.

design of analog computer circuits for simulation. The block diagram of Equations (8.23) is indicated in Figure 8.13. Construction of this block diagram is a convenient intermediate step in synthesis or simulation procedures.

The phase space variables are also discussed in Section 2.10 [see Equations (2.99) and (2.100)].

Example 8.3

Consider the mechanical system for which $ms^2C + bsC + kC = kR$. Notice that for this case X_1 is displacement and X_2 is velocity. Using Figure 8.13 as a guide, the operational amplifier circuit shown in Figure 8.14 could be constructed. Notice the convention for indicating initial conditions used in Figure 8.14.

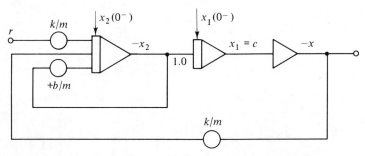

Figure 8.14. Simulation of a mechanical system using the phase space representation.

*   *   *

An advantage of using the phase space representation in the synthesis of a transfer function is that the initial conditions, $x_i(0^-)$, are easily related to physical quantities.

8.4 A Second Representation for Linear Constant Coefficient Systems Without Numerator Dynamics

Again consider the input-output relation

$$a_n s^n C + a_{n-1} s^{n-1} C + \cdots + a_0 C = kR.$$

A second type of representation is developed by proceeding in the following manner:

1. Divide by $a_n s^n$ and solve for C:

$$C = -\frac{a_{n-1}}{a_n}\frac{1}{s}C - \frac{a_{n-2}}{a_n}\frac{1}{s^2}C - \cdots - \frac{a_0}{a_n}\frac{1}{s^n}C + \frac{k}{a_n s^n}R.$$

2. Collect like terms in powers of s:

$$C = -\frac{a_{n-1}}{a_n}\frac{1}{s}C - \frac{a_{n-2}}{a_n}\frac{1}{s^2}C - \cdots + \left[-\frac{a_0 C}{a_n} + \frac{kR}{a_n}\right]\frac{1}{s^n}.$$

3. Nest integrations:

$$C = \frac{1}{s}\left[-\frac{a_{n-1}}{a_n}C + \frac{1}{s}\left[-\frac{a_{n-2}}{a_n}C + \cdots + \frac{1}{s}\left[\frac{k}{a_n}R - \frac{a_0}{a_n}C\right]\right]\right].$$

4. Define the state space variables by working *into* the nesting:

$$\boxed{X_n(s) = C(s).} \tag{8.24}$$

$$X_n(s) = \frac{1}{s}\left[-\frac{a_{n-1}}{a_n}X_n + X_{n-1}\right],$$

$$X_{n-1}(s) = \frac{1}{s}\left[-\frac{a_{n-2}}{a_n}X_n + X_{n-2}\right],$$

$$\vdots \tag{8.25}$$

$$X_2(s) = \frac{1}{s}\left[-\frac{a_1}{a_n}X_n + X_1\right],$$

$$X_1(s) = \frac{1}{s}\left[-\frac{a_0}{a_n}X_n + \frac{k}{a_n}R\right].$$

These equations can be written in the matrix form

$$s\begin{bmatrix} X_1 \\ X_2 \\ \vdots \\ \\ X_n \end{bmatrix} = \begin{bmatrix} 0 & 0 & 0 & \cdots & & -\dfrac{a_0}{a_n} \\ 1 & 0 & 0 & \cdots & & -\dfrac{a_1}{a_n} \\ 0 & 1 & 0 & & & \vdots \\ \vdots & \vdots & \vdots & & & \\ 0 & 0 & 0 & \cdots & 1 & -\dfrac{a_{n-1}}{a_n} \end{bmatrix}\begin{bmatrix} X_1 \\ X_2 \\ \vdots \\ \\ X_n \end{bmatrix} + \begin{bmatrix} \dfrac{k}{a_n} \\ 0 \\ \vdots \\ \\ 0 \end{bmatrix} R. \tag{8.26}$$

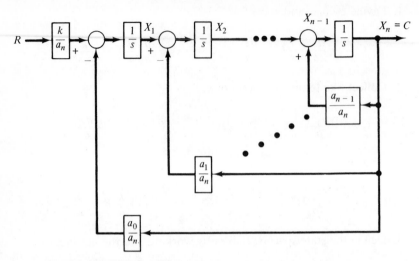

Figure 8.15. Block diagram of the synthesis of dynamic systems using the alternative state space representation.

Equations (8.24) and (8.26) should be compared to Equations (8.22) and (8.20). A block diagram of Equations (8.25) is illustrated in Figure 8.15.

Example 8.4

Again consider the system of Example 8.3 for which

$$ms^2C + bsC + kC = kR.$$

The synthesis developed using the alternative state space representation is shown in Figure 8.16.

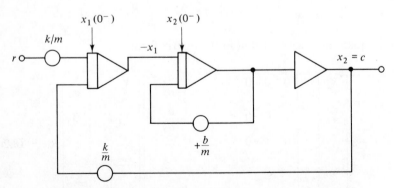

Figure 8.16. Synthesis of a dynamic system using the alternative state representation.

* * *

The advantage of this type of synthesis is that the summations are not all performed at one amplifier (this decreases the possibility of saturation). The disadvantage of the use of the alternative state space approach to simulation is that it is difficult to relate the $x_i(0^-)$ to physical quantities. A point that should be emphasized is that different operational amplifier circuits can be constructed for the *same* dynamic system. All one needs to do is define a set of variables, called the *state space variables*, such that the input-output relation reduces to a set of simultaneous first-order integrations. This procedure was demonstrated above for the phase space and alternative state space variables. For present purposes, three example representations will prove sufficient: the two discussed above and the modal representation which will be discussed in Section 8.6.

8.5 Extension of the Phase Space and Alternative Representations to the Synthesis of Transfer Functions with Numerator Dynamics[3]

Consider the synthesis and/or simulation of the general transfer function relation

$$C(s) = \frac{\{b_m s^m + b_{m-1} s^{m-1} + \cdots + b_0\} R(s)}{a_n s^n + a_{n-1} s^{n-1} + \cdots + a_0}. \tag{8.27}$$

Rewrite this equation as

$$C(s) = \left[-\frac{a_{n-1}}{a_n}\frac{1}{s} - \frac{a_{n-2}}{a_n}\frac{1}{s^2} - \cdots - \frac{a_0}{a_n}\frac{1}{s^n} \right] C(s)$$
$$+ \left[\frac{b_m}{a_n}\frac{1}{s^{n-m}} + \frac{b_{m-1}}{a_n}\frac{1}{s^{n-m+1}} + \cdots + \frac{b_0}{a_n}\frac{1}{s^n} \right] R(s). \tag{8.28}$$

For convenience, define α_k and β_k by

$$a_k = a_n \alpha_{n-k},$$
$$b_k = a_n \beta_{n-k}, \quad k = 0, 1, 2, \ldots, n-1, \tag{8.29}$$

so that Equation (8.28) can be rewritten

$$C(s) = \left[-\frac{\alpha_1}{s} - \frac{\alpha_2}{s^2} - \cdots - \frac{\alpha_n}{s^n} \right] C(s)$$
$$+ \left[\frac{\beta_{n-m}}{s^{n-m}} + \frac{\beta_{n-m+1}}{s^{n-m+1}} + \cdots + \frac{\beta_n}{s^n} \right] R(s). \tag{8.30}$$

The block diagram of this equation is provided in Figure 8.17. The state variables, X_k, are taken to be the outputs of the integration blocks. It follows

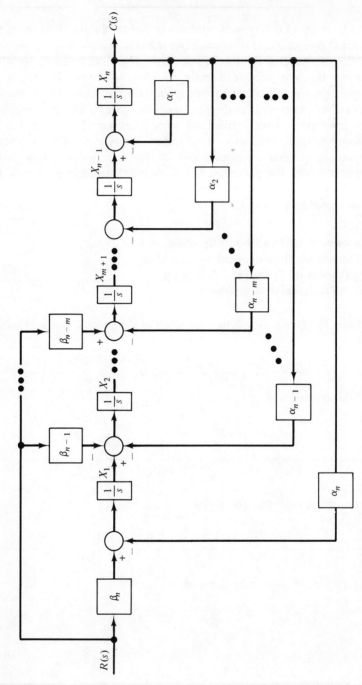

Figure 8.17. Block diagram of Equation (8.30). Compare this diagram with the diagram of the alternative state space representation shown in Figure 8.15.

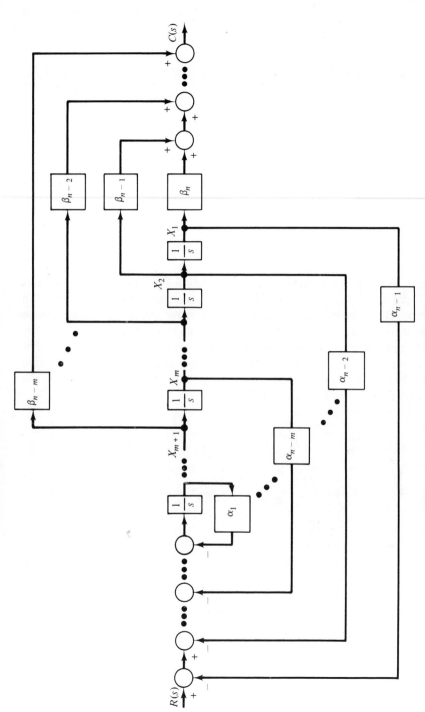

Figure 8.18. Block diagram of Equations (8.32) and (8.33). Generalization of the phase space synthesis procedure to the case of numerator dynamics.

from Figure 8.17 that

$$
s \begin{bmatrix} X_1 \\ X_2 \\ \vdots \\ X_{m+1} \\ X_{m+2} \\ \vdots \\ X_n \end{bmatrix} = \begin{bmatrix} 0 & 0 & \cdots & 0 & 0 & \cdots & 0 & -\alpha_n \\ 1 & 0 & \cdots & 0 & 0 & \cdots & 0 & -\alpha_{n-1} \\ & & & & & & & \vdots \\ 0 & 0 & \cdots & 0 & 0 & \cdots & 0 & -\alpha_m \\ 0 & 0 & \cdots & 1 & 0 & \cdots & 0 & -\alpha_{m-1} \\ & & & & & & & \vdots \\ 0 & 0 & \cdots & 0 & 0 & \cdots & 1 & -\alpha_1 \end{bmatrix} \begin{bmatrix} X_1 \\ X_2 \\ \vdots \\ X_{m+1} \\ X_{m+2} \\ \vdots \\ X_n \end{bmatrix} + \begin{bmatrix} \beta_n \\ \beta_{n-1} \\ \vdots \\ \beta_{n-m} \\ 0 \\ \vdots \\ 0 \end{bmatrix} R.
$$

$$(8.31)$$

When one compares Equations (8.31) and (8.26) and compares Figures 8.17 and 8.15 it becomes clear that the state space representation defined in Figure 8.17 is the generalization of the *alternative* representation to the case of numerator dynamics. Equations (8.29) relate transfer function coefficients to elements of the state space matrices. Since all the symbols in a block diagram are synthesized by the operational amplifier building blocks discussed at the beginning of the chapter, Figure 8.17 leads directly to a synthesis of the transfer function in Equation (8.27).

Equation (8.27) can also be rewritten as

$$C(s) = \{\beta_{n-m}s^m + \cdots + \beta_{n-1}s + \beta_n\}C_0(s), \qquad (8.32)$$

where

$$C_0(s) = \frac{1}{s^n + \alpha_1 s^{n-1} + \cdots + \alpha_{n-1}s + \alpha_n} R(s). \qquad (8.33)$$

Suppose that the transfer function relating $C_0(s)$ to $R(s)$ is synthesized using the phase space representation (see Figure 8.14); then $s^k C_0(s)$ is the input to the kth integrator to the left of $C_0(s)$. This fact is used to construct the block diagram of Equations (8.32) and (8.33) illustrated in Figure 8.18. Clearly, this diagram indicates the general phase space synthesis procedure. Note that for this synthesis the relation between $C(s)$ and the state variables is

$$C(s) = \beta_n X_1 + \beta_{n-1} X_2 + \cdots + \beta_{n-m} X_{m+1}. \qquad (8.34)$$

It will be demonstrated in the following example that Figure 8.18 and Equation (8.34) are not a unique phase space synthesis of the transfer function in Equation (8.27).

Example 8.5

Consider the system in Figure 8.19. For this system, it was shown in Chapter 4 that

$$ms^2C + bsC + kC = bsR + kR. \tag{i}$$

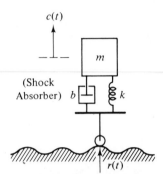

Figure 8.19. Model of an automobile (heave mode).

The phase space and the alternative state space representations will now be developed. First, to determine the alternative state space representation, one need only consult Figure 8.17. The analog computer program suggested by Figure 8.17 is illustrated in Figure 8.20a. A simplification of this circuit is shown in Figure 8.20b. Note the feed-forward lines in these circuits. Feed-forward is associated with numerator dynamics.

The phase space simulation for the automobile could be developed by translating the block diagram of Figure 8.18. However, it will prove instructive to use an alternative, analytical procedure. Define

$$X_1 \triangleq C \triangleq \frac{1}{s}[X_2 + d_1 R] \tag{ii}$$

and

$$X_2 \triangleq \frac{1}{s}\left[-\frac{k}{m}X_1 - \frac{b}{m}X_2 + d_2 R \right], \tag{iii}$$

where d_1 and d_2 are unknown constants. These constants are determined by substituting Equation (ii) into Equation (i) to find

$$\frac{1}{s}[X_2 + d_1 R] = \frac{1}{s}\left(\frac{1}{s}\left[\frac{k}{m}R - \frac{k}{m}X_1 \right] + \left[\frac{b}{m}R - \frac{b}{m}X_1 \right] \right).$$

Substitute Equation (iii) for X_2 on the left-hand side of the above equation to

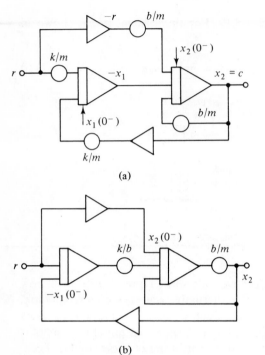

(a)

(b)

Figure 8.20. (a) Simulation of automobile (heave mode). Alternative state space, (b) simplified version of the circuit in (a). Two potentiometers have been eliminated.

find

$$\frac{1}{s}\left(\frac{1}{s}\left[-\frac{k}{m}X_1 - \frac{b}{m}X_2 + d_2R\right] + d_1R\right)$$

$$= \frac{1}{s}\left(\frac{1}{s}\left[\frac{k}{m}R - \frac{k}{m}X_1\right] + \left[\frac{b}{m}R - \frac{b}{m}X_1\right]\right).$$

Now substitute Equation (ii) for X_1 on the right-hand side of this equation so that

$$\frac{1}{s}\left(\frac{1}{s}[d_2R] + d_1R\right) = \frac{1}{s}\left(\frac{1}{s}\left[\frac{k}{m}R - \frac{b}{m}d_1R\right] + \frac{b}{m}R\right).$$

By inspection of the last equation,

$$d_1 = \frac{b}{m} \tag{iv}$$

and

$$d_2 = \frac{k}{m} - \frac{b}{m}d_1 = \frac{k}{m} - \frac{b^2}{m^2}. \tag{v}$$

The analog computer circuit follows immediately from Equations (ii)–(v). The circuit is shown in Figure 8.21. Again notice the feed-forward lines.

* * *

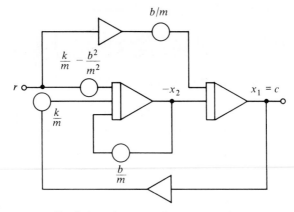

Figure 8.21. Simulation of automobile (heave mode). Phase space representation.

For the reader's convenience, the following is provided to aid in the development of this second type of generalized phase space representation of systems with numerator dynamics. For the input-output relation

$$a_n s^n C + a_{n-1} s^{n-1} C + \cdots + a_1 s C + a_0 C$$
$$= b_n s^n R + b_{n-1} s^{n-1} R + \cdots + b_0 R, \tag{8.35}$$

define

$$X_1 \triangleq C - \frac{b_n}{a_n} R = C - d_0 R,$$

$$X_1 = \frac{1}{s}[X_2 + d_1 R],$$

$$X_2 = \frac{1}{s}[X_3 + d_2 R], \tag{8.36}$$

$$\vdots$$

$$X_n = \frac{1}{s}\left[-\sum_{k=1}^{n} \frac{a_k}{a_n} X_k + d_n R\right]$$

where the constants d_i are given by[3]

$$d_1 = \frac{1}{a_n}\left[b_{n-1} - a_{n-1}\frac{b_n}{a_n}\right],$$

$$d_2 = \frac{1}{a_n}\left[b_{n-2} - a_{n-2}\frac{b_n}{a_n} - a_{n-1}d_1\right], \tag{8.37}$$

$$\vdots$$

$$d_k = \frac{1}{a_n}\left[b_{n-k} - a_{n-k}\frac{b_n}{a_n} - a_{n-k+1}d_{k-1} - \cdots - a_{n-1}d_{k-1}\right].$$

Or, the general form is

$$d_k = \frac{1}{a_n}\left[b_{n-k} - \sum_{\alpha=1}^{k} d_{\alpha-1} a_{n-k-1+\alpha} \right]$$

if one takes $d_0 = b_n/a_n$. In matrix form

$$s\begin{bmatrix} X_1 \\ X_2 \\ \cdot \\ \cdot \\ \cdot \\ X_n \end{bmatrix} = \begin{bmatrix} 0 & 1 & 0 & \cdots & 0 \\ 0 & 0 & 1 & \cdots & 0 \\ \cdot & & & & \\ \cdot & & & & \\ 0 & 0 & 0 & \cdots & 1 \\ \dfrac{-a_{n-1}}{a_n} & & \cdots & & \dfrac{-a_0}{a_n} \end{bmatrix} \begin{bmatrix} X_1 \\ X_2 \\ \cdot \\ \cdot \\ \cdot \\ X_n \end{bmatrix} + \begin{bmatrix} d_1 \\ d_2 \\ \cdot \\ \cdot \\ \cdot \\ d_n \end{bmatrix} R. \qquad (8.38)$$

Notice that this representation is distinct from that of Figure 8.18.

8.6 The Modal Representation: Transformation from Phase Space to the Modal State Space

Still another state space representation of a dynamic system is described in this section. Again, the representation could prove useful in a synthesis procedure or in the development of an analog computer simulation. The representation will be referred to as the *modal* representation and will be closely related to the calculation of poles and residues (see Chapter 2).

A transformation from the phase space representation to the modal state space, introduced in this chapter, will also be discussed. This type of transformation will be generalized in Section 8.7.

As was the case for the discussion of the phase space and alternative state space representations, the analysis begins with Equation (8.27), which may be rewritten

$$\frac{C(s)}{R(s)} = G(s) = \frac{(1/a_n)(b_m s^m + \cdots + b_1 s + b_0)}{(s + p_1)(s + p_2) \cdots (s + p_n)}, \qquad (8.39)$$

where, obviously, the p_i are the *poles* of the transfer function. In this section, the case where the p_i are *real* and *distinct* will be considered first.

To find a state space representation to aid in the construction of a computer simulation, one must define a set of variables such that relation (8.39) reduces to a set of first-order simultaneous integrations. To achieve this end, begin by performing a heavyside expansion of the transfer function of Equation (8.39):

$$G(s) = \frac{r_1}{(s + p_1)} + \frac{r_2}{(s + p_2)} + \cdots + \frac{r_3}{(s + p_n)}. \qquad (8.40)$$

As was shown in Chapter 2, the *residues*

$$r_i = (s + p_i)G(s)|_{s = -p_i}. \tag{8.41}$$

The r_i may be interpreted as the residues of the *impulse response* of the system in question (see Chapter 2).

Define the *modes* (*normal coordinates*) by

$$Z_1 = \frac{r_1 p_1}{s + p_1} R,$$

$$Z_2 = \frac{r_2 p_2}{s + p_2} R,$$

$$\cdot$$
$$\cdot$$
$$\cdot$$

$$Z_n = \frac{r_n p_n}{s + p_n} R, \tag{8.42}$$

or by

$$Z_1 = \frac{r_1}{T_1 s + 1} R,$$

$$Z_2 = \frac{r_2}{T_2 s + 1} R,$$

$$\cdot$$
$$\cdot$$
$$\cdot$$

$$Z_n = \frac{r_3}{T_n s + 1} R, \tag{8.43}$$

where the *time constants*

$$T_i = \frac{1}{p_i}. \tag{8.44}$$

It follows from Equations (8.40) and (8.42) that

$$C(s) = \frac{1}{p_1} Z_1(s) + \frac{1}{p_2} Z_2(s) + \cdots + \frac{1}{p_n} Z_n(s). \tag{8.45}$$

To summarize, the definition of the set of modal states,

$$Z_1 = \frac{1}{s}[r_1 p_1 R - p_1 Z_1],$$

$$Z_2 = \frac{1}{s}[r_2 p_2 R - p_2 Z_2],$$

$$\cdot$$
$$\cdot$$
$$\cdot$$

$$Z_n = \frac{1}{s}[r_n p_n R - p_n Z], \tag{8.46}$$

where the p_i are poles of $G(s)$ and the r_i satisfy Equation (8.41), leads to the relationship of Equation (8.45).

In matrix form, the s-domain Equations (8.45) satisfied by the modes are

$$
s\begin{bmatrix} Z_1 \\ Z_2 \\ Z_3 \\ \cdot \\ \cdot \\ \cdot \\ Z_n \end{bmatrix} = \begin{bmatrix} -p_1 & 0 & 0 & \cdots & 0 \\ 0 & -p_2 & 0 & \cdots & 0 \\ 0 & 0 & -p_3 & \cdots & 0 \\ \cdot & \cdot & \cdot & & \cdot \\ \cdot & \cdot & \cdot & & \cdot \\ \cdot & \cdot & \cdot & & \cdot \\ 0 & 0 & 0 & \cdots & -p_n \end{bmatrix} \begin{bmatrix} Z_1 \\ Z_2 \\ Z_3 \\ \cdot \\ \cdot \\ \cdot \\ Z_n \end{bmatrix} + \begin{bmatrix} r_1 p_1 \\ r_2 p_2 \\ r_3 p_3 \\ \cdot \\ \cdot \\ \cdot \\ r_n p_n \end{bmatrix} R. \quad (8.47)
$$

Note the diagonal form of the matrix in this equation; this fact indicates that the modes are independent of one another; i.e., the modal state variables are *uncoupled.*

The synthesis of the ith mode with an operational amplifier circuit is shown in Figure 8.22a. Figure 8.22b indicates a simplification of this circuit which is convenient because the residues and pole are set on separate potentiometers. The complete synthesis, which provides the output variable, $c(t)$, is shown in Figure 8.23. The decoupled nature of the modal state variables is apparent from this circuit. It is useful to derive a relationship between the modal and phase space representations. First, consider the case of no numerator dynamics. From Equations (8.20a) and (8.45) it follows that the first phase space variable

$$
X_1 = \frac{1}{p_1}Z_1 + \frac{1}{p_2}Z_2 + \cdots + \frac{1}{p_n}Z_n. \quad (8.48)
$$

Occasionally, it is desired to relate the other phase space variables, X_i, to the modal state variables, Z_i. To do this, note first that

$$
X_1 = C = G(s)R.
$$

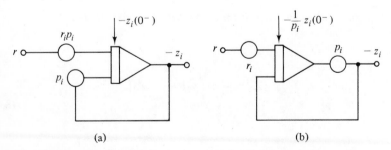

(a)   (b)

Figure 8.22. (a) Synthesis of the ith modal state, (b) simplified modal synthesis.

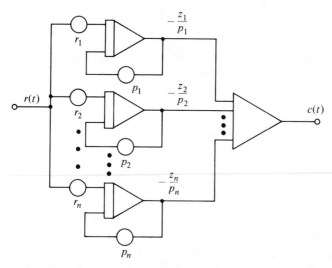

Figure 8.23. Synthesis of a dynamic system (real, distinct poles) using the modal state space representation. *Note:* Notice that the poles and impulse response residues are set on independent potentiometers.

Thus,

$$X_2 \triangleq sX_1 = sG(s)R = \left(\frac{r'_1}{s + p_1} + \frac{r'_2}{s + p_2} + \cdots + \frac{r'_n}{s + p_n}\right)R, \quad (8.49)$$

where

$$r'_i = (s + p_i)sG(s)|_{s = -p_i}$$
$$= -p_i r_i. \quad (8.50)$$

It follows immediately from Equations (8.45), (8.49), and (8.50) that

$$X_2 = -Z_1 - Z_2 - \cdots - Z_n. \quad (8.51)$$

Similarly, using the fact that $X_3 \triangleq sX_2$, it is easy to show that

$$X_3 = p_1 Z_1 + p_2 Z_2 + \cdots + p_n Z_n. \quad (8.52)$$

Proceeding in a similar manner for the remaining phase space variables, one finds

$$X_1 = \frac{1}{p_1}Z_1 + \frac{1}{p_2}Z_2 + \cdots + \frac{1}{p_n}Z_n,$$
$$X_2 = -Z_1 - Z_2 - \quad \cdots - Z_n,$$
$$X_3 = p_1 Z_1 + p_2 Z_2 + \quad \cdots + p_n Z_n, \quad (8.53)$$
$$\vdots \qquad\qquad \vdots$$
$$X_n = (-1)^{n+1}[p_1^{n-2}Z + \cdots + p_n^{n-2}Z_n].$$

In matrix form and in the time domain, Equations (8.53) may be rewritten

$$\mathbf{x}(t) = \mathfrak{I}\mathbf{z}(t), \tag{8.54}$$

where

$$\mathfrak{I} \triangleq \begin{bmatrix} \dfrac{1}{p_1} & \dfrac{1}{p_2} & \cdot\cdot & \dfrac{1}{p_n} \\[2mm] -1 & -1 & \cdot\cdot & -1 \\ \cdot & & & \cdot \\ \cdot & & & \cdot \\ \cdot & & & \cdot \\ -1^{n+1}p_1^{n-2} & \cdots & & -1^{n+1}p_n^{n-2} \end{bmatrix}. \tag{8.55}$$

$\mathfrak{I}$ is called the *phase space synthesizer matrix* because, as Equation (8.54) indicates, it "synthesizes", or reconstructs, the phase space variables from the modal state variables. Notice that the columns of this matrix are proportional to the eigenvectors of the *companion matrix* [see Equation (2.101)].

Multiply Equation (8.54) by $\mathfrak{I}^{-1}$ and find

$$\mathbf{z}(t) = \mathfrak{I}^{-1}\mathbf{x}(t). \tag{8.56}$$

$\mathfrak{I}^{-1}$ is called the *phase space analyzer matrix* because it "analyzes," or resolves, phase space variables into the modes or normal coordinates. In particular, Equation (8.56) must be satisfied at $t = 0^-$, i.e.,

$$\mathbf{z}(0^-) = \mathfrak{I}^{-1}\mathbf{x}(0^-). \tag{8.57}$$

Example 8.6

Consider the overdamped mechanical system (Figure 8.24) for which

$$m = 1.0 \text{ slug}, \quad k = 2 \text{ lb/ft}, \quad b = 3 \text{ lb/sec/ft}.$$

The transfer function is

$$\frac{C(s)}{R(s)} = \frac{2}{s^2 + 3s + 2} = \frac{2}{(s + 1)(s + 2)}, \tag{i}$$

so that

$$\begin{aligned} p_1 &= 1.0, \quad r_1 = 2 \\ p_2 &= 2.0, \quad r_2 = -2. \end{aligned} \tag{ii}$$

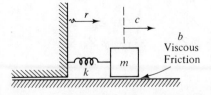

Figure 8.24. Overdamped mechanical system.

Thus the synthesizer matrix is

$$\mathcal{J} = \begin{bmatrix} 1 & \frac{1}{2} \\ -1 & -1 \end{bmatrix},$$

from which the analyzer is found to be

$$\mathcal{J}^{-1} = \begin{bmatrix} 2 & 1 \\ -2 & -2 \end{bmatrix}.$$

To check this result, notice that

$$\mathcal{J}\mathcal{J}^{-1} = \begin{bmatrix} 1 & 0 \\ 0 & 1 \end{bmatrix}.$$

It follows from Equation (8.54) that

$$\begin{bmatrix} z_1 \\ z_2 \end{bmatrix} = \begin{bmatrix} 2 & 1 \\ -2 & -2 \end{bmatrix}\begin{bmatrix} x_1 \\ x_2 \end{bmatrix}$$

or

$$z_1 = 2x_1 + x_2,$$
$$z_2 = -2x_1 - 2x_2.$$

In particular,

$$z_1(0^-) = 2x_1(0^-) + x_2(0^-),$$
$$z_2(0^-) = -x_2(0^-) - x_2(0^-),$$

where

$$x_1(0^-) = \text{Initial displacement},$$
$$x_2(0^-) = \text{Initial velocity}.$$

* * *

The analyzer and synthesizer matrices do not change form for a system with numerator dynamics. This assertion is difficult to prove in general: the assertion will be demonstrated by considering a particular example.

Example 8.7

Consider the dynamic system for which the output is related to the input by

$$C = \left[\frac{s^3 + 8s^2 + 17s + 8}{(s + 1)(s + 2)(s + 3)}\right]R. \tag{i}$$

A heavyside expansion of the transfer function yields

$$\frac{s^3 + 8s^2 + 17s + 8}{(s + 1)(s + 2)(s + 3)} = r_0 + \frac{r_1}{s + 1} + \frac{r_2}{s + 2} + \frac{r_3}{s + 3}.$$

The r_0 term must be included because the order of the numerator is equal to (not less than) the order of the denominator. In the usual manner,

$$r_1 = -1,$$

$$r_2 = \frac{-8 + 32 - 34 + 8}{(-1)(1)} = 2,$$

$$r_3 = \frac{-27 + 72 - 51 + 8}{(-2)(-1)} = 1,$$

and

$$r_0 = G(s)|_{s=\infty} = 1.$$

Define the modes by

$$Z_1 = \frac{-1}{s+1}R,$$

$$Z_2 = \frac{4}{s+2}R,$$

$$Z_3 = \frac{3}{s+3}R,$$

so that

$$C = R + [Z_1 + \tfrac{1}{2}Z_2 + \tfrac{1}{3}Z_3]. \tag{ii}$$

Compare Equation (ii) with the first row of the synthesizer matrix for a system with no numerator dynamics [Equation (8.55)].

The phase space variables satisfy the equations

$$X_1 = C - R = \left[Z_1 + \frac{1}{2}Z_2 + \frac{1}{3}Z_3\right] = \frac{(2s^2 + 6s + 2)}{(s+1)(s+2)(s+3)}R,$$

$$X_2 = sX_1 - d_1R,$$

where, by Equation (8.37),

$$d_1 = \frac{1}{a_n}[b_{n-1} - a_{n-1}d_0] = [8 - 6] = 2.$$

Thus,

$$X_2 = \left[\frac{2s^2 + 6s + 2s}{s^3 + 6s^2 + 11s + 6} - 2\right]R = -\frac{6s^2 - 20s - 12}{(s+1)(s+2)(s+3)}R$$

$$= +\frac{1}{(s+1)} + \frac{-1}{(s+2)} + \frac{-1}{(s+3)}$$

$$= -Z_1 - Z_2 - Z_3.$$

Also,

$$X_3 = sX_2 - d_2R.$$

Using Equation (8.37) again,

$$d_2 = \frac{1}{a_n}[b_{n-2} - a_{n-2}d_0 - a_{n-1}d_1]$$

$$= [17 - 11 - 6(2)] = -6.$$

Also,

$$X_3 = \left[\frac{-6s^3 - 20s^2 - 12s}{s^3 + 6s^2 + 11s + 6} + 6\right]R = \frac{16s^2 + 54s + 36}{(s + 1)(s + 2)(s + 3)}R$$

$$= \left[\frac{-10}{s + 1} + \frac{+8}{s + 2} + \frac{+9}{s + 3}\right]R$$

$$= 1Z_1 + 2Z_2 + 3Z_3.$$

To summarize,

$$\begin{bmatrix} x_1 \\ x_2 \\ x_3 \end{bmatrix} = \begin{bmatrix} 1 & \frac{1}{2} & \frac{1}{3} \\ -1 & -1 & -1 \\ 1 & 2 & 3 \end{bmatrix} \begin{bmatrix} z_1 \\ z_2 \\ z_3 \end{bmatrix}.$$

Synthesizer

Thus, the synthesizer matrix has the same form as a system without numerator dynamics.

The heavyside expansion technique illustrated above is quite useful for finding synthesizer matrices for state spaces other than the phase space.

* * *

The modal representation of dynamic systems with repeated real poles presents some difficulties which will be illustrated for the particular transfer function

$$\frac{C(s)}{R(s)} = G(s) = \frac{1}{(s + p_1)^2(s + p_2)}. \tag{8.58}$$

Perform a heavyside expansion of this transfer function to find

$$G(s) = \frac{r_{12}}{s + p_1} + \frac{r_{11}}{(s + p_1)^2} + \frac{r_2}{s + p_2}. \tag{8.59}$$

The residues are found using the techniques of Chapter 2, i.e.,

$$r_2 = (s + p_2)G(s)|_{s=-p_2} = \frac{1}{(p_1 - p_2)^2}, \tag{8.60}$$

$$r_{11} = (s + p_1)^2 G(s)|_{s=-p_1} = \frac{1}{p_2 - p_1}, \tag{8.61}$$

$$r_{12} = \left[\frac{d}{ds}\{(s + p_1)^2 G(s)\}\right]_{s=-p_1} = -\frac{1}{(s + p_2)^2}\bigg|_{s=-p_1} = -\frac{1}{(p_2 - p_1)^2}. \tag{8.62}$$

Define the modes by

$$Z_{12} = \frac{r_{12}p_1}{s + p_1}R,$$

$$Z_{11} \triangleq \frac{r_{11}p_1^2}{(s + p_1)^2}R = \frac{(r_{11}/r_{12})p_1}{(s + p_1)}Z_{12},$$ (8.63)

$$Z_2 = \frac{r_2p_2}{s + p_2},$$

or, in matrix form,

$$s\begin{bmatrix} Z_{11} \\ Z_{12} \\ Z_2 \end{bmatrix} = \begin{bmatrix} -p_1 & \dfrac{r_{11}}{r_{12}p_1} & 0 \\ 0 & -p_1 & 0 \\ 0 & 0 & -p_2 \end{bmatrix}\begin{bmatrix} Z_{11} \\ Z_{12} \\ Z_2 \end{bmatrix} + \begin{bmatrix} 0 \\ r_{12}p_1 \\ r_2p_2 \end{bmatrix}R.$$ (8.64)

The relation between the output and the modes is found from Equations (8.58) and (8.59) to be

$$C = \left[\frac{1}{p_1}Z_{12} + \frac{1}{p_1^2}Z_{11} + \frac{1}{p_2}Z_2\right].$$ (8.65)

It is apparent from Equation (8.64) that the modes Z_{11} and Z_{12} are coupled. This fact is also apparent from an examination of Figure 8.25, which is a schematic of the analog computer circuit for this system.

Analyzer and synthesizer matrices for the case of repeated poles are also found using the heavyside expansion technique illustrated above for distinct real poles.

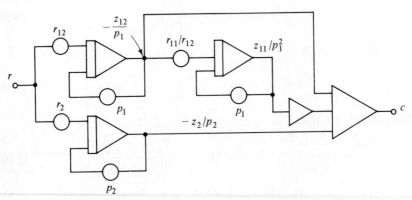

Figure 8.25. Analog computer simulation of a system with a repeated real pole: modal representation.

The modal representation of dynamic systems with complex conjugate poles in the transfer functions will be illustrated for the particular case

$$C = \frac{1}{(s + p_1)[(s + \sigma)^2 + \omega^2]} R. \tag{8.66}$$

Assume that $\omega^2 > 0$ so that the complex poles are at

$$s = -\sigma \pm i\omega. \tag{8.67}$$

Perform a heavyside expansion of the transfer function to find

$$C = \left[\frac{r_1}{s + p_1} + \frac{r_3 s + r_2}{(s + \sigma)^2 + \omega^2}\right] R. \tag{8.68}$$

As before,

$$r_1 = (s + p_1)G(s)|_{s=-p_1}. \tag{8.69}$$

r_2 and r_3 can be found from an inspection of

$$\frac{r_3 s + r_2}{[(s + \sigma)^2 + \omega^2]} = G(s) - \frac{r_1}{s + p_1}. \tag{8.70}$$

Define the mode for the real pole by

$$Z_1 \triangleq \frac{r_1 p_1}{s + p_1} R. \tag{8.71}$$

For the complex poles define the modes by[5]

$$\begin{bmatrix} sZ_2 \\ sZ_3 \end{bmatrix} \triangleq \begin{bmatrix} -\sigma & \omega \\ -\omega & -\sigma \end{bmatrix} \begin{bmatrix} Z_2 \\ Z_3 \end{bmatrix} + \begin{bmatrix} 0 \\ r_3 \end{bmatrix} R. \tag{8.72}$$

To find expressions for $Z_2(s)$ and $Z_3(s)$ in terms of $R(s)$, take the first term on the right-hand side of Equation (8.72) to the left-hand side to find

$$\begin{bmatrix} s + \sigma & -\omega \\ \omega & s + \sigma \end{bmatrix} \begin{bmatrix} Z_2 \\ Z_3 \end{bmatrix} = \begin{bmatrix} 0 \\ r_3 \end{bmatrix} R$$

or

$$\begin{bmatrix} Z_2 \\ Z_3 \end{bmatrix} = \begin{bmatrix} \dfrac{s + \sigma}{d} & \dfrac{\omega}{d} \\ -\dfrac{\omega}{d} & \dfrac{s + \sigma}{d} \end{bmatrix} \begin{bmatrix} 0 \\ r_3 \end{bmatrix} R = \begin{bmatrix} \dfrac{\omega r_3}{d} \\ \dfrac{(s + \sigma) r_3}{d} \end{bmatrix} R,$$

where

$$d = (s + \sigma)^2 + \omega^2.$$

It follows that

$$Z_2 = \frac{\omega r_3}{(s + \sigma)^2 + \omega^2} R,$$

$$Z_3 = \frac{r_3(s + \sigma)}{(s + \sigma)^2 + \omega^2} R. \tag{8.73}$$

Thus, from Equations (8.68), (8.71), and (8.73), find that

$$C = \frac{1}{p_1}Z_1 + Z_3 + kZ_2, \tag{8.74}$$

where

$$k = -\frac{\sigma}{\omega} + \frac{r_2}{r_2\omega}. \tag{8.75}$$

The analog computer circuit which is a synthesis of the above equations can be constructed using Equations (8.70) and (8.72). This circuit is shown in Figure 8.26.

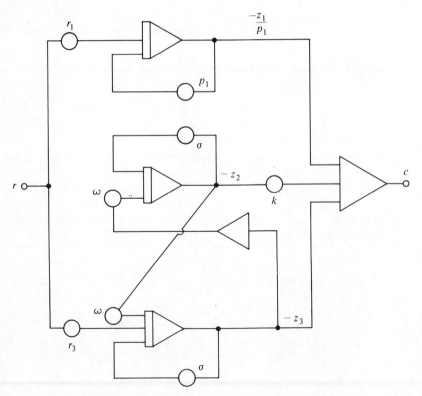

Figure 8.26. Analog computer simulation of a system with complex poles: modal representation.

Combine Equations (8.71) and (8.72) to show that

$$s\begin{bmatrix} Z_1 \\ Z_2 \\ Z_3 \end{bmatrix} = \begin{bmatrix} -p_1 & 0 & 0 \\ 0 & -\sigma & \omega \\ 0 & -\omega & -\sigma \end{bmatrix}\begin{bmatrix} Z_1 \\ Z_2 \\ Z_3 \end{bmatrix} + \begin{bmatrix} r_1 \\ 0 \\ r_3 \end{bmatrix}. \tag{8.76}$$

Note the coupling between Z_2 and Z_3.

8.7 General Methods for Finding Synthesizer and Analyzer Matrices: Controllability and Observability (Distinct Poles)

While the notions discussed in this section do not relate to the problems of synthesis or simulation, they are fundamental to modern control theory and will be referred to in sections of advanced material in succeeding chapters. As will be seen, the notions of *controllability* and *observability* are related to the cancellation of a pole of a transfer function by a zero of that transfer function.

Begin with the basic relation of Equation (8.27). Suppose that one has defined a state space vector $\mathbf{X}(s)$ such that

$$s\mathbf{X}(s) = \mathbf{\alpha}\mathbf{X}(s) + \mathbf{b}R(s) \tag{8.77}$$

and

$$C(s) = \mathbf{k}'\mathbf{X}(s) \tag{8.78}$$

for some matrices $\mathbf{\alpha}$, $\mathbf{b}$, and $\mathbf{k}$. Equations (8.31) and (8.38) are examples of the general-form Equation (8.77) and Equations (8.24) and (8.34) are examples of the general-form Equation (8.78).

Let $\mathbf{e}_k$ be an eigenvector of $\mathbf{\alpha}$ [see Equation (2.91)], associated with the eigenvalue λ_k, i.e.,

$$\mathbf{\alpha}\mathbf{e}_k = \lambda_k\mathbf{e}_k. \tag{8.79}$$

Only the particular case of distinct eigenvalues will be considered in this text. Take the matrix $\mathbf{\mathfrak{I}}$ to be the matrix whose columns are the eigenvectors, i.e.,

$$\mathbf{\mathfrak{I}} = [\mathbf{e}_1 \quad \mathbf{e}_2 \quad \cdots \quad \mathbf{e}_n]. \tag{8.80}$$

Notice that Equation (8.79) indicates that

$$\mathbf{\alpha}\mathbf{\mathfrak{I}} = [\lambda_1\mathbf{e}_1 \quad \lambda_2\mathbf{e}_2 \quad \cdots \quad \lambda_n\mathbf{e}_n]. \tag{8.81}$$

Consider the inverse of $\mathfrak{I}$ partitioned into rows

$$\mathfrak{I}^{-1} = \begin{bmatrix} \mathbf{v}'_1 \\ \mathbf{v}'_2 \\ \cdot \\ \cdot \\ \cdot \\ \mathbf{v}'_n \end{bmatrix}. \tag{8.82}$$

It follows from the definition of an inverse matrix that

$$\mathbf{v}'_i \mathbf{e}_k = \begin{cases} 1, & k = l, \\ 0, & k \neq l. \end{cases} \tag{8.83}$$

It follows from Equations (8.81)–(8.83) that if the inverse exists, then

$$\mathfrak{I}^{-1}\mathfrak{Q}\mathfrak{I} = \begin{bmatrix} \lambda_1 & 0 & \cdots & 0 \\ 0 & \lambda_2 & \cdots & 0 \\ \cdot & & & \cdot \\ \cdot & & & \cdot \\ 0 & 0 & \cdots & \lambda_n \end{bmatrix}. \tag{8.84}$$

The above result is used in the following way: Define the vector $\mathbf{Z}(s)$ by

$$\mathbf{X}(s) = \mathfrak{I}\mathbf{Z}(s). \tag{8.85}$$

Equation (8.77) can be rewritten

$$s\mathfrak{I}\mathbf{Z}(s) = \mathfrak{Q}\mathfrak{I}\mathbf{Z}(s) + \mathbf{b}R(s)$$

or

$$s\mathbf{Z}(s) = \mathfrak{I}^{-1}\mathfrak{Q}\mathfrak{I}\mathbf{Z}(s) + \mathfrak{I}^{-1}\mathbf{b}R(s). \tag{8.86}$$

Denote the diagonal matrix on the right-hand side of Equation (8.84) by $\mathbf{\Lambda}$ so that Equation (8.86) becomes

$$s\mathbf{Z}(s) = \mathbf{\Lambda}\mathbf{Z}(s) + \mathfrak{I}^{-1}\mathbf{b}R(s). \tag{8.87}$$

Because this equation has the same form as Equation (8.47), the state vector, $\mathbf{Z}(s)$, defined by Equation (8.85), is the modal state vector.

Notice that Equation (8.78) may be rewritten

$$C(s) = \mathbf{k}'\mathfrak{I}\mathbf{Z}(s). \tag{8.88}$$

Compare Equations (8.60) and (8.85) and Equations (8.47) and (8.87). Clearly, the matrix, $\mathfrak{I}$, defined in Equation (8.80), is the general *synthesizer* matrix and, therefore, $\mathfrak{I}^{-1}$ is the general *analyzer* matrix.

It will be useful to present at least one method for finding eigenvectors for an arbitrary $\mathcal{C}$-matrix. Only the case of distinct eigenvalues (poles) will be discussed, and the reader is referred to the literature for a more general discussion.[4] By definition, the kth eigenvalue, λ_k, satisfies the characteristic equation

$$
\det \begin{bmatrix}
a_{11} - \lambda_k & a_{12} & \cdots & a_{1n} \\
a_{21} & a_{22} - \lambda_k & \cdots & a_{2n} \\
\cdot & \cdot & & \cdot \\
\cdot & \cdot & & \cdot \\
\cdot & \cdot & & \cdot \\
a_{n1} & a_{n2} & \cdots & a_{nn} - \lambda_k
\end{bmatrix} = 0. \tag{8.89}
$$

Denote the l—m cofactor of the above matrix by $A_{(k)lm}$; then, if the above determinant is expanded by the first row,

$$
a_{11}A_{(k)11} + a_{12}A_{(k)12} + \cdots + a_{1n}A_{(k)1n} = \lambda_k A_{(k)11}. \tag{8.90}
$$

An even more general result can be obtained. Replace the first row of the matrix in Equation (8.89) by the mth row and notice that the determinant must still be zero (see Theorem B.3 in Appendix B). If the determinant of the resulting matrix is expanded in terms of its first row, one finds that

$$
a_{m1}A_{(k)11} + a_{m2}A_{(k)12} + \cdots + a_{mm}A_{(k)1m} + \cdots + a_{mn}A_{(k)1n} = \lambda_k A_{(k)1m}. \tag{8.91}
$$

When one compares Equations (8.79) and (8.91), it becomes apparent that *one value* of the kth eigenvector is given by

$$
\mathbf{e}_k = \begin{bmatrix}
A_{(k)11} \\
A_{(k)12} \\
\cdot \\
\cdot \\
\cdot \\
A_{(k)1n}
\end{bmatrix}. \tag{8.92}
$$

This formula is very convenient for computation. It will be left as an exercise for the reader to show that a more general formula for the kth eigenvector is

$$
\mathbf{e}_k = \begin{bmatrix}
A_{(k)l1} \\
A_{(k)l2} \\
\cdot \\
\cdot \\
\cdot \\
A_{(k)ln}
\end{bmatrix} \tag{8.93}
$$

for any l. It is important to note that one calculates a different e_k for a different l; however, all the e_k satisfy Equation (8.79) and they will be proportional to one another for the same value of k. It is usually convenient to take $l = 1$ except that on certain rare occasions $A_{(k)1r} = 0$ for all r so that a different value of l must be used.[4]

Example 8.8

Determine the synthesizer and analyzer matrices for the system of equations

$$\dot{x}_1 = 2x_1 + x_2,$$
$$\dot{x}_2 = -3x_1 - 2x_2.$$

The characteristic equation

$$\det \begin{bmatrix} 2 - \lambda & 1 \\ -3 & -(2 + \lambda) \end{bmatrix} = \lambda^2 - 1 = 0. \tag{i}$$

Thus, $\lambda_1 = 1$ and $\lambda_2 = -1$. Thus,

$$\begin{bmatrix} 2 - \lambda_1 & 1 \\ -3 & -(2 + \lambda_1) \end{bmatrix} = \begin{bmatrix} 1 & 1 \\ -3 & -3 \end{bmatrix} \text{ and }$$

$$\begin{bmatrix} 2 - \lambda_2 & 1 \\ -3 & -(2 + \lambda_2) \end{bmatrix} = \begin{bmatrix} 3 & 1 \\ -3 & -1 \end{bmatrix}. \tag{ii}$$

Take $l = 1$ in Equation (8.93) and find from Equations (ii) that

$$e_1 = \begin{bmatrix} -3 \\ 3 \end{bmatrix} \text{ and } e_2 = \begin{bmatrix} -1 \\ 3 \end{bmatrix}.$$

It follows from Equation (8.80) that the synthesizer matrix

$$\mathcal{T} = \begin{bmatrix} -3 & -1 \\ 3 & 3 \end{bmatrix}, \tag{iii}$$

so that the analyzer matrix

$$\mathcal{T}^{-1} = \begin{bmatrix} -\frac{1}{2} & -\frac{1}{6} \\ \frac{1}{2} & \frac{1}{2} \end{bmatrix}. \tag{iv}$$

To check this result, notice that

$$\mathcal{T}^{-1} \mathcal{Q} \mathcal{T} = \begin{bmatrix} -\frac{1}{2} & -\frac{1}{6} \\ \frac{1}{2} & \frac{1}{2} \end{bmatrix} \begin{bmatrix} 2 & 1 \\ -3 & -2 \end{bmatrix} \begin{bmatrix} -3 & -1 \\ 3 & 3 \end{bmatrix} = \begin{bmatrix} 1 & 0 \\ 0 & -1 \end{bmatrix},$$

a result which is in agreement with Equation (8.84).

$$* \quad * \quad *$$

Consider Equation (8.88). Occasionally, one finds that the ith component of the row vector $\mathbf{k}'\mathfrak{I}$ is zero. Obviously, this means that the ith mode, $z_i(t)$, has no influence on the output, $c(t)$. In this case the system is said to have an *unobservable mode*[3] or, more simply, the system is called *unobservable*. The significance of the concept is illustrated in the next example.

Example 8.9

Check the observability of the system for which a state space representation is

$$\dot{x}_1 = 2x_1 + x_2,$$

$$\dot{x}_2 = -3x_1 - 2x_2,$$

$$c = (1 \quad 1)\begin{bmatrix} x_1 \\ x_2 \end{bmatrix}.$$

Obviously, $\mathbf{k}' = (1 \quad 1)$ so that using the results of Example 8.8 leads to

$$\mathbf{k}'\mathfrak{I} = (1 \quad 1)\begin{bmatrix} -3 & -1 \\ 3 & 3 \end{bmatrix} = (0 \quad 2),$$

so that the first mode is unobservable. It follows from Equation (8.87) that

$$z_1(t) = z_1(0^-)e^t$$

and that

$$z_2(t) = z_2(0^-)e^{-t}.$$

From Equation (8.88), it follows that

$$c(t) = 2z_2(0^-)e^{-t}.$$

Notice that the output is stable (decays with time) even though the unobservable unstable mode is present. This would be a highly undesirable effect in the operation of a feedback system since the system output would give no indication of the instability in the feedback information to the controller (compensator). The first indication of the instability would be the saturation of some other system variable which would result in a malfunction of the entire loop.

$$* \quad * \quad *$$

Consider Equation (8.87). If the ith component of the column vector $\mathfrak{I}^{-1}\mathbf{b}$ is zero, the input, $r(t)$, has no influence on the ith mode, and the system is said to have an *uncontrollable mode* or the system is called *uncontrollable*. The discussion is more complicated in the case of nondistinct eigenvalues (poles).[1,3,4,5] The significance of the concept is illustrated by the next example.

Example 8.10

Check the controllability of the system for which a state space representation is

$$\dot{x}_1 = 2x_1 + x_2 - r,$$
$$\dot{x}_2 = -3x_1 - 2x_2 + 3r,$$
$$c = (1 \quad -1)\begin{bmatrix} x_1 \\ x_2 \end{bmatrix}. \tag{i}$$

First check observability: Since, for this case, $\mathbf{k}' = (1 \quad -1)$,

$$\mathbf{k}'\mathfrak{I} = (1 \quad -1)\begin{bmatrix} -3 & -1 \\ 3 & 3 \end{bmatrix} = (-6 \quad -4).$$

Thus, both modes are observable. In particular, if $c(t)$ were a feedback variable, the controller would be receiving information about the instability. To check controllability, notice that Equations (i) indicate that

$$\mathbf{b} = \begin{bmatrix} -1 \\ 3 \end{bmatrix},$$

so that using the results of Example 8.8 leads to

$$\mathfrak{I}^{-1}\mathbf{b} = \begin{bmatrix} -\frac{1}{2} & -\frac{1}{6} \\ \frac{1}{2} & \frac{1}{2} \end{bmatrix}\begin{bmatrix} -1 \\ 3 \end{bmatrix} = \begin{bmatrix} 0 \\ 1 \end{bmatrix}.$$

Thus, the first (and unstable) mode is uncontrollable. In summary, even though the controller is receiving information about the instability, it is unable to take corrective action because any regulation of the variable $r(t)$ will have no influence on the instability.

* * *

8.8   Synthesizer and Analyzer Matrices for Symmetric Coefficient Matrices

In succeeding chapters it will be useful to use results associated with symmetric $\mathfrak{C}$ matrices which will now be derived.

First, consider some theorems on the eigenvalues of some symmetric matrix $\mathbb{Q}$; i.e., a matrix for which

$$\mathbb{Q} = \mathbb{Q}'. \tag{8.94}$$

Theorem:

The eigenvalues of a real symmetric matrix, Q, are real.   (8.95)

Proof. Assume the hypothesis is not true; i.e., there exists a complex number λ such that

$$Qe = \lambda e \tag{i}$$

for some e. Take the complex conjugate (denoted by overbar) of this equation to find

$$Q\bar{e} = \bar{\lambda}\bar{e}. \tag{ii}$$

It follows from these equations that

$$\bar{e}'Qe = \lambda\bar{e}'e \tag{iii}$$

and

$$e'Q\bar{e} = \bar{\lambda}e'\bar{e}. \tag{iv}$$

The transpose of a scalar is equal to the scalar, so that

$$\bar{e}'Qe = e'Q'\bar{e}$$
$$= e'Q\bar{e}. \tag{v}$$

The last equality follows from Equation (8.94). It follows from (iii), (iv) and (v) that

$$(\lambda - \bar{\lambda})e'\bar{e} = 0,$$

which can be satisfied only if $\lambda = \bar{\lambda}$. This is a contradiction of the original assumption so that the theorem must be true.

Theorem:

Any two eigenvectors associated with distinct eigenvalues of a real symmetric matrix are orthogonal.

Proof. Take λ_k, λ_l to be eigenvalues so that

$$Qe_k = \lambda_k e_k \tag{i}$$

and

$$Qe_l = \lambda_l e_l \tag{ii}$$

for some e_k and e_l. Thus

$$e_k'Qe_l = \lambda_l e_k'e_l \tag{iii}$$

and

$$e_l'Qe_k = \lambda_k e_l'e_k. \tag{iv}$$

But, since the transpose of a scalar is equal to the scalar,

$$\mathbf{e}'_k \mathbb{Q} \mathbf{e}_l = (\mathbf{e}'_k \mathbb{Q} \mathbf{e}_l)'$$
$$= \mathbf{e}'_l \mathbb{Q}' \mathbf{e}_k$$
$$= \mathbf{e}'_l \mathbb{Q} \mathbf{e}_k.$$

Whence

$$(\lambda_l - \lambda_k) \mathbf{e}'_k \mathbf{e}_l = 0,$$

so that

$$\mathbf{e}'_k \mathbf{e}_l = 0 \quad \text{if } k \neq l, \tag{8.96}$$

and the theorem is established.

Take the synthesizer matrix $\mathfrak{T}$ to be defined as in Equation 8.80 and consider the product

$$\mathfrak{T}'\mathfrak{T} = \begin{bmatrix} \mathbf{e}'_1 \\ \mathbf{e}'_2 \\ \cdot \\ \cdot \\ \cdot \\ \mathbf{e}'_n \end{bmatrix} \begin{bmatrix} \mathbf{e}_1 & \mathbf{e}_2 & \cdots & \mathbf{e}_n \end{bmatrix}.$$

If $\mathbb{Q}$ is symmetric, it follows from 8.96 that

$$\mathfrak{T}'\mathfrak{T} = \mathscr{I}$$

so that

$$\mathfrak{T}' = \mathfrak{T}^{-1}. \tag{8.97a}$$

Whence, the analyzer matrix for a symmetric $\mathbb{Q}$ matrix is the transpose of the synthesizer matrix. It follows that Equation (8.84) can be rewritten

$$\mathfrak{T}'\mathbb{Q}\mathfrak{T} = \Lambda \tag{8.97b}$$

if $\mathbb{Q}$ is symmetric and where Λ is the diagonal matrix of eigenvalues.

8.9 A Final State Space Representation Which Is More Closely Associated with the Dynamics of a Physical System[2]

In this section, a state space representation more closely related to the discussion of previous chapters is described. This representation has no application in synthesis procedures but is quite useful in simulation studies. The representation is also quite useful in the design of multivariable feedback systems, which is discussed in Chapter 11.

Notice that Equations (4.63) and (4.64) and (the more general) Equations (5.64) and (5.65) define sets of first-order, coupled, ordinary differential equations which are state space representations of dynamic systems. As has been indicated in the present chapter, an operational amplifier circuit is easily constructed from such a set of equations in order to provide a simulation of the dynamic system. When the techniques of state variable analysis, which were discussed in Chapters 4 through 6, are applied, the results are the matrix equations

$$\dot{x} = \mathfrak{A}x + \mathfrak{B}u, \tag{8.98}$$

$$c = \mathfrak{D}x. \tag{8.98a}$$

The x vector is a vector of effort and flow variables, the u vector is a vector of input effort and flows, and the $\mathfrak{A}$, $\mathfrak{B}$, and $\mathfrak{D}$ matrices have elements which are functions of resistances, compliances, inertias, transformer constants, and gyrator constants.

Since the outputs of integrators in a simulation circuit are the state variables, it follows that all integrator outputs of simulations developed from Equations (8.98) and (8.98a) have direct physical interpretations (i.e., are interpreted as physical efforts and flows). This fact can be quite useful to the analyst.

Equations (8.98) and (8.98a) are slightly more general versions of Equations (8.77) and (8.78); however, all of the techniques presented in Section 8.7 may be applied to the state space representation currently being discussed except that, now, the kth mode is called *uncontrollable* if the entire kth row of $\mathfrak{J}^{-1}\mathfrak{B}$ vanishes and it is called *unobservable* if the entire kth row of $\mathfrak{D}\mathfrak{J}$ vanishes.

Example 8.11[1]

Check the controllability of the circuit analyzed in Example 4.9 for the special case

$$CR_c = \frac{L}{R_L}. \tag{i}$$

The state variable equations are

$$\begin{bmatrix} \dot{e}_c \\ \dot{f}_L \end{bmatrix} = \begin{bmatrix} 0 & -\dfrac{1}{C} \\ \dfrac{1}{L} & \dfrac{-(R_L + R_c)}{L} \end{bmatrix} \begin{bmatrix} e_c \\ f_L \end{bmatrix} + \begin{bmatrix} \dfrac{1}{C} \\ \dfrac{R_c}{L} \end{bmatrix} f_0, \tag{ii}$$

so that the characteristic equation

$$\det \begin{bmatrix} \lambda & \dfrac{1}{C} \\ -\dfrac{1}{L} & \lambda + \dfrac{R_L + R_c}{L} \end{bmatrix} = \lambda^2 + \frac{R_L + R_c}{L}\lambda + \frac{1}{CL} = 0.$$

Use the quadratic formula to find

$$\lambda_{1,2} = -\frac{R_L + R_c}{2L} \pm \frac{1}{2}\sqrt{\frac{(R_L + R_c)^2}{L^2} - \frac{4}{CL}}.$$

Use Equation (i) to show that

$$\lambda_{1,2} = -\frac{R_L + R_c}{2L} \pm \frac{R_L - R_c}{2L},$$

so that

$$\lambda_1 = -\frac{R_L}{L}, \qquad \lambda_2 = -\frac{R_c}{L}. \tag{iii}$$

Thus,

$$\begin{bmatrix} -\lambda_1 & -\dfrac{1}{C} \\[2ex] \dfrac{1}{L} & \dfrac{-(R_L + R_c)}{L} - \lambda_1 \end{bmatrix} = \begin{bmatrix} \dfrac{R_L}{L} & -\dfrac{1}{C} \\[2ex] \dfrac{1}{L} & -\dfrac{R_c}{L} \end{bmatrix}$$

and

$$\begin{bmatrix} -\lambda_2 & -\dfrac{1}{C} \\[2ex] \dfrac{1}{L} & \dfrac{-(R_L + R_c)}{L} - \lambda_2 \end{bmatrix} = \begin{bmatrix} \dfrac{R_c}{L} & -\dfrac{1}{C} \\[2ex] \dfrac{1}{L} & -\dfrac{R_L}{L} \end{bmatrix},$$

so that the synthesizer matrix [see Equation (8.92)]

$$\mathfrak{T} = \begin{bmatrix} -\dfrac{R_c}{L} & -\dfrac{R_L}{L} \\[2ex] -\dfrac{1}{L} & -\dfrac{1}{L} \end{bmatrix}$$

and the analyzer matrix

$$\mathfrak{T}^{-1} = \begin{bmatrix} \dfrac{-L}{R_c - R_L} & \dfrac{R_L L}{R_c - R_L} \\[2ex] \dfrac{L}{R_c - R_L} & \dfrac{-R_c L}{R_c - R_L} \end{bmatrix}. \tag{iv}$$

It follows from Equations (ii) and (iv) that

$$\mathfrak{T}^{-1}\mathbf{b} = \begin{bmatrix} \dfrac{R_c R_L - L/C}{R_c - R_L} \\[2ex] \dfrac{L/C - R_c^2}{R_c - R_L} \end{bmatrix}. \tag{v}$$

It follows from Equation (i) that the first mode is uncontrollable. In this example, the uncontrollable mode is stable since $\lambda_1 = -R_L/L$. Refer to Problem 3.5 for an important insight into condition (i).

*   *   *

8.10   A Comment on Digital Simulation

There are many fine texts written on the subject of digital simulation, and there is no need to go into a detailed discussion of this topic in the present text. However, for the sake of completeness, one digital simulation technique will be presented.*

It was shown in Chapter 2 [Equation (2.88)] that the solution to the state space Equation (8.98a) is

$$\mathbf{x}(t) = \boldsymbol{\phi}(t)\mathbf{x}(0^-) + \int_0^t \boldsymbol{\phi}(t - \xi)\mathfrak{B}\mathbf{u}(\xi)\, d\xi, \qquad (8.99a)$$

where $\boldsymbol{\phi}(t)$ is the *fundamental solution*, which, as shown in Chapter 2, can be determined by Laplace transform techniques. It will now be shown that

$$\boldsymbol{\phi}(t) = \exp\{\mathfrak{A}t\}, \qquad (8.99b)$$

where, as is indicated in Appendix B [Equation (B.63)],

$$\exp\{\mathfrak{A}t\} = \boldsymbol{\mathcal{I}} + \mathfrak{A}t + \frac{1}{2!}\mathfrak{A}^2t^2 + \frac{1}{3!}\mathfrak{A}^3t^3 + \frac{1}{4!}\mathfrak{A}^4t^4 + \cdots. \qquad (8.99c)$$

To prove this assertion, consider the unforced case for which Equation (8.99a) reduces to

$$\mathbf{x}(t) = \exp\{\mathfrak{A}t\}\mathbf{x}(0^-) \qquad (8.100)$$

if Equation (8.99b) is true. The assertion is proved if the proposed solution, in Equation (8.100), satisfies the initial conditions and the differential equation. First notice that Equation (8.99c) can be used to show that

$$\mathbf{x}(0) = \exp\{\mathfrak{A}0\}\mathbf{x}(0^-) = \boldsymbol{\mathcal{I}}\mathbf{x}(0^-) = \mathbf{x}(0^-).$$

Thus, the proposed solution satisfies the proper initial conditions. It also follows from Equations (8.99c) and (8.100) that

$$\dot{\mathbf{x}}(t) = \left\{ \mathfrak{A} + \mathfrak{A}^2t + \frac{1}{2!}\mathfrak{A}^3t^2 + \frac{1}{3!}\mathfrak{A}^4t^3 + \cdots \right\}\mathbf{x}(0^-)$$

$$= \mathfrak{A}\left\{ \boldsymbol{\mathcal{I}} + \mathfrak{A}t + \frac{1}{2!}\mathfrak{A}^2t^2 + \frac{1}{3!}\mathfrak{A}^3t^3 + \cdots \right\}\mathbf{x}(0^-)$$

$$= \mathfrak{A}\exp\{\mathfrak{A}t\}\mathbf{x}(0^-) = \mathfrak{A}\mathbf{x}(t),$$

and the assertion is proved.

---

* The method to be presented was devised by Professor H. Paynter of M.I.T.

Consider now the numerical computation of the fundamental solution $\phi(T)$ using Equation (8.99a). For convenience, denote

$$\mathcal{P} \triangleq \mathcal{Q}T. \tag{8.101}$$

Then, by Equations (8.99b) and (8.99c),

$$\phi(T) = \mathcal{I} + \mathcal{P} + \frac{1}{2!}\mathcal{P}^2 + \frac{1}{3!}\mathcal{P}^3 + \cdots + \frac{1}{(q-1)!}\mathcal{P}^{q-1} + \mathcal{R}_q, \tag{8.102a}$$

where, obviously, the remainder matrix, $\mathcal{R}_q$, is given by

$$\mathcal{R}_q = \frac{1}{q!}\mathcal{P}^q + \frac{1}{(q+1)!}\mathcal{P}^{q+1} + \cdots. \tag{8.102b}$$

Let w, the *maximum element*, be defined by

$$w = \max_{i,j} |\mathcal{P}_{ij}|. \tag{8.103}$$

Consider the matrix $\mathcal{W}$ defined by

$$(\mathcal{W})_{ij} = w \quad \text{for all } i, j. \tag{8.104}$$

It follows that the elements of $\mathcal{W}^2$ are given by

$$(\mathcal{W}^2)_{ij} = nw^2 \quad \text{for all } i, j,$$

where $n =$ the order of the matrix $\mathcal{P}$. Then, for $\mathcal{W}^3$

$$(\mathcal{W}^3)_{ij} = n^2 w^3 \quad \text{for all } i, j$$

and in general

$$(\mathcal{W}^N)_{ij} = n^{N-1} w^N \quad \text{for all } i, j. \tag{8.105}$$

From Equations (8.102a) and (8.103) it follows that

$$|R_{ij}| \leq \left[\frac{n^{q-1}w^q}{q!} + \frac{n^q w^{q+1}}{(q+1)!} + \cdots \right]$$

$$= \frac{n^{q-1}w^q}{q!}\left[1 + \frac{nw}{q+1} + \frac{(nw)^2}{(q+1)(q+2)} + \cdots \right].$$

Since $q > 1$, it follows that

$$|R_{ij}| < \frac{n^{q-1}w^q}{q!}\left[1 + nw + \frac{(nw)^2}{2!} + \cdots \right].$$

Recognizing the series in brackets as the Taylor series for the exponential function (see Appendix B), it follows that

$$|R_{ij}| < \frac{n^{q-1}w^q e^{nw}}{q!}.$$

Finally, since $n^{q-1} < n^q$, it follows that

$$|R_{ij}| < \frac{(nw)^q e^{nw}}{q!} \quad \text{for all } i, j. \tag{8.106}$$

Equation (8.106) provides an upper bound for all elements of the remainder (or error) matrix associated with truncating the series in Equation (8.102) after q terms. Thus, one can discontinue machine computation of $\phi(T)$ after q steps if

$$\frac{(nw)^q}{q!} e^{nw}$$

is "negligibly small." Whether or not this term is negligibly small depends on some predetermined accuracy that must be prescribed for the elements of $\phi(T)$.

A FORTRAN IV program to determine $\phi(T)$ is shown in Figure 8.27. In this program, the following notation is used:

ACRACY $\triangleq$ Specified accuracy of elements of ϕ_{ij}.

$$\text{POWER (I,J)} = \frac{(\mathbf{a})_{ij}^L}{L!} \quad \text{for some integer } L.$$

The results of the program can be extended to the forced case. In Equation (8.99a), let

$$\eta = nT - \xi;$$

then in the time interval $(n-1)T \le t \le nT$

$$\int_{(n-1)T}^{nT} \phi(nT - \xi)\mathbf{B}\mathbf{u}(\xi)\,d\xi = -\int_{T}^{0} \phi(\eta)\mathbf{B}\mathbf{u}(nT - \eta)\,d\eta$$

or

$$\int_{(n-1)T}^{nT} \phi(nT - \xi)\mathbf{B}\mathbf{u}(\xi)\,d\xi = \int_{0}^{T} \phi(\eta)\mathbf{B}\mathbf{u}(nT - \eta)\,d\eta. \tag{8.107}$$

One way to proceed is to make the approximation that

$$\mathbf{u}(\eta - nT) \simeq \mathbf{u}([n-1]T), \quad 0 < \eta < T. \tag{8.108}$$

Then

$$\int_{(n-1)T}^{nT} \phi(nT - \xi)\mathbf{B}\mathbf{u}(\xi)\,d\xi = \int_{0}^{T} \phi(\eta)\,d\eta\,\mathbf{B}\mathbf{u}([n-1]T). \tag{8.109}$$

```
                    W = A(1, 1) * T              {input: A, ACRACY, T, N,
                    DO  2  I=1, N
                    DO  1  J=1, N
   1                IF (ABS(A(I, J)) * T . GT . W) W = ABS (A(I, J)) * T
   2                CONTINUE                      ⎧ Note:
                    WN = W * FLOAT (N)            ⎨ might choose
                    Q = 1.                        ⎩ ACRACY = I.E-6* W ⎭
                    TEST = EXP (WN)
   3                TEST = TEST * WN/Q
                    IF (TEST . LE . ACRACY) GO TO 4
                    Q = Q + 1
                    GO TO 3
   4                NQ = Q
                    DO  6  I = I, N
                    DO  5  J = I, N
                    POWER (I, J) = A(I, J) * T
   5                PHI (I, J) = A (I, J) * T
   6                PHI (I, I) = PHI (I, I) + 1.
                    DO  10  L=2, NQ
                    FACT = FLOAT (L)
                    DO  9  I = 1, N
                    DO  8  J = 1, N
                    TEMP (J) = 0
                    DO  7  K = 1, N
   7                TEMP (J) = POWER (I, K) * A(K,J) * T/FACT + TEMP(J)
   8                PHI (I,J) = PHI(I,J) + TEMP(J)
                    DO  100  J = 1, N
  100               POWER (I,J) = TEMP(J)
   9                CONTINUE
  10                CONTINUE
                    STOP
                    END                           {output: PHI (I, J)
```

Figure 8.27. FORTRAN IV program for the machine computation of the fundamental solution.

It will be left as an exercise to show that if $\mathbf{\alpha}^{-1}$ exists, then

$$\int_0^T \phi(\eta)\, d\eta = \mathbf{\alpha}^{-1}[\phi(T) - \mathbf{\mathit{I}}]. \tag{8.110}$$

Finally,

$$\int_{(n-1)T}^{nT} \phi(nT - \xi)\mathbf{\mathcal{B}}\mathbf{u}(\xi)\, d\xi \simeq \mathbf{\alpha}^{-1}[\phi(T) - \mathbf{\mathit{I}}]\mathbf{\mathcal{B}}\mathbf{u}([n-1]T). \tag{8.111}$$

Define the matrix

$$\mathbf{\mathcal{F}}(T) \triangleq \mathbf{\alpha}^{-1}[\phi(T) - \mathbf{\mathit{I}}]\mathbf{\mathcal{B}}. \tag{8.112}$$

It is left as an exercise to show that

$$\mathbf{x}(nT) \simeq \phi(T)\mathbf{x}([n-1]T) + \mathbf{\mathcal{F}}(T)\mathbf{u}([n-1]T) \tag{8.113}$$

for any integer n. This equation is easily programmed on a digital computer.

8.11 Magnitude and Time Scaling

As has been shown, analog and digital computer simulations are easily developed from the state space equation

$$sX(s) = \mathbf{\alpha}X(s) + \mathbf{b}R(s) \tag{8.114}$$

and the output equation

$$c(t) = \mathbf{k}'\mathbf{x}(t). \tag{8.115}$$

Sometimes it is desired to scale the state variables, x_i (e.g., to avoid saturation of operational amplifiers). The scaled variables are defined by

$$x_{1s} \triangleq \frac{1}{\omega_1}x_1,$$

$$\cdot$$
$$\cdot \tag{8.116}$$
$$\cdot$$

$$x_{ns} \triangleq \frac{1}{\omega_n}x_n,$$

where the ω_i are arbitrarily chosen scale factors.

In matrix form, Equation (8.116) can be written

$$\mathbf{x}_s = \begin{bmatrix} \frac{1}{\omega_1} & 0 & 0 & \cdots & 0 \\ 0 & \frac{1}{\omega_2} & 0 & \cdots & 0 \\ \cdot & \cdot & & & \cdot \\ \cdot & \cdot & & & \cdot \\ \cdot & \cdot & & & \cdot \\ 0 & 0 & & \cdots & \frac{1}{\omega_n} \end{bmatrix} \mathbf{x} = \mathbf{\Omega}\mathbf{x}. \tag{8.117}$$

$\mathbf{\Omega}$ will be referred to as the *scaling matrix*. Notice that

$$s\mathbf{X}_s = s\mathbf{\Omega}\mathbf{X} = \mathbf{\Omega}\mathbf{\alpha}\mathbf{X} + \mathbf{\Omega}\mathbf{b}R$$

or

$$s\mathbf{X}_s = \mathbf{\Omega}\mathbf{\alpha}\mathbf{\Omega}^{-1}\mathbf{X} + \mathbf{\Omega}\mathbf{b}R, \tag{8.118}$$

and also that

$$C(s) = \mathbf{k}'\mathbf{\Omega}^{-1}\mathbf{X}_s(s). \tag{8.119}$$

Comparing this equation with Equation (8.115), it is seen that magnitude scaling modifies the coefficient matrix and the input and output vectors in a rather simple manner.

Notice that

$$\Omega^{-1} = \begin{bmatrix} \omega_1 & 0 & 0 & \cdots & 0 \\ 0 & \omega_2 & 0 & \cdots & 0 \\ \cdot & \cdot & & & \cdot \\ \cdot & \cdot & & & \cdot \\ \cdot & \cdot & & & \cdot \\ 0 & 0 & & \cdots & \omega_n \end{bmatrix}. \tag{8.120}$$

Finally, suppose it is desired to scale time; i.e., it is desired to relate real time and machine time by

$$t_{real} = \epsilon t_{machine}, \tag{8.121}$$

where ϵ is an arbitrarily chosen scale factor.

Since s implies d/dt, it follows that

$$s_{real} = \frac{1}{\epsilon} s_{machine}. \tag{8.122}$$

The last equation leads to the conclusion that to scale time by a factor ϵ, one need only scale s by a factor $1/\epsilon$, and the state space equations become

$$s_M X = \epsilon \alpha X + \epsilon b R; \tag{8.123}$$

$$c(t) = k' x(t). \tag{8.124}$$

Hence, one need only modify the coefficient matrix, α, and the input vector, b, in order to scale time.

References

1. KALMAN, R. E., "Mathematical Description of Linear Dynamical Systems," *J.S.I.A.M. Control*, Series A, 1, No. 2, 1963, 152–192.

2. KARNOPP, D., and ROSENBERG, C., *Analysis and Simulation of Multiport Systems*, M.I.T. Press, Cambridge, Mass., 1968.

3. LOSCUTOFF, W., *Class Notes for Courses M.E. 171 and M.E. 271*, Dept. of Mechanical Engineering, Univ. of California, Davis (unpublished).

4. OGATA, K., *State Space Analysis of Control Systems*, Prentice-Hall, Englewood Cliffs, N.J., 1967.

5. TAKAHASHI, Y., et al., *Control*, Addison-Wesley, Reading, Mass., 1970.

Problems

8.1. Show that integrators will attenuate high-frequency noise signals.

8.2. Establish the validity of Equation (8.16).

8.3. Plot the ratio E_2/E_1 vs. x using first Equation (8.16) and then the approximate equation (8.17) and the typical parameter values

$$R = 50 \times 10^3 \text{ ohms,}$$

$$R_1 = 10^6 \text{ ohms.}$$

Repeat the calculations for $R_1 = 10^3$ ohms.

8.4. Consider the seismometer shown in Figure 8.28. Let $x(t)$ be the absolute displacement of the mass and define the output as the relative displacement

$$c(t) \triangleq x(t) - r(t).$$

Then, it can be shown that

$$ms^2 C + bsC + kC = ms^2 R.$$

Develop an operational amplifier circuit to simulate the seismometer response, $c(t)$. Use the phase space representation.

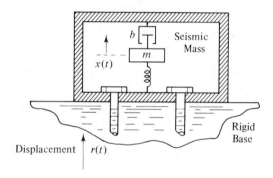

Figure 8.28. Seismometer.

8.5. Repeat Problem 8.4, but use the alternative state space representation.

8.6. Devise an operational amplifier circuit which is a synthesis of a *Padé filter* for which

$$\frac{C(s)}{R(s)} = \frac{1 - s}{1 + s}. \tag{8.125}$$

8.7. The differential equation for the unforced mechanical system shown in Figure 8.29 (see next page) is

$$m\ddot{x} + b\dot{x} + kx = 0.$$

Using a phase space representation, develop an analog computer circuit to simulate this system. Determine the simulated responses for the two cases shown on the next page.

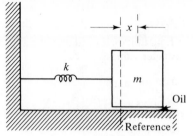

(Equilibrium Position)   **Figure 8.29.** Mechanical system.

Run 1	Run 2
$m = 1.0$ slug	$m, k, x(0^-)$ as in run 1
$k = 16.0$ lb/ft	b such that system is
$b = 4$ lb-sec/ft	critically damped
$x(0^-) = -\frac{1}{2}$ ft	Released from rest
Released from rest	

For both runs, plot $x(t)$ vs. t and x vs. $\dot{x}$.

8.8. Consider the two-tank system shown in Figure 8.30. The appropriate input-output relation is

$$(1 + A_1 R_1 s)(1 + A_2 R_2 s)C = R_2 F.$$

Start with the phase variable representation and determine the analyzer and the synthesizer matrices for this system.

8.9. Simulate the system in Problem 8.8 on an analog computer using the modal state space representation. For constants use

$$A_1 R_1 = 1.0 \text{ min}^{-1},$$
$$A_2 R_2 = 10.0 \text{ min}^{-1},$$
$$R_2 = 1.0 \text{ min/sq ft}.$$

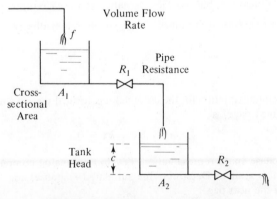

Figure 8.30. Two-tank system (cascade).

Repeat the simulation for the following cases:

Case	$c(0^-)$	$\dot{c}(0^-)$
1	0	1
2	0	-1
3	1	0
4	-1	0
5	1	-1

Plot c vs. t and z_1 vs. z_2.

8.10. Repeat Problem 8.9, but use the phase space representation. "Observe" the modes by simulating the matrix transformation

$$z = \mathcal{T}^{-1}x$$
Analyzer matrix

on the analog computer. Plot x_1 vs. x_2 and z_1 vs. z_2. On the x_1-x_2 plane, plot the lines

$$t_{11}x_1 + t_{12}x_2 = 0,$$

$$t_{21}x_1 + t_{22}x_2 = 0,$$

where t_{ij} are elements of the analyzer matrix.

8.11. Derive Equation (8.55) from Equation (8.80).

8.12. *Decoupling feedback.* Consider the velocity servo shown in Figure 8.31. σ_m and f_m are the stress and velocity at the point shown. It can be shown that

$$s\begin{bmatrix} \Sigma_m \\ F_m \end{bmatrix} = \begin{bmatrix} -\dfrac{b}{m} & \dfrac{k}{A} \\ -\dfrac{1}{m} & 0 \end{bmatrix}\begin{bmatrix} \Sigma_m \\ F_m \end{bmatrix} + \begin{bmatrix} \dfrac{b}{A_m} & -\dfrac{k}{A} \\ \dfrac{1}{m} & 0 \end{bmatrix}\begin{bmatrix} R_1 \\ R_2 \end{bmatrix}. \qquad \text{(i)}$$

It is desired to *uncouple* (i.e., eliminate dynamic interactions between) Σ_m and F_m by using feedback. Show that this task can be accomplished by using the feedback "law"

$$\begin{bmatrix} R_1 \\ R_2 \end{bmatrix} = \begin{bmatrix} A & -m\alpha_2 \\ \dfrac{\alpha_1 A}{k} & -\dfrac{b}{k}\alpha_2 + 1 \end{bmatrix}\begin{bmatrix} \Sigma_m \\ F_m \end{bmatrix} \qquad \text{(ii)}$$

for arbitrary α_1 and α_2.

Procedure 1. Simulate the open loop system on the analog computer using Equation (i). Hold r_1 and $r_2 = 0$ and observe the transient response in σ_m and f_m. Plot the results in the σ_m-f_m plane for several sets of initial conditions in σ_m and f_m.

Procedure 2. Repeat procedure 1, but program the feedback Equation (ii) into your computer circuit. Use the following values:

$$m = 2.0, \qquad b = 0.707,$$
$$k = 1.0, \qquad A = 1.0.$$

What is the significance of the constants α_1 and α_2?

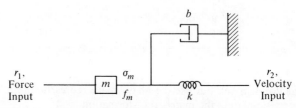

Figure 8.31. Two-input velocity servo.

8.13. Show that for the system described by the state space Equations (8.77) and (8.78)

$$G(s) = \frac{C(s)}{R(s)} = \mathbf{k}'(s\boldsymbol{\mathscr{I}} - \boldsymbol{\mathcal{R}})^{-1}\mathbf{b}.$$

Error Analysis of Simulation Studies* 9

This chapter contains an introduction to topics which may be applied to the error analysis of analog or digital simulation studies. The chapter begins with a discussion of the error, or *sensitivity*, analysis of the simulation of systems described by linear ordinary differential equations with constant coefficients and ends with a short description of the sensitivity techniques applicable to the nonlinear case.

9.1 The Source of Errors in Simulation Studies[1]

Let $c(t)$ be the output of a system for which the Laplace domain input-output relationship is

$$C(s) = \frac{(b_m s^m + \cdots + b_0)R(s)}{a_n s^n + a_{n-1}s^{n-1} + \cdots + a_0}. \tag{9.1}$$

As was indicated in earlier chapters on dynamic systems analysis, the coefficients a_i and b_i depend on physical constants which change with operating level and/or which are not known precisely during analysis. Also, $c(t)$ depends on the initial conditions $c(0^-)$, $Dc(0^-)$, ..., $D^{n-1}c(0^-)$, which also are often not known precisely. Thus, coefficients of transfer functions and initial conditions are sources of error in simulation studies.

---

* The discussion in this chapter is referred to in only the advanced-material sections of succeeding chapters. This chapter may be omitted during the first reading of the text.

The next several sections present methods for determining, quantitatively, the effect of errors in transfer function coefficients and of initial conditions on simulated outputs; the methods are remarkably simple to apply.

9.2 Sensitivity of the Simulated Outputs to the Denominator Coefficients, a_i[1]

The quantity

$$\frac{\partial c(t)}{\partial a_i} \tag{9.2}$$

is often used as a quantitative measure of sensitivity of simulated output to error in denominator coefficients. An alternative, normalized measure of the effect of such an error is

$$v_i \triangleq a_i \frac{\partial c}{\partial a_i} = \frac{\partial c}{\partial \ln (a_i)}. \tag{9.3}$$

Notice that this *sensitivity coefficient*

$$v_i = v_i(t) \tag{9.4}$$

and is interpreted as a variation in the simulated output with respect to a 100% error in a_i.

It will be left as an exercise to show that the use of a Taylor series expansion leads to the conclusion that the difference in actual output and simulated output is approximated by

$$\Delta c \simeq \frac{(a_i - a_{i0})}{a_{i0}} v_i, \tag{9.5}$$

where a_{i0} is the value of the coefficient used in the simulation and a_i is the true value of the coefficient.

Delay the discussion of systems with numerator dynamics until the next section so that a simplified version of Equation (9.1) is

$$\sum_{k=0}^{n} a_k s^k C = R. \tag{9.6}$$

Differentiate this equation with respect to a_i and multiply by a_i to find

$$\sum_{k=0}^{n} a_k s^k V_i + a_i s^i C = 0,$$

where $V_i(s) = \mathcal{L}[v_i(t)]$. Thus,

$$\sum_{k=0}^{n} a_k s^k V_i = -a_i s^i C. \tag{9.7}$$

Notice that Equation (9.7) is a transfer function relation between the sensitivity, V_i, and the simulated output, C. Notice that $a_i s^i C$ is a pseudo forcing function for this relationship.

Equation (9.7) indicates a method for determining the partial derivative $v_i(t)$: Simply use the phase space, the alternative state space, or the modal state space representation to simulate relationship (9.7) (here the quantity $s^i C$ is obtained from the original simulation). This technique would yield the single sensitivity coefficient v_i. However, as will now be shown, it is possible to obtain *all* of the sensitivities, $v_i(t)$, *simultaneously.*

Consider the quantity $q(t)$, whose Laplace transform, $Q(s)$, is arbitrarily defined by

$$\sum_{r=0}^{n} a_r s^r Q = C, \tag{9.8}$$

where $c(t)$ is the simulated output. Differentiate Equation (9.8) i times with respect to time (i.e., multiply by s^i) to find

$$\sum_{r=0}^{n} a_r s^{r+i} Q = s^i C. \tag{9.9}$$

Now consider the "modified" phase space variables, $U_i(s)$, defined by

$$U_i \triangleq -a_i s^i Q, \tag{9.10}$$

which, by Equation (9.9), satisfy the equation

$$\sum_{r=0}^{n} a_r s^r U_i = -a_i s^i C. \tag{9.11}$$

Comparing this equation with Equation (9.7), one discovers the remarkable result that the sensitivity coefficients

$$V_i(t) = U_i(t). \tag{9.12}$$

The manner in which Equation (9.12) is exploited is illustrated in Figure 9.1. Simultaneous with the original simulation, one simulates $q(t)$ as defined by Equation (9.8) with a *sensitivity circuit* (notice that the input to the sensitivity circuit is the output of the original simulation circuit). *All* of the sensitivity coefficients are then obtained as indicated in Figure 9.1.

The phase space representation *must* be used to obtain $q(t)$; however, no such restriction is made concerning $c(t)$.

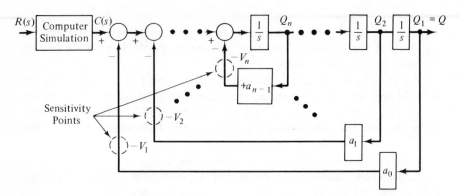

Figure 9.1. Schematic diagram of the sensitivity circuit. Sensitivity to denominator coefficients.[1]

Most often, one is not directly interested in a variation of a simulated output with respect to changes in a transfer function coefficient, a_i, but rather with respect to changes in some physical parameter, $p(t)$; i.e., the quantity $p[\partial c(t)/\partial p]$ is the quantity desired. For this case, use the above technique with the chain rule formula:

$$p\frac{\partial c}{\partial p} = p \sum_{i}^{n} \frac{\partial c}{\partial a_i}\frac{\partial a_i}{\partial p} = \sum_{i}^{n} \frac{p}{a_i}\frac{\partial a_i}{\partial p}v_i(t). \qquad (9.13)$$

9.3 Sensitivity of the Simulated Output of a Dynamic System to Errors in the Numerator Coefficients, b_i[1]

Consider the transfer function relation (9.1), which may be rewritten

$$\sum_{k}^{n} a_k s^k C = \sum_{k}^{n} b_k s^k R. \qquad (9.14)$$

Define the numerator sensitivity coefficient

$$w_i(t) = b_i\frac{\partial c}{\partial b_i} = \frac{\partial c(t)}{\partial \ln b_i}. \qquad (9.15)$$

Differentiate Equation (9.14) with respect to b_i and then multiply by b_i to find

$$b_i \sum_{k}^{n} a_k s^k \frac{\partial c}{\partial b_i} = b_i s^i R$$

or

$$\sum_{k}^{n} a_k s^k W_i(s) = b_i s^i R. \tag{9.16}$$

Equation (9.16) defines a transfer function between the numerator sensitivity, W_i, and the ith derivative of the *input* [compare Equation (9.16) with Equation (9.7)]. Consider $q_b(t)$ defined by the Laplace domain relation

$$\sum_{k}^{n} a_k s^k Q_b = R. \tag{9.17}$$

Multiply Equation (9.17) by s^i to find

$$\sum_{k}^{n} a_k s^{k+i} Q_b = s^i R. \tag{9.18}$$

Define

$$U_i \triangleq b_i s^i R. \tag{9.19}$$

Then from Equation (9.18), one finds that the U_i satisfy the equation

$$\sum_{k}^{n} a_k s^k U_i \triangleq b_i s^i R. \tag{9.20}$$

Comparing Equation (9.20) with Equation (9.16), one finds

$$u_i(t) = w_i(t) \tag{9.21}$$

when U_i is defined by Equation (9.19).

Figure 9.2 indicates the method in which Equations (9.17), (9.19), and

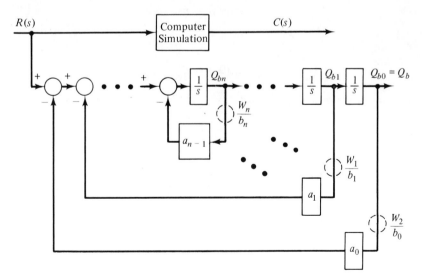

Figure 9.2. Sensitivity circuit for finding the sensitivity to errors in numerator coefficients.[1]

(9.21) are used to determine the numerator sensitivity coefficients, $w_i(t)$. Comparing this figure with Figure 9.1, one notices that the sensitivity circuit for the numerator sensitivities, w_i, is in *parallel* with the original simulation circuit, while the sensitivity circuit for the denominator sensitivities, v_i, is in *series*.

The extension of Equation (9.13) is

$$p\frac{\partial c}{\partial p} = \sum_{i=1}^{n} \frac{p}{a_i} \frac{\partial a_i}{\partial p} v_i(t) + \sum_{i=1}^{n} \frac{p}{b_i} \frac{\partial b_i}{\partial p} w_i(t). \qquad (9.22)$$

9.4 Sensitivity of Simulation Outputs to Errors in Initial Conditions

Consider the solution, $c(t)$, to a simulation constructed from the state variable equations

$$\dot{\mathbf{x}} = \mathbf{\alpha x} + \mathbf{b}r(t) \qquad (9.23)$$

and

$$c(t) = \mathbf{k'x}(t). \qquad (9.24)$$

Consider also the initial conditions

$$x_1(0^-) = \eta_1,$$
$$x_2(0^-) = \eta_2, \qquad (9.25)$$
$$\vdots$$
$$x_n(0^-) = \eta_n.$$

A quantitative measure of sensitivity might be taken to be

$$v_k(t) \triangleq \frac{\partial c(t)}{\partial \eta_k}, \qquad (9.26)$$

or, alternatively, the normalized sensitivity

$$v_k(t) \triangleq \eta_k \frac{\partial c(t)}{\partial \eta_k}. \qquad (9.27)$$

Very often $\eta_k = 0$ for several k, in which case the definition of Equation (9.26) should be used. In what follows, expressions for both definitions of sensitivity will be derived.

It was shown in Chapter 2 that the solution to Equations (9.23)–(9.25) is

$$c(t) = \mathbf{k}'\boldsymbol{\phi}(t)\boldsymbol{\eta} + \int_0^t \mathbf{k}'\boldsymbol{\phi}(t - \xi)\mathbf{b}r(\xi)\,d\xi, \tag{9.28a}$$

where

$$\boldsymbol{\phi}(t) = \mathcal{L}^{-1}\{[s\boldsymbol{\mathcal{I}} - \boldsymbol{\mathcal{C}}]^{-1}\}. \tag{9.28b}$$

The first term can be written in terms of indices

$$\mathbf{k}'\boldsymbol{\phi}(t)\boldsymbol{\eta} = \sum_l \sum_m k_l \phi_{lm} \eta_m.$$

Thus,

$$\frac{\partial c}{\partial \eta_k} = \sum_l k_l \phi_{lk} = \sum_l \phi'_{kl} k_l;$$

in vector form

$$\frac{\partial c(t)}{\partial \boldsymbol{\eta}} = \boldsymbol{\phi}'(t)\mathbf{k}. \tag{9.29}$$

The above vector is the vector of sensitivities if definition (9.26) is used. If the normalized form of Equation (9.27) is adopted, the correct modification of Equation (9.29) is

$$\mathbf{v}(t) = \boldsymbol{\mathcal{R}}\boldsymbol{\phi}'(t)\mathbf{k}, \tag{9.30}$$

where

$$\boldsymbol{\mathcal{R}} = \begin{bmatrix} \eta_1 & 0 & 0 & \cdots & 0 \\ 0 & \eta_2 & 0 & \cdots & 0 \\ 0 & 0 & \eta_3 & \cdots & 0 \\ \cdot & \cdot & \cdot & & \cdot \\ \cdot & \cdot & \cdot & & \cdot \\ \cdot & \cdot & \cdot & & \cdot \\ 0 & 0 & 0 & \cdots & \eta_n \end{bmatrix}. \tag{9.31}$$

The $\boldsymbol{\phi}(t)$ matrix can be determined using the Laplace transform technique implied in Equation (9.28b) or by the method of Paynter described in Chapter 8 (Figure 8.27).

9.5 Generalization of the Results for Parameter Sensitivities[2]

The sensitivities of a single output variable, $c(t)$, to numerator and denominator coefficients were discussed in Sections 9.2 and 9.3. In this section, the sensitivity of the *entire state vector* to model parameters is calculated.

Begin with the phase space representation

$$\dot{x} = \mathbf{\alpha}_0 x + \mathbf{b}_0 r, \quad x(0) = \mathbf{0}, \tag{9.32}$$

where

$$\mathbf{\alpha}_0 = \begin{bmatrix} 0 & 1 & 0 & \cdots & 0 \\ 0 & 0 & 1 & \cdots & 0 \\ \cdot & \cdot & \cdot & & \cdot \\ \cdot & \cdot & \cdot & & \cdot \\ 0 & 0 & 0 & \cdots & 1 \\ -\gamma_1 & -\gamma_2 & -\gamma_3 & \cdots & -\gamma_n \end{bmatrix}, \tag{9.33}$$

$$\mathbf{b}_0 = \begin{bmatrix} 0 \\ 0 \\ \cdot \\ \cdot \\ 1 \end{bmatrix}. \tag{9.34}$$

More general representations are considered in the basic reference.[2]
Define the Jacobian matrix $\mathcal{J}$ by the component equation

$$J_{kl} = \frac{\partial x_k}{\partial \gamma_l}. \tag{9.35}$$

Clearly, this matrix is the matrix of the nonnormalized sensitivities. First it will be shown that

$$J_{kl} = J_{k-1,l+1}. \tag{9.36}$$

This remarkable property may be interpreted to mean that $\mathcal{J}$ has the following form:

$$\mathcal{J} = \begin{bmatrix} J_{11} & J_{12} & J_{13} & J_{14} & \cdots & J_{1n} \\ J_{12} & J_{13} & J_{14} & & & J_{2n} \\ J_{13} & J_{14} & & & & \\ J_{14} & & & & & \cdot \\ \cdot & & & & & \cdot \\ \cdot & & & & & \cdot \\ J_{1n} & J_{2n} & & \cdots & & J_{nn} \end{bmatrix}, \tag{9.37}$$

where the dashed lines are drawn through all elements which have the same value. Thus there are *only 2n — 1 independent elements* in the Jacobian matrix. This fact means that far fewer sensitivities must be calculated than would be expected.

To prove the validity of Equation (9.36), notice that Equation (9.33) can be used to show that

$$\frac{\partial \dot{x}_k}{\partial \gamma_l} = \frac{\partial x_{k+1}}{\partial \gamma_l}; \quad k = 1, 2, \ldots, n - 1, \tag{9.38a}$$

and that

$$\frac{\partial \dot{x}_n}{\partial \gamma_l} = -\sum_{i=1}^{n} \gamma_i \frac{\partial x_i}{\partial \gamma_l} - x_l, \qquad \frac{\partial \mathbf{x}(0)}{\partial \gamma_l} = \mathbf{0}. \tag{9.38b}$$

Similarly, for any $l < n$,

$$\frac{\partial \dot{x}_k}{\partial \gamma_{l+1}} = \frac{\partial x_{k+1}}{\partial \gamma_{l+1}}, \quad k = 1, 2, \ldots, n - 1, \tag{9.39a}$$

and

$$\frac{\partial \dot{x}_n}{\partial \gamma_{l+1}} = -\sum_{i=1}^{n} \gamma_i \frac{\partial x_i}{\partial \gamma_{l+1}} - x_{l+1}, \qquad \frac{\partial \mathbf{x}(0)}{\partial \gamma_{l+1}} = \mathbf{0}. \tag{9.39b}$$

Integrate the last equation with respect to time to find

$$\int_0^t \frac{\partial \dot{x}_n}{\partial \gamma_{l+1}} \, d\xi = \frac{\partial x_n}{\partial \gamma_{l+1}} = -\sum_{i=1}^{n} \gamma_i \int_0^t \frac{\partial x_i}{\partial \gamma_{l+1}} \, d\xi - \int_0^t x_{l+1} \, d\xi.$$

Substitute Equations (9.39a) and (9.32) into the last equation to find

$$\frac{\partial \dot{x}_{n-1}}{\partial \gamma_{l+1}} = -\gamma_1 \int_0^t \frac{\partial x_1}{\partial \gamma_{l+1}} \, d\xi - \sum_{i=2}^{n} \gamma_i \frac{\partial x_{i-1}}{\partial \gamma_{l+1}} - x_l. \tag{9.40}$$

Now define the vector $\mathbf{v}(t)$ by

$$v_1 = \int_0^t \frac{\partial x_1}{\partial \gamma_{l+1}} \, d\xi, \tag{9.41a}$$

$$v_k = \frac{\partial x_{k-1}}{\partial \gamma_{l+1}}, \quad k = 2, 3, \ldots, n. \tag{9.41b}$$

Thus, using Equations (9.39a) and (9.40), one finds that

$$\dot{v}_k = v_{k+1}, \quad k = 1, 2, \ldots, n - 1,$$

$$\dot{v}_n = -\sum_{i=1}^{n} \gamma_i v_i - x_l. \tag{9.42}$$

Comparing Equations (9.42) with Equations (9.38), it follows that

$$v_k(t) = \frac{\partial x_k}{\partial \gamma_l}, \tag{9.43}$$

i.e., from Equation (9.41b)

$$\frac{\partial x_{k-1}}{\partial \gamma_{l+1}} = v_k = \frac{\partial x_k}{\partial \gamma_l}, \quad k = 2, \ldots, n; \quad l = 1, \ldots, n-1.$$

Thus, Equation (9.36) has been established.

Still another useful property of the state variable sensitivities will be obtained. It will now be shown that *all* of the coefficients of the Jacobian matrix can be obtained from the single simulation-sensitivity circuit illustrated in Figure 9.1 if the simulation is constructed from the representation of Equation (9.32).

First notice that the outputs of the integrators are

$$\frac{\partial c}{\partial \gamma_l} = \frac{\partial x_1}{\partial \gamma_l}, \tag{9.44}$$

so that these sensitivities are obtained quite naturally from the sensitivity circuit. It will now be shown that the other sensitivities can be obtained from a linear combination of these sensitivities *and* the state variables. First notice that Equations (9.37) and (9.44) indicate that the first n diagonals are determined immediately as the outputs of the integrators of the sensitivity circuits. An inspection of Equation (9.37) also indicates that the remaining elements of the Jacobian matrix can be determined by finding equations for J_{kn}, $k = 2, 3, \ldots, n$.

It follows from Equation (9.32) that

$$\frac{\partial \dot{x}_1}{\partial \gamma_n} = \frac{\partial x_2}{\partial \gamma_n}.$$

It then follows from Equation (9.37) that

$$\frac{\partial x_2}{\partial \gamma_n} = J_{2n} = \frac{\partial \dot{x}_n}{\partial \gamma_1}. \tag{9.45a}$$

Use Equation (9.32) again to show that

$$J_{2n} = -x_1 - \sum_{l=1}^{n} \gamma_l \frac{\partial x_l}{\partial \gamma_1} = -x_1 - \sum_{l=1}^{n} \gamma_l \frac{\partial x_1}{\partial \gamma_l}. \tag{9.45b}$$

Thus, J_{2n} (and therefore an entire diagonal of $\mathcal{J}$) is determined as a linear combination of x_1 and the outputs of the integrators of the sensitivity circuit.

Similarly,

$$J_{3n} = \frac{\partial x_3}{\partial \gamma_n} = \frac{\partial \dot{x}_2}{\partial \gamma_n} = \dot{J}_{2n}.$$

Use Equation (9.45b) to show that

$$J_{3n} = -\dot{x}_1 - \sum_{l=1}^{n} \gamma_l \frac{\partial \dot{x}_l}{\partial \gamma_1}$$

or, from Equations (9.33) and (9.45a)

$$J_{3n} = -x_2 - \sum_{l=1}^{n-1} \gamma_l \frac{\partial x_{l+1}}{\partial \gamma_1} - \gamma_n \frac{\partial x_2}{\partial \gamma_n}.$$

Finally,

$$J_{3n} = -x_2 - \sum_{l=1}^{n-1} \gamma_l \frac{\partial x_{l+1}}{\partial \gamma_1} - \gamma_n J_{2n}. \tag{9.46}$$

After much algebra, it can be shown that the generalization of Equation (9.46) is

$$J_{kn} = -x_{k-1} - \sum_{l=1}^{n+2-k} \gamma_l \frac{\partial x_{l+k-2}}{\partial \gamma_1} - \sum_{l=n+3-k}^{n} \gamma_l J_{l+k-n-1,n}, \quad k \geq 3 \tag{9.47}$$

This equation proves the assertion that all of the elements of the Jacobian matrix can be found from a linear combination of the phase space variables and the outputs of the integrators of the sensitivity circuit illustrated in Figure 9.1.

9.6 Interpretation of Sensitivity Coefficients: Definition of the Covariance Matrix

The purpose of *simulation* is to predict the response of a dynamic system. The purpose of *sensitivity analysis* is to evaluate the accuracy of such predictions. The use of Equation (9.5) to estimate the accuracy is a rough but useful procedure which is sufficient for many purposes. Equation (9.5) can be used to answer, in an approximate manner, two types of questions:

1. What magnitude of error, $\Delta c(t)$, is associated with a given percentage error in the value of a physical parameter and/or an initial condition?

2. To what percentage accuracy must a physical parameter or an initial condition be known in order to meet a specified constraint on the error, Δc?

Occasionally, a more quantitative description of the error in a simulation study is required. A procedure which may prove satisfactory in such a case will now be presented. The procedure is based on the quantification of certain *statistics*, which will now be defined.

Assume that n similar dynamic systems are subjected to similar inputs. One might think of the n responses in a single variable, $x(t)$, as if they were graphed in a three-dimensional diagram, as is illustrated in Figure 9.3. The

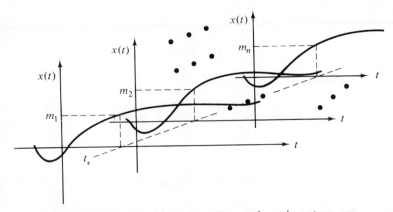

Figure 9.3. Ensemble of responses from n dynamic systems constructed in a similar manner and subjected to similar inputs. At the instant t_*, n measurements of $x(t)$ are made.

set of responses is called an *ensemble*. Suppose that at some fixed time, t_*, $x(t)$ is measured simultaneously in all n systems. Denote the measurement taken in the kth system by m_k. Because of the random nature of physical systems, the measurements will not all be of the same value. *Statistics* are measures of the degree of randomness of the measurements. An example of a statistic is the center of gravity of the measurements

$$= \frac{1}{n} \sum_{l=1}^{n} m_l.$$

One can imagine the special case where the ensemble contains an infinite number of responses, $x(t)$. For this case, the center of gravity is called the *mean* or *expected value* of $x(t_*)$ and is denoted

$$\mathcal{E}[x(t_*)] = \lim_{n \to \infty} \frac{1}{n} \sum_{l=1}^{n} m_l. \tag{9.48}$$

It is a remarkable fact of nature that in many cases this limit actually exists.

Another "statistic" is the moment of inertia of the measurements about the center of gravity

$$= \frac{1}{(n-1)} \sum_{l=1}^{n} (m_l - \mathcal{E}[x(t^*)])^2.$$

This statistic is a quantitative measure of the spread of the measurement. Notice that if all the measurements are close to the expected value, this statistic is relatively small. Conversely, if the measurements often differ greatly from the expected value, the *moment of inertia* statistic is very large.

The *variance*, D^2, is the limit value of the moment statistic, i.e.,

$$D^2 = \lim_{n \to \infty} \frac{1}{(n-1)} \sum_{l=1}^{n} (m_l - \mathcal{E}[x(t_*)])^2. \tag{9.49}$$

The square root of the variance is called the *standard deviation*. Notice that

$$D^2 = \mathcal{E}\{(x(t_*) - \mathcal{E}[x(t_*)])^2\}. \tag{9.50}$$

In general, one needs to define many other statistics in order to provide a quantitative description of randomness. For present purposes, the two statistics defined above are nearly sufficient. The only other statistic that need be introduced is a measure of the statistical interaction between two variables.

Suppose now that for *each* dynamic system, in the hypothetical measurement experiment, measurements are made of two variables, $x_k(t)$ and $x_i(t)$. Expected values and variances could be calculated for both variables using formulas similar to Equations (9.48) and (9.49). Define the *covariance*:

$$D_{ki} = \lim_{n \to \infty} \frac{1}{(n-1)} \sum_{l=1}^{n} \{(m_k)_l - \mathcal{E}[x_k(t_*)]\}\{(m_i)_l - \mathcal{E}[x_i(t_*)]\} \tag{9.51}$$

or

$$D_{ki} = \mathcal{E}[\{x_k(t_*) - \mathcal{E}[x_k(t_*)]\}\{x_i(t_*) - \mathcal{E}[x_i(t_*)]\}]. \tag{9.52}$$

Notice that

$$D_{ki} = D_{ik}. \tag{9.53}$$

Notice that if $i = k$, the covariance is the variance. Notice also that a variance is always positive, while a covariance may be positive or negative.

It is important to gain some intuitive insight into the assertion that covariance is a measure of statistical interaction between two variables. Suppose that n is large; then a large number of measurements of $x_k(t_*)$ will fall in any feasible region; for example, a large number of measurements will fall in a small region one standard deviation above $\mathcal{E}[x_k(t_*)]$. For all of the measurements in this subset of measurements, $(m_k)_l - \mathcal{E}[x_k(t_*)]$ is positive (and about equal to one standard deviation). Associated with each element of this subset of measurements, there is a corresponding measurement of $x_i(t_*)$ denoted $(m_i)_l$. If $x_k(t_*)$ and $x_i(t_*)$ are *uncorrelated* (i.e., if the randomness in the two sets of measurements is independent), then $(m_i)_l - \mathcal{E}[x_i(t_*)]$ is equally likely to be negative as positive. Thus, when one calculates the contribution to the sum in Equation (9.51) from this subset of measurements, a great deal of cancellation of positive and negative terms will occur and the contribution to the sum will be near zero. Since the same statement can be made about any subset of measurements of $x_k(t_*)$ in any small region, one concludes that in the case of absolute absence of any statistical interaction between $x_k(t_*)$ and $x_i(t_*)$,

$$D_{ki} = 0. \tag{9.54}$$

It should be clear that the covariance attains its maximum magnitude as a positive number when no cancellations occur in the sum of Equation (9.51), i.e., when the interaction is such that whenever $x_k(t_*)$ is above its expected value, $x_i(t_*)$ is also found to be above its expected value *and* whenever $x_k(t_*)$ is below its expected value, $x_i(t_*)$ is below its expected value. Similarly, the covariance attains its maximum magnitude as a negative number if the statistical interactions are such that whenever $x_k(t_*)$ is above its expected value, $x_i(t_*)$ is found to be *below* its expected value and vice versa.

Finally, suppose that measurements are made on all the state variables of each dynamic system. The expected value of the state vector

$$\mathcal{E}[\mathbf{x}(t_*)] = \begin{bmatrix} \mathcal{E}[x_1(t_*)] \\ \mathcal{E}[x_2(t_*)] \\ \cdot \\ \cdot \\ \cdot \\ \mathcal{E}[x_n(t_*)] \end{bmatrix}. \qquad (9.55)$$

One may also calculate variances for each state variable and a covariance for each pair of state variables. These last statistics may be thought of by defining a *covariance matrix*, $\mathfrak{D}$, by the component equation

$$(\mathfrak{D})_{ki} = D_{ki}, \qquad (9.56)$$

where D_{ki} is defined by Equation (9.52). It is important that the reader verify that the definition of the covariance matrix can be written in vector notation as

$$\mathfrak{D}(t_*) = \mathcal{E}[\{\mathbf{x}(t_*) - \mathcal{E}[\mathbf{x}(t_*)]\}\{\mathbf{x}(t_*) - \mathcal{E}[\mathbf{x}(t_*)]\}']. \qquad (9.57)$$

Knowledge of the covariance matrix will often satisfy the requirements of a sensitivity analysis, because the analyst can find a measure of the expected spread of an actual system's response about the simulation by consulting the appropriate element of this matrix. For this reason, an analytical link between the covariance matrix and the Jacobian matrix of the previous section will now be developed.

9.7   A Useful Approximate Relationship Between the Covariance Matrix and the Jacobian Matrix

Suppose that $\mathbf{x}$ and $\mathbf{p}$ are two vector quantities related by the nonlinear algebraic relationship

$$\mathbf{x} = \mathbf{f}(\mathbf{p}). \qquad (9.58)$$

Expand $\mathbf{f(p)}$ in a Taylor series about $\mathcal{E}(\mathbf{p})$ [see Equation (B.56) of Appendix B] to find

$$\mathbf{x} \simeq \mathbf{f}[\mathcal{E}(\mathbf{p})] + \mathcal{g}(\mathbf{p} - \mathcal{E}(\mathbf{p})), \tag{9.59}$$

where the k, l component of the $\mathcal{g}$ matrix

$$= \frac{\partial f_k}{\partial p_l}\Big|_{\mathcal{E}(\mathbf{p})}. \tag{9.60}$$

Approximation (9.59) is valid *as long as* $\mathbf{p}$ *is close to its expected value.*

Take the average or expected value of both sides of Equation (9.59) to find

$$\mathcal{E}[\mathbf{x}] \simeq \mathcal{E}\{\mathbf{f}(\mathcal{E}[\mathbf{p}])\} + \mathcal{g}\mathcal{E}(\mathbf{p} - \mathcal{E}[\mathbf{p}])$$
$$= \mathbf{f}(\mathcal{E}[\mathbf{p}]) + \mathcal{g}0,$$

so that

$$\mathcal{E}[\mathbf{x}] \approx \mathbf{f}(\mathcal{E}[\mathbf{p}]). \tag{9.61}$$

The surprising thing is that this relationship is an approximation which is valid only when $\mathbf{p}$ is narrowly distributed about its expected or average value.

Denote

$$\mathfrak{D}_x = \text{covariance matrix of the } \mathbf{x} \text{ variables}, \tag{9.62}$$

$$\mathfrak{D}_p = \text{covariance matrix of the } \mathbf{p} \text{ variables}. \tag{9.63}$$

Begin the derivation of a relationship between these two matrices by combining Equations (9.56) and (9.61) to show that

$$\mathbf{x} - \mathcal{E}[\mathbf{x}] \simeq \mathcal{g}(\mathbf{p} - \mathcal{E}[\mathbf{p}]).$$

Thus,

$$\{\mathbf{x} - \mathcal{E}[\mathbf{x}]\}\{\mathbf{x} - \mathcal{E}[\mathbf{x}]\}' \simeq \mathcal{g}\{\mathbf{p} - \mathcal{E}[\mathbf{p}]\}\{\mathbf{p} - \mathcal{E}[\mathbf{p}]\}'\mathcal{g}'. \tag{9.64}$$

Take the expected value of both sides of this equation and use definition (9.57) to show that

$$\mathfrak{D}_x \simeq \mathcal{g}\mathfrak{D}_p\mathcal{g}'. \tag{9.65}$$

If $\mathbf{x}$ is to be the state vector and $\mathbf{p}$ is the vector of mathematical or physical parameters, then the $\mathcal{g}$ matrix is calculated using one of the methods presented in earlier sections. Obviously, one must know the values of $\mathfrak{D}_p$ in order to make the calculation. This *input* covariance matrix will usually be a diagonal matrix of parameter variances since, in general, the parameter variations will be uncorrelated.

Example 9.1

Suppose that

$$\mathfrak{D}_p = \begin{bmatrix} 0.01 & 0 & 0 \\ 0 & 0.01 & 0 \\ 0 & 0 & 0.02 \end{bmatrix}$$

and that

$$\mathfrak{J} = \begin{bmatrix} 1 & 2 & 1 \\ 1 & -1 & -1 \end{bmatrix}.$$

Then

$$\mathfrak{D}_x \simeq \begin{bmatrix} 1 & 2 & 1 \\ 1 & -1 & -1 \end{bmatrix} \begin{bmatrix} 0.01 & 0 & 0 \\ 0 & 0.01 & 0 \\ 0 & 0 & 0.02 \end{bmatrix} \begin{bmatrix} 1 & 1 \\ 2 & -1 \\ 1 & -1 \end{bmatrix}$$

$$= \begin{bmatrix} 0.07 & -0.03 \\ -0.03 & 0.04 \end{bmatrix}.$$

Notice that the state variables have a covariance of relatively large magnitude (compared to the variances) even though the input parameters are uncorrelated.

*   *   *

9.8 The Sensitivity of Unforced Nonlinear Systems of Differential Equations

Consider the simulation of the nonlinear system of equations

$$\dot{\mathbf{x}} = \mathbf{f}(\mathbf{x}, p), \qquad (9.66)$$

where p is some physical parameter.

Denote the result of the simulation

$$\mathbf{x}_* = \mathbf{x}_*(t). \qquad (9.67)$$

Denote the sensitivity vector

$$\mathbf{v}(t) = \begin{bmatrix} \dfrac{\partial x_1}{\partial p} \\[2mm] \dfrac{\partial x_2}{\partial p} \\[1mm] \cdot \\ \cdot \\ \cdot \\ \dfrac{\partial x_n}{\partial p} \end{bmatrix}_{\mathbf{x} = \mathbf{x}_*} \qquad (9.68)$$

Differentiate Equation (9.66) partially with respect to p to find

$$\frac{\partial \dot{\mathbf{x}}}{\partial p} = \frac{\partial \mathbf{f}}{\partial p} + \mathcal{J}\frac{\partial \mathbf{x}}{\partial p},$$ (9.69)

where

$$J_{kl} = \frac{\partial f_k}{\partial x_l}\bigg|_{\mathbf{x}_*}.$$ (9.70)

Thus,

$$\dot{\mathbf{v}} = \mathcal{J}(\mathbf{x}_*(t))\mathbf{v} + \frac{\partial \mathbf{f}}{\partial p}(\mathbf{x}_*(t), p).$$ (9.71)

This system of differential equations must, in general, also be solved using a computer method. Note that the sensitivity equations are a system of forced linear equations with time-varying coefficients.

If the system is described by an output equation

$$c(t) = \mathbf{k}'\mathbf{x}(t),$$ (9.72)

then

$$\frac{\partial c}{\partial p}(t) = \mathbf{k}'\mathbf{v}(t).$$ (9.73)

References

1. Tomovic, R., *Sensitivity Analysis of Dynamic Systems*, McGraw-Hill, New York, 1960.

2. Wilkie, D. F., and Perkins, W. R., "Generation of Sensitivity Functions for Linear Systems Using Low Order Models," *I.E.E.E. Transactions on Automatic Control*, AC-14, No. 2, Apr. 1969.

Problems

9.1. Establish Equation (9.5).

9.2. What are the proper initial conditions on the integrators of the sensitivity circuits illustrated in Figures 9.1 and 9.2?

9.3. Determine, analytically, the sensitivities

$$\frac{\partial c(t)}{\partial T}, \qquad \frac{\partial c(t)}{\partial K}$$

when

$$\frac{C(s)}{R(s)} = \frac{K}{Ts + 1}$$

and

$$R(s) = \frac{1}{s}.$$

9.4. Determine, analytically, the sensitivity

$$\frac{\partial c(t)}{\partial \omega}$$

when

$$\frac{C(s)}{R(s)} = \frac{\omega^2}{s^2 + \omega^2}$$

and

$$R(s) = \frac{1}{s}.$$

9.5. Determine, analytically, an approximate time history of the variance of $c(t)$ for the system of Problem 9.3 when K and T are uncorrelated and

$$K = 2.0,$$
$$T = 1.0 \text{ sec},$$

and the variances

$$D_K^2 = 0.02,$$
$$D_T^2 = 0.01 \text{ sec}^2.$$

9.6. Generalize Equations (9.13) and (9.22) for the case when both the a_i and the b_i depend on the same parameter p.

9.7. Consider Problem 8.9. Construct a computer circuit or program and determine the accuracy to which R_1 and R_2 must be known so that the simulation results will be accurate to $\pm 5\%$.

Feedback Control: Design Criteria and Elementary Methods for Stability Analysis

10

To this point, the discussion has emphasized the analysis, synthesis, and simulation of dynamic systems. In the present chapter, attention shifts to the design of dynamic systems. In particular, the design of feedback systems (see Chapter 1) will be studied. The theory of the design of feedback systems is often referred to as *automatic control* theory.

A rational discussion of any type of design must begin with a description of the design criteria which an engineer uses to quantify, and thereby compare, alternative designs. The main purpose for including the present chapter is to provide just such a description. Feedback system design criteria will be defined and discussed in sufficient detail to encourage the reader to develop an intuitive understanding of the goals of feedback design.

The remaining chapters of the text are then devoted to the description of techniques which the designer uses to develop an automatically controlled system which meets the design criteria.

10.1 Negative Feedback Control of Scalar Variable Systems: Terminology and the Basic Transfer Functions

A schematic of the negative feedback control concept is provided in Figure 10.1. Notice that the prime mover *plus* controlled object are called the *plant* or *process*.

Since only a single output and a single set point are indicated, the feedback

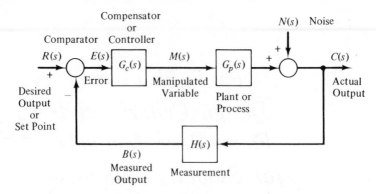

Figure 10.1. Scalar variable feedback control: conventional terminology.

system illustrated in Figure 10.1 is called a *scalar* feedback system. Such systems will receive the major emphasis in this text, although an introduction to multivariable system design will also be provided.

Notice that the output

$$C(s) = E(s)G_c G_p$$

and that the error

$$E(s) = R(s) - C(s)H(s).$$

Thus, the closed loop transfer function between input and output is

$$\frac{C(s)}{R(s)} = \frac{G_c G_p}{1 + G_c G_p H}. \tag{10.1}$$

Similarly, it can be shown that other closed loop transfer functions are given by

$$\frac{E(s)}{R(s)} = \frac{1}{1 + G_c G_p H} \tag{10.2}$$

and

$$\frac{M(s)}{R(s)} = \frac{G_c}{1 + G_c G_p H}. \tag{10.3}$$

The *closed loop* poles, denoted s_p, are those values of s for which the denominators of the closed loop transfer functions vanish. It follows from Equations (10.1)–(10.3) that the closed loop poles satisfy the equation

$$\boxed{1 + G_c(s_p)G_p(s_p)H(s_p) = 0.} \tag{10.4}$$

This equation will be referred to as the *characteristic equation*.

The characteristic equation shows the relationship between *closed loop* poles and the *open loop* transfer functions G_c, G_p, and H. Notice that all of the closed loop transfer functions have the same poles. Only the numerator dynamics are different for the different transfer functions. The transfer functions and the characteristic equation will be referred to in the description of feedback system design criteria.

Several particular examples of the general system, illustrated in Figure 10.1, are provided in Chapters 1 and 7 (see Figures 7.14 and 7.22). A temperature control of a fluid stream is illustrated in Figures 10.2a and 10.2b. A position control system (a *servomechanism*) is illustrated in Figure 10.3. A block diagram of this feedback system is provided in Figure 10.4.

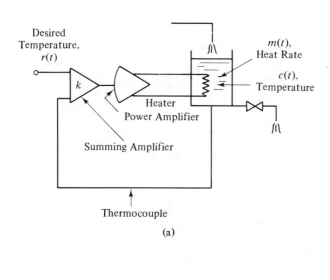

(a)

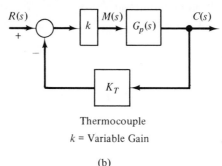

Thermocouple

k = Variable Gain

(b)

Figure 10.2. (a) Automatic control of the temperature of a fluid stream, (b) block diagram of the system illustrated in (a).

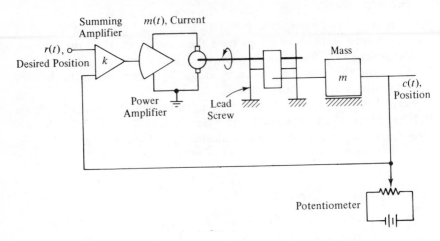

Figure 10.3. Automatic control of position; a servomechanism.

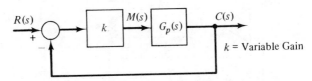

Figure 10.4. Block diagram of the servosystem illustrated in Figure 10.3.

10.2 Speed and Damping of Response as Design Criteria; Contribution of Individual Closed Loop Poles to the Dynamic Response of an Automatically Controlled System[3]

Suppose that the closed loop transfer function, defined in Equation (10.1), can be rewritten in the form

$$\frac{C(s)}{R(s)} = \frac{G_c G_p}{1 + G_c G_p H} = \frac{A(s)}{B(s)(s + \alpha)}, \tag{10.5}$$

where $A(s)$ and $B(s)$ are polynomials in s. Then a pole zero diagram of a portion of the closed loop transfer function will be as illustrated in Figure 10.5. That is, a real closed loop pole exists at $s_p = -\alpha$.

It follows from Equation (10.5) and the discussion in Chapter 2 that the contribution to the transient closed loop response from the pole at $s_p = -\alpha$ is

$$R_0 e^{-\alpha t} u(t), \tag{10.6}$$

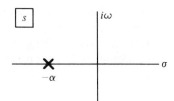

Figure 10.5. Real closed loop pole.

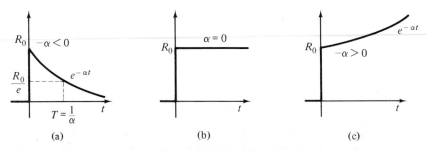

Figure 10.6. (a) Contribution to closed loop response of a real pole
in the left half s plane, (b) contribution to closed loop response of
a real pole at the origin of the s plane, (c) contribution to closed
loop response of a real pole in the right half s plane.

where

$$R_0 = \frac{A(-\alpha)}{B(-\alpha)}R(-\alpha). \qquad (10.7)$$

Plots of Equation (10.6) for three possible ranges of α are illustrated in Figures
10.6a, 10.6b, and 10.6c.

It follows from Equation (10.6) that the contribution of a real pole in the
left half s plane to the closed loop transient response falls to $R_0/e\,(= 0.368R_0)$
at the time

$$t = T \triangleq \frac{1}{\alpha}. \qquad (10.8)$$

T is called the *time constant*. If the closed loop transfer function contains only
a single real pole, the time constant can be measured quite simply by measuring the length of time necessary for the system to come to 63.2% of its
final value after being subjected to a step input. For obvious reasons, a
closed loop system described by a transfer function with poles in the left half
s plane is called *stable*.

The time constant is a measure of the *speed of response* of a dynamic
system. Obviously, one would like to design a control system so that the time
constants associated with the poles are as small as possible.

That closed loop real pole which is *nearest* the origin in the s plane is described by the value of α with the smallest magnitude—thus, by Equation (10.8), by the largest time constant. This real pole is called the *dominant real pole* because it often "dominates" the transient response since its contribution to the response decays the least rapidly. A careful analysis of residues is, however, required in order to determine the truly dominant pole.

Suppose now that the closed loop transfer function can be rewritten

$$\frac{G_c G_p}{1 + G_p G_c H} = \frac{A(s)}{B(s)[(s + \alpha)^2 + \beta^2]}. \tag{10.9}$$

The constants α and β are given the names

$$\alpha \triangleq \text{ damping constant}, \tag{10.10}$$

$$\beta \triangleq \text{ natural frequency}. \tag{10.11}$$

Assume that the system is *underdamped*; i.e., $\beta > 0$ and real. Then, the contribution to the closed loop transient response is determined by the techniques of Chapter 2 to be

$$A_0 e^{-\alpha t} \cos (\beta t + \phi), \tag{10.12}$$

where the amplitude

$$A_0 = \left| \frac{A(-\alpha + i\beta) R(-\alpha + i\beta)}{B(-\alpha + i\beta) i\beta} \right| \tag{10.13}$$

and the phase angle, ϕ, is the angle of the complex number ratio

$$\frac{A(-\alpha + i\beta) R(-\alpha + i\beta)}{B(-\alpha + i\beta) i\beta}. \tag{10.14}$$

The pole-zero diagram for a portion of the transfer function described by Equation (10.9) is illustrated in Figure 10.7.

The quantity $T \triangleq (1/\alpha)$ again is a measure of the speed of response of the closed loop pole, but now this time constant must be associated with the envelope of the oscillatory response (see Figures 10.8).

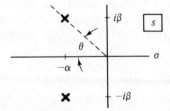

Figure 10.7. Complex closed loop poles.

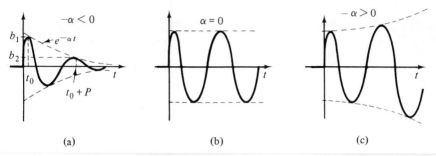

(a) (b) (c)

Figure 10.8. (a) Contribution from left half plane complex poles, (b) contribution from complex poles on the imaginary axis, (c) contribution from right half plane complex poles.

When the closed loop transfer function has complex poles, the speed of response is not the only dynamic design criterion. The designer must also consider the rate of damping of transient terms. In other words, it is not sufficient to design a system for which the exponential envelopes (see the dashed lines in Figures 10.8) decay quickly since the number of significant oscillations are also an important factor. To quantify rate of damping rewrite the quadratic factor in the alternate form

$$\frac{1}{(s + \alpha)^2 + \beta^2} = \frac{1}{s^2 + 2\zeta\omega_n s + \omega_n^2} \tag{10.15}$$

where the constants ζ and ω_n are given the names

$$\zeta \triangleq \text{damping ratio}, \tag{10.16}$$

$$\omega_n \triangleq \text{undamped natural frequency}. \tag{10.17}$$

It is left as an exercise for the reader to show that

1.

$$\zeta = \frac{\alpha}{\omega_n}. \tag{10.18}$$

2.

$$\beta = \omega_n\sqrt{1 - \zeta^2}. \tag{10.19}$$

3. The *underdamped* condition ($\beta > 0$ and real) may be written

$$0 \leq \zeta \leq 1. \tag{10.20}$$

4. The *critically* damped condition ($\beta = 0$) may be written

$$\zeta = 1.0. \tag{10.21}$$

5. The stability limit condition ($\alpha = 0$) may be written

$$\xi = 0$$

or

$$\beta = \omega_n. \tag{10.22}$$

Let θ be defined in the manner indicated in Figure 10.7. It follows from Equations (10.18) and (10.19) that

$$\cot \theta = \frac{\alpha}{\beta} = \frac{\xi \omega_n}{\omega_n \sqrt{1 - \xi^2}} = \frac{\xi}{\sqrt{1 - \xi^2}}. \tag{10.23}$$

This equation leads to the conclusion that *all poles on the same radial line from the origin have the same damping ratio, ξ.*

An important design criterion for automatic control systems is the *damping per cycle* defined by

$$\frac{b_1}{b_2} = \frac{e^{-\alpha t_0} \cos (\beta t_0 + \phi) u(t)}{e^{-\alpha (t_0 + P)} \cos [\beta(t_0 + P) + \phi] u(t)}, \tag{10.24}$$

where b_1 and b_2 are as illustrated in Figure 10.8a and P is the period of oscillation given by

$$P = \frac{2\pi}{\beta}. \tag{10.25}$$

Notice that

$$\cos (\beta t_0 + \phi) = \cos (\beta(t_0 + P) + \phi),$$

so that

$$\frac{b_1}{b_2} = e^{\alpha P} = e^{\xi \omega_n 2\pi / \omega_n \sqrt{1 - \xi^2}}$$

or

$$\frac{b_1}{b_2} = e^{2\pi \xi / \sqrt{1 - \xi^2}}. \tag{10.26}$$

Combine Equations (10.23) and (10.26) to find that

$$\boxed{\frac{b_1}{b_2} = e^{2\pi \cot \theta}.} \tag{10.27}$$

Obviously, damping per cycle is a quantitative measure of rate of damping, i.e., to the number of oscillations in the transient response. Clearly, the designer will usually wish to reduce the number of significant oscillations in the transient period in order to reduce equipment wear.

It follows from Equations (10.26) and (10.27) that ξ or θ could be used as equivalent design criteria for the damping per cycle, b_1/b_2.

The damping per cycle for different β, at constant α, is illustrated in Figure 10.9.

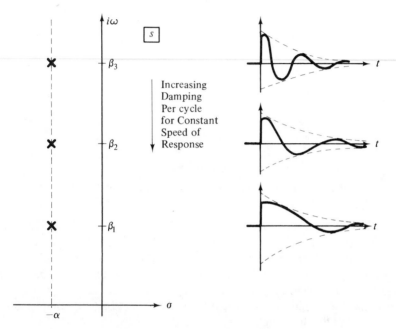

Figure 10.9. Damping per cycle for constant α. All complex pole pairs have equal speed of response.

To understand why the term *damping ratio* is associated with the symbol ξ, let α denote the damping constant of an underdamped system and let α_c denote the value of the damping constant for the critical damping condition, $\beta = 0$. It follows from Equation (10.15) that $\alpha_c = \omega_n$ at critical damping. It then follows from Equation (10.18) that

$$\frac{\alpha}{\alpha_c} = \frac{\omega_n \xi}{\omega_n} = \xi. \tag{10.28}$$

Hence, the name damping ratio.

Equations (10.26) and (10.27) were used to construct the graph of Figure 10.10. Notice that complex poles for which θ is $65°$ or less contribute transient response terms with damping per cycle in excess of 19.1 to 1. A damping per cycle of 5 to 1 is often satisfactory. As indicated in Figure 10.10, such a value

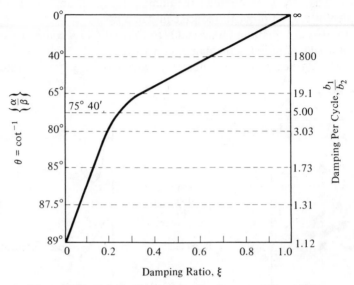

Figure 10.10. Relationship between the three quantifiers of the rate of damping.

of damping per cycle is achieved for $\theta \leq 75°$ ($\xi \geq 0.3$). It is surprising that θ can be so large.

A conventional procedure is to determine an allowed region for the location of closed loop poles in the s plane for a given set of design specifications. For instance, suppose that a designer is confronted with the specification that the closed loop transient response should be 63% completed in 0.5 sec and that the damping per cycle should be in excess of 5 to 1. The specifica-

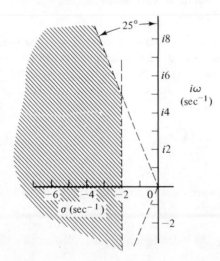

Figure 10.11. Allowed region for closed loop poles for design specifications of a 0.5-sec time constant and a 5 to 1 damping per cycle.

tion on the speed of response can be satisfied if the real part, α, of all closed loop poles lies to the left of a vertical line through the -2 point on the real axis of the s plane. The damping criterion will be met if all poles lie below the radial line which makes an angle of 75° with the negative real axis. Thus, to meet both specifications, the designer must develop a feedback system for which the closed loop poles lie in the shaded region in Figure 10.11.

10.3 Other Design Criteria for the Transient Performance of Feedback Systems[7]

A designer's customer will usually provide design criteria for both speed of response *and* damping. As indicated in the previous section, these criteria can be stated in terms of the location of closed loop poles. However, there exist many other ways of stating such criteria. A few of the alternative statements are described in this section. Other alternatives will be discussed in succeeding chapters.

Consider the closed loop step response illustrated in Figure 10.12. A tolerance band about steady state may be specified. Usually, the band is taken to be $\pm 2\%$ or $\pm 5\%$ of the steady state value. The *rise time* is defined to be the instant when the response first enters the tolerance band, and the *settling time* is defined to be that instant when the response enters the band for the final time.

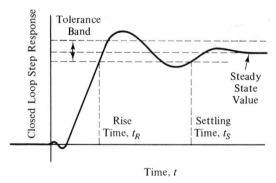

Figure 10.12. Definition of rise time and settling time.

Rise time is a measure of the speed of response, and the ratio of settling time to rise time is a measure of damping. The design criteria for the transient response of a feedback system are often stated in terms of these two measures. Of course, the statement of the criteria in terms of closed loop pole location has meaning only if the feedback system can be modeled by a transfer function or by linear differential equations with constant coefficients. In that sense, the rise time–settling time criteria are of a more general nature than are statements about allowed closed loop pole location.

It is possible to devise single criteria which quantify both the speed of response *and* the damping. One such criterion is the *integral-absolute deviation* criterion defined by

$$I_a = \int_0^\infty |c(t) - c(\infty)| \, dt. \tag{10.29}$$

Here $c(t)$ is the step response of the closed loop system and $c(\infty)$ is the steady state value of $c(t)$. I_a may be interpreted as the shaded area indicated in Figure 10.13. The criterion is most valuable when used to determine the relative value of competing designs; i.e., the "best" design is that which minimizes I_a. It is instructive to speculate on the nature of the step responses that will result if the designer accepts this definition of the *best* design. It is clear from an examination of Figure 10.13 that the step response of the best design will be damped since continuous cycling will cause I_a to approach infinity [notice that this statement would not be true if the deviation, $c(t) - c(\infty)$, were integrated instead of its absolute value]. Notice, however, that a large contribution to I_a is made before $c(t)$ first crosses $c(\infty)$. To minimize this first contribution, the best design must respond very quickly; thus, I_a does indeed quantify both speed of response and damping.

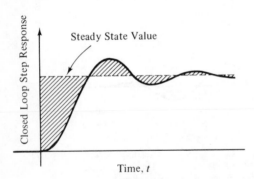

Figure 10.13. Graphical representation of the integral-absolute deviation criterion.

When a designer uses I_a as a criterion, he will often find that some of the closed loop poles of the best design have a damping ratio of about 0.3 ($\theta \simeq 75°$), which means that overshoot and some oscillation will occur. The reason for this result is that the best system responds so quickly to reduce the initial contribution to I_a that overshoot ensues. If the designer finds that this *optimum* response is unsatisfactory, he could adopt the alternative criterion

$$I_{at} = \int_0^\infty t \, |c(t) - c(\infty)| \, dt. \tag{10.30}$$

Notice that this criterion places a stronger emphasis on the contribution of oscillations to I_{at} than on the initial contribution of deviations. The design that minimizes I_{at} will display a less oscillatory response than the design which minimizes I_a.

Another alternative criterion is the *integral-squared-deviation* criterion defined by[6]

$$I_s = \int_0^\infty \{c(t) - c(\infty)\}^2 \, dt. \tag{10.31}$$

The optimum system associated with this criterion will have much the same properties as the system defined by I_a. However, the analysis problems associated with minimizing I_s are less difficult to solve than those associated with I_a. In fact, in many modern applications, analysts prefer to work with I_s, and its generalizations, rather than with I_a.[1,5,8]

10.4 Design Criteria for the Steady State Performance of a Feedback System

Suppose that the feedback system illustrated in Figure 10.1 is subjected to a step input of magnitude a; i.e., suppose that

$$R(s) = \frac{a}{s}. \tag{10.32}$$

It then follows from the final value theorem and Equation (10.2) that the *steady state error*

$$e(\infty) = \lim_{s \to 0} \frac{a}{(1 + G_c G_p H)}$$

or

$$\frac{e(\infty)}{a} = \lim_{s \to 0} \left[\frac{1}{1 + G_c(s)G_p(s)H(s)} \right]. \tag{10.33}$$

Define the gains of open loop transfer functions:

$$K_c = \lim_{s \to 0} G_c(s),$$

$$K_p = \lim_{s \to 0} G_p(s), \tag{10.34}$$

$$K_H = \lim_{s \to 0} H(s).$$

It then follows that

$$\frac{e(\infty)}{a} = \frac{1}{1 + K_c K_p K_H} \tag{10.35}$$

where $e(\infty)/a$ is called the *normalized steady state error*. Inasmuch as the error, $e(t)$, is the difference between the *actual* output and the *desired* output, one often would like to make the normalized steady state error as small as possible. It is clear from Equation (10.35) that one way to accomplish this end is to use a compensator with a high gain, i.e., make K_c a large number. Unfortunately, one often finds that large values of K_c produce highly oscillatory (even unstable) transient response.

Thus, if the designer must satisfy both transient and steady state criteria, he will often find that these criteria are competing goals. In other words, to reduce steady state error, the designer may have to be satisfied with increased oscillation and vice versa.

Another way to achieve zero steady state error is to include a pole at the origin of the s plane in the compensator transfer function, $G_c(s)$. This statement can, of course, be proved mathematically, but an intuitive argument will suffice for present purposes. If the transfer function, $G_c(s)$, has a pole at the origin, the compensator will behave much as an integrator does. However, an integrator can have a constant steady output only if its input is zero. Since the input to the compensator is the error signal, it follows that zero steady state error will occur if $G_c(s)$ has a pole at the origin.

It will be shown in succeeding chapters that if $G_c(s)$ is synthesized in such a manner that a pole exists at the origin, the stability and speed of response of the overall feedback system will often be degraded. For such cases, the designer will again have to make a trade-off between steady state error and transient response.

It must be understood that the designer of feedback systems will not *always* be concerned with steady state error. In some cases, steady state error cannot be tolerated, while in other cases it is either quite acceptable or easily accounted for. For instance, consider the feedback control of the temperature of a furnace wherein the desired output (set point) is input by turning a dial face on a potentiometer. The potentiometer voltage is compared with the temperature feedback signal (which is another voltage) to produce an error signal. In this case the dial of the potentiometer can be calibrated to correspond to the actual furnace temperature and the existance of a nonzero error signal has no significance whatsoever.

Equation (10.35) applies only to the case of a step input. On occasion it is useful to derive equations for the steady state error from ramp or parabolic inputs.[3] These derivations are left to the reader as exercises at the end of the chapter.

10.5   The Equivalent Open Loop System: Parameter Variation and Disturbance Signals: The Main Reason for Using Feedback[2,4,9]

In Figure 10.1 the *plant* refers to that segment of the dynamic system over which the designer often has little influence. This segment contains the prime mover (chief supplier of energy) and other devices necessary to produce the output. The designer adds the measuring device, $H(s)$, the comparator, and the compensator, $G_c(s)$, in order to synthesize the feedback and thereby realize the design objectives.

It is important to note that, theoretically, a system which is structurally different but mathematically equivalent to the negative feedback system can be constructed. This equivalent system is illustrated in Figure 10.14. This is an open loop compensation because there is no feedback signal and comparator.

Figure 10.14. Open loop compensation.

Suppose that G_{c0} is designed and synthesized so that

$$G_{c0} = \frac{G_c(s)}{1 + G_c G_p H},\tag{10.36}$$

where G_c and H are the transfer functions of the compensator and measuring device that the designer might have used in the synthesis of a feedback system. It follows from Equation (10.36) that the relationship between $R(s)$ and $C(s)$ is

$$C(s) = \left[\frac{G_c G_p}{1 + G_c G_p H}\right] R(s).\tag{10.37}$$

Comparing Equations (10.1) and (10.37), it is obvious that the feedback and open loop compensation systems have identical input-output relations. It then follows that both the static and transient properties of the two compensating systems are identical.

The question that now arises is why use feedback when the open loop system is simpler and perhaps less expensive to construct? The answer to this question is that the so-called *sensitivity* properties of the two alternative designs differ greatly. A system is said to be insensitive if an unforeseen change in the value of a parameter does not significantly alter the performance of the system. Most often (but not always) the sensitivity properties of the feedback system are the more desirable.

The analysis and the exploitation of the sensitivity properties of dynamic systems will receive much discussion in succeeding chapters on feedback design. For present purposes, it suffices to demonstrate, by example, the assertion that the sensitivity properties of feedback systems are often superior to those of open loop systems.

Example 10.1

Consider the two equivalent systems shown in Figures 10.15a and 10.15b. The unit step response for both systems is

$$c_{\text{nom}} = \tfrac{10}{11}[1 - e^{-11t}]u(t).\tag{10.38}$$

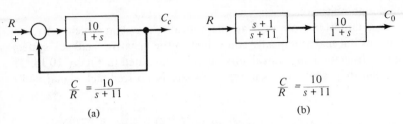

Figure 10.15. (a) Feedback system, (b) open loop system which is equivalent, *in a mathematical sense*, to the feedback system.

Suppose that in the course of operation, a parameter variation occurs in such a manner that the plant transfer function becomes

$$G_p(s) = \frac{10}{s + 0.5}.$$ (10.39)

(This change might be associated with a change in operating level for a linearized transfer function.) The transfer functions for the two systems become

$$\frac{C_c}{R} = \frac{10}{s + 10.5}$$ (10.40)

for the closed loop system and

$$\frac{C_0}{R} = \frac{10(s + 1)}{(s + 11)(s + 0.5)}$$ (10.41)

for the open loop system. The unit step response for the two cases (after parameter variation) becomes

$$c_c = \frac{10}{10.5}[1 - e^{-10.5t}]u(t)$$ (10.42)

for the closed loop system and

$$c_0 = \frac{20}{11}[1 - 0.52e^{-0.5t} - 0.48e^{-11t}]u(t)$$ (10.43)

for the open loop system.

Notice that the residue associated with the slower decaying exponential in Equation (10.43) is large. Equations (10.38), (10.42), and (10.43) are plotted in Figure 10.16.

Obviously, the closed loop system is less sensitive to parameter change.

* * *

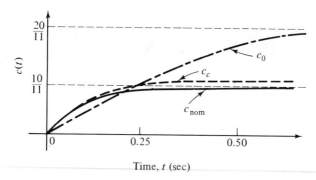

Figure 10.16. Step responses before and after parameter variation for both of the systems illustrated in Figure 10.15.

The insensitivity of feedback systems was emphasized in Chapter 1 and will be emphasized in succeeding chapters since it is this property which truly distinguishes feedback systems from open loop compensation systems.

There exist quantifiers of the sensitivity properties of dynamic systems which can be useful in design.[4] Some of these measures will be discussed in the present section and others will be discussed in later chapters.

Define the *loop gain*:

$$K = K_c K_p K_H. \tag{10.44}$$

Equation (10.35) can be rewritten

$$\frac{e(\infty)}{a} = \frac{1}{1 + K}. \tag{10.45}$$

A measure of the sensitivity of steady state error to changes in K is the normalized partial derivative

$$K \frac{\partial}{\partial K} \left[\frac{e(\infty)}{a} \right], \tag{10.46}$$

which is called the *sensitivity coefficient*.

It follows from Equation (10.45) that

$$K \frac{\partial}{\partial K} \left[\frac{e(\infty)}{a} \right] = - \frac{K}{(1 + K)^2} = -K \left[\frac{e(\infty)}{a} \right]^2. \tag{10.47}$$

This sensitivity is plotted against K in Figure 10.17. Notice that the sensitivity coefficient attains a maximum magnitude at a loop gain of 1.0. More importantly, it should be noted that the sensitivity approaches zero as the loop gain becomes large. Thus, not only does high gain decrease steady state error, but

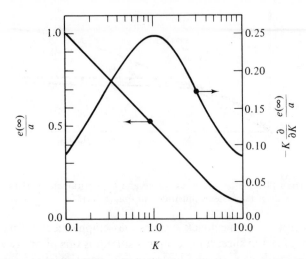

Figure 10.17. Sensitivity of the steady state error to changes in loop gain, K. The input is a step of magnitude a.

it also desensitizes the steady state properties of the feedback system to changes in open loop gains.

It is easily shown that if steady state error due to a step input is eliminated by incorporating an integrator in the compensator, the feedback system's steady state response is perfectly insensitive to parameter change.

Sensitivity coefficients (normalized partial derivatives) can also be calculated for transient performance measures. The sensitivity coefficients for closed loop pole location will now be discussed.

Rewrite the characteristic equation [Equation (10.4)] as a polynomial in s_p, i.e.,

$$\sum_{\beta=0}^{n} b_\beta s_p^\beta = 0. \tag{10.48}$$

The coefficients b_β depend on both the numerator and denominator coefficients of G_c, G_p, and H. These coefficients, in turn, are related to physical parameters such as inertances, resistances, compliances, and gyrator constants. Very often the values of these constants cannot be determined during design or will change during the operation of the system. Denote one of these physical parameters by d_k.

It follows from Equation (10.48) that

$$\sum_{\beta=0}^{n} \frac{\partial b_\beta}{\partial d_k} s_p^\beta + \sum_{\beta=1}^{n} \beta b_\beta s_p^{\beta-1} \frac{\partial s_p}{\partial d_k} = 0,$$

where s_p is a particular closed loop pole. Thus,

$$\frac{\partial s_p}{\partial d_k} = -\frac{\sum\limits_{\beta=0}^{n} (\partial b_\beta / \partial d_k) s_p^\beta}{\sum\limits_{\beta=1}^{n} \beta b_\beta s_p^{\beta-1}} \tag{10.49}$$

and the sensitivity coefficient

$$d_k \frac{\partial s_p}{\partial d_k} = -\frac{\sum\limits_{\beta=0}^{n} d_k (\partial b_\beta / \partial d_k) s_p^\beta}{\sum\limits_{\beta=1}^{n} \beta b_\beta s^{\beta-1}}. \tag{10.50}$$

Notice that only the closed loop poles, s_p, need be known in order to calculate these sensitivities. If s_p is complex, the sensitivity coefficient will, in general, also be complex.

The reader who finds himself becoming interested in sensitivity analysis of dynamic systems might find the advanced material in Chapter 9 valuable.

The other chief reason for using feedback is to continually correct the influences on plant output of unwanted disturbance (*noise*) signals. This point was discussed at length in Chapter 1. In a later chapter on frequency domain methods, the quantification of the noise insensitivity properties of feedback systems will be explored.

10.6 Summary of the Discussion of Design Criteria for Feedback System Synthesis

As has been noted in previous sections, the following list of classes of design criteria are considerations which enter into the design of almost all feedback systems:

1. *Transient state criteria:*
   a. Stability.
   b. Speed of response.
   c. Damping.
2. *Equilibrium state criteria:*
   d. Steady state error. $\qquad(10.51)$
3. *Sensitivity criteria:*
   e. Sensitivity to noise.
   f. Sensitivity to parameter variation.
4. *Effect of nonlinearities:*
   g. Stability.

This list of criteria appears to be the one invariant of control theory. That is, the techniques and methodologies are changing quite rapidly, while the goals of the application of the techniques do not.

It is this author's bias that the sensitivity criteria are the most important ones since they quantify the truly unique properties of feedback. It is interesting to note that control system theorists sometimes lose sight of this fact during periods of rapid growth of the field. These periods are frequent and dramatic for system theory, and it is no wonder that system theorists temporarily lose sight of their goals in their excitement.

10.7   An Elementary Method of Stability Analysis: The $s = i\omega$ Method

In many design environments, the development of the feedback control system is the last task that is performed. Early design efforts are usually devoted to the selection of the prime mover and the controlled object from among several alternative conceptual designs. A reasonable starting point for control system design is the determination of that feedback which will provide stable performance. The discussion of mathematical techniques of stability analysis begins in the present section.

Assume that all components in a feedback system are adequately described by linear ordinary differential equations with constant coefficients—hence, by transfer functions. It follows from the discussion in Section 10.2 that the basic stability criterion is that all closed loop poles lie in the left half of the s plane.

The present section is a discussion of an elementary method for checking the characteristic equation [Equation (10.4)] to determine if any roots are on the imaginary axis, i.e., to check if the feedback system is at the so-called *stability limit*. Notice that if $s = i\omega_u$ is a closed loop pole [i.e., $s = i\omega_u$ satisfies Equation (10.4)] for some ω_u, the system is at the stability limit. This fact can be used in the design of automatic control systems, as is illustrated by the next example.

Example 10.2

For the system in Figure 10.18,

$$G_c = K,$$

$$G_p = \frac{1}{(1 + s)^3},$$

and

$$H = 1.0.$$

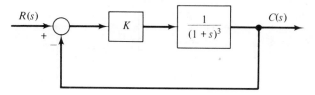

Figure 10.18. Feedback system.

Thus, from Equation (10.4), the characteristic equation is

$$1 + G_c G_p H = 1 + \frac{K}{(1 + s_p)^3} = 0$$

or

$$K + (1 + s_p)^3 = 0. \qquad (i)$$

To find that value of K (denoted K_u) for which $s_p = i\omega_u$ (stability limit condition), substitute $s_p = i\omega_u$ into Equation (i) to find

$$K_u + 1 + 3i\omega_u + 3(-\omega_u^2) + i(-\omega_u^3) = 0.$$

This equation between complex quantities transforms to

$$K + 1 - 3\omega_u^2 = 0 \quad \text{(real part)} \qquad (ii)$$

and

$$3\omega_u - \omega_u^3 = 0 \quad \text{(imaginary part)}, \qquad (iii)$$

since for a complex number to vanish, both its real and imaginary parts must be zero. It follows from Equation (iii) that

$$\omega_u = \sqrt{3},$$

where ω_u is called the *hunting* frequency at the stability limit. Substituting for ω_u in Equation (ii) leads to

$$K_u = 8.$$

The stability limit value of K is called the *ultimate sensitivity*.

* * *

Example 10.3

The $s = i\omega$ method may be used when open loop transfer functions are transcendental functions of s. This is the unique advantage of this method.

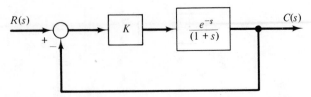

Figure 10.19. Feedback system described by a transfer function which is a transcendental function of s.

The characteristic equation for the system illustrated in Figure 10.19 is

$$1 + \frac{Ke^{-s}}{1 + s} = 0.$$

At the stability limit $s = i\omega_u$ and $K = K_u$ so that

$$1 + \frac{K_u e^{-i\omega_u}}{1 + i\omega_u} = 0$$

or

$$1 + i\omega_u + K_u \cos \omega_u - iK_u \sin \omega_u = 0. \tag{i}$$

Splitting this complex equation into its real and imaginary parts,

$$1 + K_u \cos \omega_u = 0$$

and

$$\omega_u - K_u \sin \omega_u = 0.$$

Solve these equations for K_u and find that

$$K_u = \frac{-1}{\cos \omega_u}. \tag{ii}$$

Thus,

$$-\omega_u = \tan \omega_u. \tag{iii}$$

The second equation can be solved graphically in the manner illustrated in Figure 10.20. In this figure, both sides of Equation (iii) are plotted vs. ω. The points of intersection are the values of ω for which Equation (iii) is satisfied, i.e., the values of ω_u.

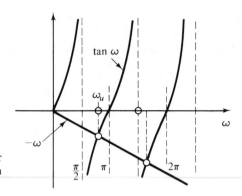

Figure 10.20. Graphical solution for hunting frequencies of the system illustrated in Figure 10.19.

Examination of this figure leads one to conclude that Equations (ii) and (iii) have an infinite number of solutions. In general one is interested in that solution which corresponds to the lowest value of K_u, although this statement is not always true. This uncertainty is one of the weaknesses of the method.

* * *

To summarize, then, the $s = i\omega$ method of stability analysis is

1. Determine the characteristic equation.
2. Substitute $s = i\omega_u$ and $K = K_u$ into the characteristic equation.
3. Separate the resulting equation into two equations for the real and imaginary parts.
4. Solve the two equations for the two unknowns ω_u and K_u.

These values of ω_u and K_u can be used with Table 7.1 to determine recommended P.I.D. controller settings.

10.8 An Elementary Method of Stability Analysis: The Routh Array Method

There is another procedure for determining whether or not the characteristic Equation (10.4) has any roots in the right half of the s plane. The method can be applied only to the case where G_c, G_p, and H are the ratio of polynomials so that the characteristic equation can be rewritten

$$b_n s^n + b_{n-1} s^{n-1} + \cdots + b_1 s + b_0 = 0. \tag{10.52}$$

The method, called the Routh *array*, is given without proof.

The Routh array method is far easier to apply than is the $s = i\omega$ method when the characteristic equation has the form indicated in Equation (10.52). However, if transcendental transfer functions are used to describe distributed parameter components, the Routh method is not applicable. Another advantage of the Routh method is that the uncertainty mentioned at the end of Example 10.3 does not exist during the application of the Routh array.

Rule 1: The polynomial Equation (10.52) has roots on the imaginary axis or in the right half of the s plane if the b_k do not all have the same sign or if any b_k is zero.   (10.53)

Example 10.4

The following polynomials have unstable roots:

$$s^5 + 3Ks^4 + 2s^3 + s^2 + Ks - 1 = 0,$$

$$s^5 + 3s^3 + 2Ks^2 + s + K = 0.$$

$$* \quad * \quad *$$

The inverse of Rule 1 is not true. One next forms the Routh array associated with the polynomial of Equation (10.52). The first two rows of the array are formed in the manner indicated below:

$$
\begin{array}{ccccc}
b_n & b_{n-2} & \cdots & b_2 & b_0 \\
b_{n-1} & b_{n-3} & \cdots & b_1 & 0
\end{array}
$$
                                                 (10.54)

if n is an even integer. If n is an odd integer, the last two columns of the first two rows will have the form

$$
\begin{array}{cc}
\cdots\, b_3 & b_1 \\
\cdots\, b_2 & b_0
\end{array}
$$
                                                 (10.55)

Let $A_{k,l}$ be the Routh array element in the kth row and lth column. The remainder of the Routh array is formed by using the formula

$$A_{k,l} = \frac{A_{k-2,\,l+1}A_{k-1,\,1} - A_{k-2,\,1}A_{k-1,\,l+1}}{A_{k-1,\,1}}$$

$$= A_{k-2,\,l+1} - \left[\frac{A_{k-2,\,1}}{A_{k-1,\,1}}\right]A_{k-1,\,l+1} \qquad \text{for } k = 3, 4, \ldots, n + 1.$$
                                                 (10.56)

The second rule is applied to the first column of the resulting array.

Rule 2: The number of sign changes in the first column of the Routh array of a polynomial is equal to the number of roots of that polynomial in the right half of the s plane.

Example 10.5

The polynomial

$$s^4 + 2s^3 + 3s^2 + 8s + 2 = 0$$

is associated with the Routh array, which from statements (10.54) and (10.56) is

$$\begin{array}{ccc}
1 & 3 & 2 \\
2 & 8 & \\
-1 & 2 & \\
12 & & \\
2 & &
\end{array}$$

The first column of the array has two sign changes; thus, by Rule 2, two right half-plane roots of the characteristic equation exist.

* * *

Example 10.6

Consider again the system illustrated in Figure 10.18 for which the characteristic equation is

$$1 + G_c G_p H = 1 + \frac{K}{(1+s)^3} = 0$$

or

$$s^3 + 3s^2 + 3s + 1 + K = 0.$$

The corresponding Routh array is determined from Equations (10.55) and (10.56) to be

$$\begin{array}{cc}
1 & 3 \\
3 & 1+K \\
\dfrac{8-K}{3} & \\
1+K &
\end{array}$$

For the feedback system to be stable, there must be no sign changes in the first column; thus, the gain, K, is restricted to

$$-1 \leq K \leq 8.$$

This inequality defines the range of stable operation for the automatic control system. Compare this result with that obtained in Example 10.2.

* * *

The computer program IROUTH, described and listed in Appendix C, outputs the first column of the Routh array when the coefficients of the characteristic equation are input to it.

There are other rules which simplify calculations and/or are used in special cases.

Rule 3: Any row of the Routh array can be divided by a positive constant without changing the signs of the first column.

The validity of this rule is indicated by an examination of Equation (10.56). Notice that if the $(k - 1)$st row is divided by α, the values of the kth row are unchanged. If the $(k - 2)$nd row is divided by α, the kth row is also divided by α. These two facts are sufficient to establish Rule 3.

Example 10.7

Reconstruct the Routh array of Example 10.5, but after each step, divide the kth row by a positive constant such that the first element in the kth row has a magnitude of 1. The result is

$$
\begin{array}{rrr}
1 & 3 & 2 \\
1 & 4 & \\
-1 & 2 & \\
1 & & \\
1 & &
\end{array}
$$

The calculations are simplified and the conclusions are the same.

* * *

If for any k, $A_{k,1} = 0$, one cannot determine the $(k + 2)$nd row [see Equation (10.56)], one can merely stop here if it is sufficient to know that the system is not stable. If it is desired to also know how many closed loop poles are on the imaginary axis or are in the right half-plane, one could continue the construction of the Routh array by using Rule 4.

Rule 4. If $A_{k,1} = 0$ for any k, substitute $A_{k,1} = \delta$ and complete the array and then apply Rule 2 by assuming that δ is very small.

Example 10.8

Construct the Routh array for

$$s^5 + 2s^4 + 2s^3 + 4s^2 + s + 1 = 0.$$

The result is

$$
\begin{array}{ccc}
1 & 2 & 1 \\
2 & 4 & 1 \\
\delta & 0.5 & \\
\dfrac{4\delta - 1}{\delta} & 1 & \\
0.5 & &
\end{array}
$$

δ was substituted for $A_{3,1}$, which would be zero otherwise. $A_{4,1}$ is a large negative number for small δ so that two sign changes occur in the first column of the Routh array; hence, two of the roots of the polynomial lie in the right half s plane.

$$* \quad * \quad *$$

10.9 An Elementary Design Technique Based on the Elementary Methods of Stability Analysis [3]

One can extend the usefulness of the methods discussed in the last two sections in order to determine if the characteristic equation of the closed loop system has roots which satisfy constraints on closed loop poles. These constraints could be the result of transient performance design specifications (see Figure 10.11).

These extended methods are based on transformations of the s variable in the characteristic equation. For instance, suppose that the analyst wishes to make a quick check of the characteristic equation to determine whether or not the time constants, associated with the closed loop poles, are all less than some specified time constant, T_0. Make the transformation in variables

$$s = s_0 - \frac{1}{T_0}, \tag{10.57}$$

so that the characteristic equation becomes

$$1 + G_c\left(s_0 - \frac{1}{T_0}\right)G_p\left(s_0 - \frac{1}{T_0}\right)H\left(s_0 - \frac{1}{T_0}\right) = 0. \tag{10.58}$$

A graphical interpretation of Equation (10.57) is indicated in Figure 10.21. If the $s_0 = i\omega$ method or the Routh array method is applied to Equation (10.58), the analyst will determine whether or not all of the closed loop poles are in the *left half of the s_0 plane*. It follows from Equation (10.57) that he will

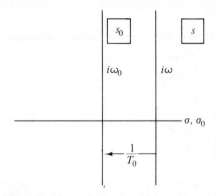

Figure 10.21. Interpretation of the transformation of Equation (10.57) as a shift in the imaginary axis.

then have determined whether or not all closed loop poles lie to the left of a line through $-1/T_0$ in the s plane, i.e., whether or not all closed loop poles are associated with a time constant less than T_0.

Example 10.9

For the system illustrated in Figure 10.18, determine that value of gain, K, so that all closed loop poles have a time constant less than 2.0 sec. The original characteristic equation is

$$s^3 + \tfrac{3}{2}s^2 + 3s + 1 + K = 0.$$

Make the transformation of Equation (10.57) to find

$$s_0^3 + \tfrac{3}{2}s_0^2 + \tfrac{3}{4}s_0 + \tfrac{1}{8} + K = 0.$$

The Routh array for this last equation is

$$
\begin{array}{cc}
1 & \tfrac{3}{4} \\[2mm]
\tfrac{3}{2} & \tfrac{1}{8} + K \\[2mm]
\dfrac{2(1-K)}{3} & \\[2mm]
\tfrac{1}{8} + K &
\end{array}
$$

No changes occur in the signs of the first column if

$$-\tfrac{1}{8} \le K \le 1.0.$$

Hence, for this range of K, all of the closed loop poles are associated with time constants less than 2.0 sec.

* * *

Example 10.10

Develop a method which can be used to determine the maximum time constant, for a given value of K, associated with any closed loop pole of the system illustrated in Figure 10.19. The original characteristic equation is

$$1 + s + Ke^{-s} = 0.$$

Make the transformation of Equation (10.57) to find

$$1 - \frac{1}{T_0} + s_0 + Ke^{1/T_0}e^{-s_0} = 0.$$

For a given value of K, a closed loop pole lies on the imaginary axis of the s_0 plane (hence, on a vertical line through $-1/T_0$ in the s plane) if

$$1 - \frac{1}{T_0} + i\omega_0 + Ke^{1/T_0}\{\cos \omega_0 - i \sin \omega_0\} = 0.$$

Separate this equation into real and imaginary parts to find

$$1 - \frac{1}{T_0} + Ke^{1/T_0} \cos \omega_0 = 0$$

and

$$\omega_0 - Ke^{1/T_0} \sin \omega_0 = 0.$$

These equations can be combined to show that

$$T_0 = \frac{1}{\ln [\omega_0/K \sin \omega_0]} \tag{10.59}$$

and that

$$1 + \ln\left[\frac{K \sin \omega_0}{\omega_0}\right] + \frac{\omega_0}{\tan \omega_0} = 0. \tag{10.60}$$

Equation (10.60) can be solved, for a given value of K, by Newton iteration (see Appendix B) or by a modification of the graphical technique demonstrated in Example 10.3. T_0 is then found using Equation (10.59).

It follows from Equation (10.18) that the damping ratio associated with the pole with maximum time constant is

$$\xi = \frac{1/T_0}{[(1/T_0)^2 + \omega_0^2]^{1/2}} = \frac{1}{[1 + T_0^2\omega_0^2]^{1/2}}.$$

* * *

10.10   Design Criteria for State Variable
Feedback Systems: Controllability
and Observability*

The feedback system illustrated in Figure 10.1 is called a *scalar* feedback system because only a single input and a single output are associated with the system. A more general type of feedback is illustrated in Figure 10.22. This diagram illustrates the relationship between the time domain variables:

$\mathbf{x}(t) = $ vector of state variables,

$\mathbf{c}(t) = $ vector of output variables,

$\mathbf{u}(t) = $ vector of manipulated variables,   (10.61)

$\mathbf{r}(t) = $ vector of desired output variables (set points),

$\mathbf{e}(t) = \mathbf{r}(t) - \mathbf{c}(t)$.

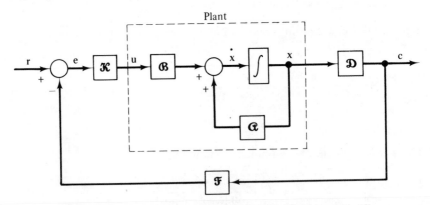

Figure 10.22. Block diagram of multivariable feedback control. All signals are time domain vector quantities.

The *plant* (prime mover and controlled object) is described by the block diagram symbols enclosed in the box drawn with dashed lines. Notice that

$$\dot{\mathbf{x}} = \boldsymbol{\mathcal{C}}\mathbf{x} + \boldsymbol{\mathcal{B}}\mathbf{u} \qquad (10.62)$$

and that

$$\mathbf{c} = \boldsymbol{\mathcal{D}}\mathbf{x}. \qquad (10.63)$$

In other words, it is assumed that the plant has been analyzed using the state variable formalism described in Section 4.8 of Chapter 4 and in Chapter 8.

---

* The discussion in this section will be referred to only in special sections of succeeding chapters. This section may be omitted during the first reading of the text.

Notice that in multivariable feedback control, an entire vector of outputs, $\mathbf{c}(t)$, is fed back and compared with set points in order to generate an error vector. *Compensator matrices*, $\mathfrak{F}$ and $\mathfrak{K}$, are added to the system in order to meet design objectives. Thus, in multivariable feedback, $\mathfrak{K}$ and $\mathfrak{F}$ play the same role as $G_c(s)$ in scalar feedback. The $\mathfrak{D}$ matrix is analogous to the transfer function, $H(s)$, in scalar feedback control.

In general, the vectors $\mathbf{r}$, $\mathbf{e}$, $\mathbf{u}$, $\mathbf{x}$, and $\mathbf{c}$ will all be of different dimension. Thus, only the $\mathfrak{A}$ matrix will, in general, be square.

Notice that

$$\mathbf{u} = \mathfrak{K}(\mathbf{r} - \mathfrak{F}\mathbf{c}) = \mathfrak{K}(\mathbf{r} - \mathfrak{F}\mathfrak{D}\mathbf{x}). \tag{10.64}$$

Combine Equations (10.62)–(10.64) to show that after the feedback loop is synthesized, the state variable equations become

$$\dot{\mathbf{x}} = \{\mathfrak{A} - \mathfrak{B}\mathfrak{K}\mathfrak{F}\mathfrak{D}\}\mathbf{x} + \mathfrak{B}\mathfrak{K}\mathbf{r},$$
$$\mathbf{c} = \mathfrak{D}\mathbf{x}. \tag{10.65}$$

These equations have the same general form as the open loop Equations (10.62) and (10.63).

As demonstrated in Section 2.10, the transient performance (speed of response and damping) of a system described by Equations (10.65) is quantified by the eigenvalues of the matrix

$$\mathfrak{A}_c = \mathfrak{A} - \mathfrak{B}\mathfrak{K}\mathfrak{F}\mathfrak{D}. \tag{10.66}$$

Thus, the discussion in Section 10.2 may be applied to these closed loop eigenvalues in order to state transient performance criteria.

At steady state, $\dot{\mathbf{x}} = \mathbf{0}$; thus, from the first of Equations (10.65)

$$\{\mathfrak{A} - \mathfrak{B}\mathfrak{K}\mathfrak{F}\mathfrak{D}\}\mathbf{x}(\infty) = -\mathfrak{B}\mathfrak{K}\mathbf{r}(\infty). \tag{10.67}$$

The steady state error

$$\mathbf{e}(\infty) = \mathbf{r}(\infty) - \mathfrak{F}\mathfrak{D}\mathbf{x}(\infty). \tag{10.68}$$

Combine these equations to show that

$$\mathbf{e}(\infty) = [\mathfrak{I} + \mathfrak{F}\mathfrak{D}\{\mathfrak{A} - \mathfrak{B}\mathfrak{K}\mathfrak{F}\mathfrak{D}\}^{-1}\mathfrak{B}\mathfrak{K}]\mathbf{r}(\infty) \tag{10.69}$$

if the inverse exists.

The reader should convince himself, by examining Figure 10.22, that $\mathbf{e}(\infty)$ cannot be $\mathbf{0}$ as long as $\mathfrak{A}\mathbf{x}$ is not $\mathbf{0}$ and $\mathfrak{K}$ and $\mathfrak{B}$ are constant matrices. If steady state error must be eliminated, then integrators must be added to appropriate diagonal elements of the $\mathfrak{K}$ matrix.

Currently, the most popular design criteria for the transient performance of a multivariable feedback system are not the locations of the closed loop eigenvalues [i.e., the eigenvalues of the matrix $\mathfrak{A}_c$ defined by Equation (10.66)]. Rather, analysts prefer to work with generalizations of the scalar criterion defined by Equation (10.31). Define the *deviation vectors*

$$\boldsymbol{\delta}(t) = \mathbf{u}(t) - \mathbf{u}(\infty), \tag{10.70}$$

$$\mathbf{d}(t) = \mathbf{c}(t) - \mathbf{c}(\infty), \tag{10.71}$$

where, from Equations (10.64), (10.65), and (10.67), it can be shown that

$$\boldsymbol{\delta}(\infty) = \mathfrak{K}\mathbf{e}(\infty), \tag{10.72}$$

$$\mathbf{c}(\infty) = -\mathfrak{D}\{\mathfrak{A} - \mathfrak{B}\mathfrak{K}\mathfrak{F}\mathfrak{D}\}^{-1}\mathfrak{B}\mathfrak{K}\mathbf{r}(\infty). \tag{10.73}$$

One example of a generalization of Equation (10.31) is the *quadratic criterion*

$$I_Q = \int_0^\infty \{\mathbf{d}'(t)\mathcal{P}\mathbf{d}(t) + \boldsymbol{\delta}'(t)\mathbf{W}\boldsymbol{\delta}(t)\}\,dt, \tag{10.74}$$

where $\mathcal{P}$ and $\mathbf{W}$ are called *weighting matrices*; they will be discussed below. The second term in the integrand of Equation (10.74) has no analog in Equation (10.31) and will also be discussed below.

Notice that the matrix of the form

$$Q(\mathbf{v}) = \mathbf{v}'(t)\mathfrak{M}\mathbf{v}(t) \tag{10.75}$$

is a scalar. This scalar is called a *quadratic form* and $\mathfrak{M}$ is called the *matrix of a quadratic form*. It is interesting to note that $\mathfrak{M}$ can always be made symmetric.

Example 10.11

Consider the quadratic form

$$Q(\mathbf{v}) = v_1^2 - v_1 v_2 + v_2^2$$

(the presence of only quadratic terms is the reason for the name quadratic form). In matrix notation

$$Q(\mathbf{v}) = (v_1 v_2)\begin{bmatrix} 1 & -1 \\ 0 & 1 \end{bmatrix}\begin{bmatrix} v_1 \\ v_2 \end{bmatrix},$$

and the matrix is not symmetric. However, notice that the form could be rewritten

$$Q(\mathbf{v}) = v_1^2 - \frac{v_1 v_2}{2} - \frac{v_2 v_1}{2} + v_2^2$$

in which case, the matrix representation

$$Q(\mathbf{v}) = (v_1 v_2)\begin{bmatrix} 1 & -\frac{1}{2} \\ -\frac{1}{2} & 1 \end{bmatrix}\begin{bmatrix} v_1 \\ v_2 \end{bmatrix}$$

and now the matrix is symmetric. It should be clear that by rewriting a form, in any number of v variables, one may use a symmetric matrix in the matrix representation of a quadratic form.

$$*\quad *\quad *$$

As in the case of integral-squared error, only the relative values of the quadratic criterion, I_Q, are significant. In other words, the designer can select the best or *optimal* feedback system from a set of conceptual designs by selecting that system which minimizes I_Q. In the discussion of the integral-absolute-error criterion, it was pointed out that this criterion quantified both speed of response and damping because its associated integrand was always positive. It seems clear that the same restriction should be imposed on the integrand of Equation (10.74). As will be shown, one need only choose $\mathcal{P}$ and $\mathcal{W}$ with some care in order to satisfy this restriction.

The integrand of Equation (10.74) is the sum of two quadratic forms. In order that the first term quantify both speed and damping, one imposes the condition

$$\mathbf{d}'(t)\mathcal{P}\mathbf{d}(t) > 0 \tag{10.76}$$

for any $\mathbf{d}$. The second term in the integrand quantifies the deviations of the manipulated variables from equilibrium values and is included in order to decrease the likelihood of saturations in the optimal system (e.g., saturation in the position of a servo spool valve). Since both negative and positive deviations could cause saturation, one imposes the condition

$$\boldsymbol{\delta}'(t)\mathcal{W}\boldsymbol{\delta}(t) > 0 \tag{10.77}$$

for all possible $\boldsymbol{\delta}$; i.e., negative and positive deviations are to be penalized equally. It is not clear, as yet, just how one could impose conditions (10.76) and (10.77).

If the quadratic form

$$Q(\mathbf{v}) = \mathbf{v}'\mathcal{M}\mathbf{v} > 0 \tag{10.78}$$

for all $\mathbf{v}$, then Q is said to be a *positive definite quadratic form* and $\mathfrak{M}$ is said to be *positive definite*. It is clear from inequalities (10.76) and (10.77) that the designer of feedback systems should impose the condition that the weighting matrices $\mathcal{P}$ and $\mathcal{W}$ should be positive definite.

An arbitrary symmetric matrix can be tested in several ways in order to determine whether or not it is positive definite. One of these tests is now presented. Make the transformation in variables

$$\mathbf{v} = \mathfrak{T}\mathbf{y}, \tag{10.79}$$

where $\mathbf{y}$ and $\mathbf{v}$ are of the same dimension. Then

$$Q(\mathbf{v}) = Q_0(\mathbf{y}) = \mathbf{y}'\mathfrak{T}'\mathfrak{M}\mathfrak{T}\mathbf{y}. \tag{10.80}$$

It was shown in section 9.8 that if $\mathfrak{M}$ is symmetric, the $\mathfrak{T}$ can be chosen so that

$$\mathfrak{T}'\mathfrak{M}\mathfrak{T} = \Lambda, \tag{10.81}$$

where Λ is a diagonal matrix with the eigenvalues of $\mathfrak{M}$ as diagonal elements. Moreover, the elements of $\mathfrak{T}$ and the eigenvalues of $\mathfrak{M}$ are real numbers if $\mathfrak{M}$ is symmetric. Make such a choice for $\mathfrak{T}$ in Equation (10.79) so that

$$Q(\mathbf{v}) = Q_0(\mathbf{y}) = \mathbf{y}'\Lambda\mathbf{y}$$
$$= \lambda_1 y_1^2 + \lambda_2 y_2^2 + \cdots + \lambda_n y_n^2. \tag{10.82}$$

Thus,

$$Q(\mathbf{v}) > 0$$

(i.e., $\mathfrak{M}$ is positive definite) if

$$\lambda_k > 0 \quad \textit{for all } k, \tag{10.83}$$

where the λ_k are the eigenvalues of $\mathfrak{M}$.

Example 10.12

Test the matrix of the quadratic form of Example 10.11 to determine whether or not it is positive definite:

$$\mathfrak{M} = \begin{bmatrix} 1 & -\frac{1}{2} \\ -\frac{1}{2} & 1 \end{bmatrix}.$$

Thus, the eigenvalues satisfy

$$\det \begin{bmatrix} 1 - \lambda & -\frac{1}{2} \\ -\frac{1}{2} & 1 - \lambda \end{bmatrix} = (1 - \lambda)^2 - \frac{1}{4} = 0$$

or

$$\lambda = 1 \pm \tfrac{1}{2}.$$

Thus

$$\lambda_1 = \tfrac{1}{2}, \qquad \lambda_2 = \tfrac{3}{2}.$$

Since both eigenvalues are positive, $\mathfrak{M}$ is positive definite.

* * *

Notice that some of the elements of $\mathfrak{M}$ in the example are negative. *The eigenvalues of a matrix and not the elements of the matrix* are indications of whether or not a matrix is positive definite.

Finally, it is concluded that only weighting matrices with all positive eigenvalues should be used in the quadratic criterion [Equation (10.74)]. This is the only restriction that is placed on these matrices. Clearly, one will define a different optimally controlled feedback system for different choices of the $\mathscr{P}$ and $\mathscr{W}$ matrices. This degree of freedom can be exploited by the designer by performing trade-off studies between speed of response and saturation limits on manipulated variables.

An important modification of the criterion of Equation (10.74) is

$$I_{EQ} = \mathbf{e}'(\infty)\mathscr{L}\mathbf{e}(\infty) + \int_0^\infty \{\mathbf{d}'(t)\mathscr{P}\mathbf{d}(t) + \boldsymbol{\delta}'(t)\mathscr{W}\boldsymbol{\delta}(t)\}\, dt, \qquad (10.84)$$

where $\mathscr{L}$ is a weighting matrix. Obviously, this criterion quantifies speed of response, damping, the likelihood of saturation, *and* steady state error. Since negative error is usually as undesirable as positive error, it seems reasonable to require that $\mathscr{L}$ be positive definite, i.e., to require that $\mathscr{L}$ have positive eigenvalues.

Finally, the role of the concepts of controllability and observability in feedback systems design are discussed. These concepts were introduced in Section 8.7. Obviously, a designer should not develop a system with an unobservable or uncontrollable pole in the right half of the s plane. However, it is not quite so obvious how one should answer the question, Is the existence of an unobservable or uncontrollable pole, in the left half of the s plane, to be tolerated? This question will be explored by considering a particular example.

For the scalar feedback system illustrated in Figure 10.23, the characteristic equation is

$$1 + \frac{2}{(s_p + 5)(s_p + 2)} = 0$$

or

$$s_p^2 + 7s_p + 12 = (s_p + 4)(s_p + 3) = 0, \qquad (10.85)$$

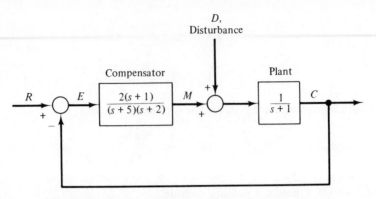

Figure 10.23. Feedback system for which a compensator is synthesized so as to cancel a plant pole.

so that closed loop time constants are $\frac{1}{4}$ and $\frac{1}{3}$ sec. Notice that the plant has an open loop time constant of 1.0 sec.

The pole zero cancellation arouses one's suspicion. The controllability and observability of the system will now be investigated by first developing a state variable representation of the feedback system. Notice that

$$sC = -C + M \qquad (10.86a)$$

and that

$$M = \left[\frac{2(s+1)}{(s+5)(s+2)} \right] E = \left[\frac{\frac{8}{3}}{s+5} - \frac{\frac{2}{3}}{s+2} \right] E. \qquad (10.86b)$$

Define the state variables

$$X_1 = \frac{\frac{8}{3}}{s+5} E = \frac{\frac{8}{3}}{s+5}(R - C),$$

$$X_2 = \frac{\frac{2}{3}}{s+2} E = \frac{\frac{2}{3}}{s+2}(R - C), \qquad (10.87)$$

$$X_3 = C.$$

Notice that

$$M = X_1 - X_3. \qquad (10.88)$$

In the time domain, these equations may be rewritten

$$\dot{\mathbf{x}} = \begin{bmatrix} -5 & 0 & -\frac{8}{3} \\ 0 & -2 & -\frac{2}{3} \\ 1 & -1 & -1 \end{bmatrix} \mathbf{x} + \begin{bmatrix} \frac{8}{3} \\ \frac{2}{3} \\ 0 \end{bmatrix} r \qquad (10.89)$$

$$c(t) = (0 \quad 0 \quad 1)\mathbf{x}. \qquad (10.90a)$$

The characteristic equation of the closed loop $\mathcal{Q}$ matrix is

$$\lambda^3 + 8\lambda^2 + 9\lambda + 12 = (\lambda + 4)(\lambda + 3)(\lambda + 1) = 0. \quad (10.90b)$$

The third root, $\lambda_3 = -1$, appears in this equation but not in Equation (10.85) because of the pole zero cancellation. Use the techniques of Chapter 8 to show that

$$\mathfrak{I} = \begin{bmatrix} \frac{16}{3} & \frac{4}{3} & -\frac{2}{3} \\ -\frac{2}{3} & -\frac{2}{3} & -\frac{2}{3} \\ -2 & -1 & 1 \end{bmatrix}. \quad (10.91)$$

It follows from Equations (10.90a) and (10.92) that the observability vector

$$\mathbf{k}'\mathfrak{I} = (-2 \quad -1 \quad 1). \quad (10.92)$$

Thus, the third mode (associated with a 1-sec time constant) *is observable* in this closed loop response.

Next show that

$$\mathfrak{I}^{-1} = \begin{bmatrix} \frac{1}{3} & \frac{1}{6} & \frac{1}{3} \\ -\frac{1}{2} & -1 & -1 \\ \frac{1}{6} & -\frac{2}{3} & \frac{2}{3} \end{bmatrix}. \quad (10.93)$$

It follows from Equations (10.89) and (10.94) that, with respect to $r(t)$, the controllability vector

$$\mathfrak{I}^{-1}\mathbf{b} = \begin{bmatrix} 1 \\ -2 \\ 0 \end{bmatrix}, \quad (10.94)$$

so that the third mode will not be excited by a set point signal, $r(t)$.

One may now ask whether it is important to worry about the slow third mode if it cannot be excited by an input to the system. The answer is yes; one must be concerned with this mode because it can be excited by disturbance signals entering at other points in the feedback loop. To understand this point, consider the disturbance signal illustrated in Figure 10.23. Equation (10.86a) can be modified to read

$$sC = -C + M + D.$$

Neglect R, and combine Equations (10.87) and (10.88) to find

$$\dot{\mathbf{x}} = \begin{bmatrix} -5 & 0 & -\frac{8}{3} \\ 0 & -2 & -\frac{2}{3} \\ 1 & -1 & -1 \end{bmatrix} \mathbf{x} + \begin{bmatrix} 0 \\ 0 \\ 1 \end{bmatrix} d. \quad (10.95)$$

Since the $\mathfrak{A}$ matrix for this equation is the same as that in Equation (10.89), the $\mathfrak{T}$ and $\mathfrak{T}^{-1}$ matrices are unchanged. Thus, with respect to the disturbance input, the controllability vector is

$$\mathfrak{T}'\mathbf{b}_d = \begin{bmatrix} \tfrac{1}{6} \\ -\tfrac{2}{3} \\ \tfrac{2}{3} \end{bmatrix}. \tag{10.96}$$

Thus, *all three modes are controllable by (i.e., can be excited by) the disturbance signal.*

The final conclusion is that the pole zero cancellation indicated in Figure 10.23 has resulted in the existence of an uncontrollable and slow mode which can be excited by a disturbance signal. The designer must now decide whether the speed of response is so seriously affected that design objectives will not be met. It is not possible to present generally applicable rules about the desirability or undesirability of uncontrollable modes. Each design situation must be judged independently.

References

1. ATHANS, M., and FALB, P. L., *Optimal Control*, McGraw-Hill, New York, 1966.

2. BREWER, J. W., *The Application of Numerical Methods to System Design*, Ph.D. Dissertation, Dept. of Mechanical Engineering, Univ. of California, Berkeley, 1966.

3. D'AZZO, J. J., and HOUPIS, C. H., *Feedback Control System Analysis and Synthesis*, McGraw-Hill, New York, 1966.

4. HOROWITZ, I. M., *Synthesis of Feedback Systems*, Academic Press, New York, 1963.

5. "Special Issue on Linear-Quadratic-Gaussian Problem," *I.E.E.E. Transactions on Automatic Control*, *AC*-16, No. 6, Dec. 1971.

6. NEWTON, G.C., GOULD, L.A., and KAISER, J.F., *Analytical Design of Feedback Controls*, Wiley, New York, 1957.

7. OGATA, K., *State Space Analysis of Control Systems*, Prentice-Hall, Englewood Cliffs, N.J., 1967.

8. SCHULTZ, D. G., and MELSA, J. L., *State Functions and Linear Control Systems*, McGraw-Hill, New York, 1967.

9. PERKINS, W. R., and CRUZ, J. B., "The Parameter Variation Problem in State Feedback Control Systems," in *Joint Automatic Control Conference Proceedings*, Stanford, California, June 1964.

Problems

10.1. For the feedback system illustrated in Figure 10.1, define the open loop transfer function

$$G_0(s) = G_c(s)G_p(s)H(s)$$

and define the closed loop transfer functions

$$G_1(s) = \frac{E(s)}{R(s)}, \qquad G_2(s) = \frac{M(s)}{R(s)}, \qquad G_3(s) = \frac{C(s)}{R(s)}$$

and show that
(a) The *zeros* of G_1 are the *poles* of G_0.
(b) The *zeros* of G_2 are the *poles* of G_p, the *poles* of H, and the *zeros* of G_c.
(c) The *zeros* of G_3 are the *poles* of H, the *zeros* of G_c, and the *zeros* of G_p.
How are the gains of G_1, G_2, and G_3 related to the gains of G_c, G_p, and H?

10.2. Construct a block diagram for the thickness control system of Problem 2.10. What is the characteristic equation for this feedback system?

10.3. It is desired to determine the value of the capacitance, C, in the network illustrated in Figure 10.24. The resistor has a resistance of 10^6 ohms. A 10-v supply is connected across the input and $e_f(t)$ is recorded. What is the value of C?

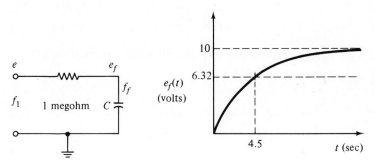

Figure 10.24. Step response for an R-C filter.

10.4. Let d_c denote the damping per cycle [Equation (10.27)]. Plot d_c and the sensitivity coefficient

$$\theta \frac{\partial d_c}{\partial \theta}$$

against θ (θ is defined in Figure 10.7).

10.5. Plot d_c and the sensitivity coefficient

$$\xi \frac{\partial d_c}{\partial \xi}$$

against ξ [ξ is the damping ratio defined in Equation (10.15)].

10.6. For the thickness control system of Problem 10.2, assume that $\tau = 1.0$ sec, and, using the $s = i\omega$ method, find ω_u and the ultimate sensitivity. What are the Ziegler-Nichols settings?

10.7. Apply Routh's criterion to the following *open* loop transfer functions to determine if the corresponding closed loop systems are stable or unstable:

(a) $G(s) = \dfrac{18s^2 + 3s + 10}{s^5 + 3s^4 + 7s^3 + 2s^2 + 3s + 5}$

(b) $G(s) = \dfrac{s^2 + 5s + 2}{s(s^3 + 2s^2 + 2s + 3)}$

(c) $G(s) = \dfrac{s^2 + s - 1}{15s^5 + 3s^4 + 2s^3 + s^2 + 2}$

where $G = G_cG_pH$.

10.8. Use Routh's criterion to determine the stability limit ranges of K, when the *loop is closed*, for the following *open loop* transfer functions:

(a) $\dfrac{K(1 - s)}{s + 5}$.   (b) $\dfrac{1}{(s + K)^3}$.   (c) $\dfrac{K(s + 2)^2}{s(s^2 + 1)}$.

10.9. This problem illustrates an example of a significant type of *open loop control*. The transfer function of a dynamic system is given by

$$\frac{C(s)}{R(s)} = \frac{1}{s^2 + 1}.$$

Find $c(t)$ if $r(t)$ is as shown in Figure 10.25 (refer to Problem 2.6).

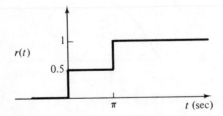

Figure 10.25. Posi-cast input.

Design of Scalar Automatic Control Systems by the Root Locus Method

11

The classical *root locus* method for the design of feedback systems is described in this chapter[1, 2, 3, 4]. This method is very useful for designing systems to meet transient response criteria which are stated in terms of closed loop pole locations (refer to Chapter 10). Graphical methods are emphasized, although a digital computer is easily used to achieve the same results. For instance, the program EIGRAD described in Appendix C can be used to determine root loci.

The method consists of the following steps:

1. Construct the locus of closed loop poles in the s plane by solving the characteristic equation for several values of some parameter. The parameter for this locus is some coefficient in the transfer function of the compensator whose value may be chosen by the designer.

2. Calibrate the locus in terms of the parameter.

3. Determine the *optimum* location of closed loop poles.

4. Determine the optimum value of the parameter (follows immediately from steps 2 and 3).

5. Finally, determine the sensitivity of the closed loop pole location to variation of some plant parameter.

Only the case of single input-single output (i.e., scalar feedback) systems are discussed in this chapter. The intuitive skill one will develop from a study of this simple case will prove invaluable to an analyst. This statement is true of even those analysts who are interested in multivariable feedback.

11.1   Angle and Magnitude Criteria for the Closed Loop Root Locus

Consider the scalar variable automatic control system shown in Figure 11.1. In this diagram, the gain of the compensator, K, is illustrated in an individual block. Thus, the gain of $G_c(s)$ is 1.0. A basic problem of design is to choose K so that the system satisfies some design criteria. In this chapter criteria associated with transient response will be considered.

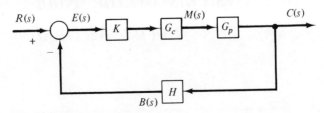

Figure 11.1. Control system. K is the gain of the compensator.

As was shown in Chapter 10,

$$\frac{C}{R} = \frac{KG_cG_p}{1 + KG_cG_pH}, \tag{11.1}$$

$$\frac{M}{R} = \frac{KG_c}{1 + KG_cG_pH}, \tag{11.2}$$

$$\frac{E}{R} = \frac{1}{1 + KG_cG_pH}, \tag{11.3}$$

so that the characteristic equation is

$$1 + KG_0 = 0, \tag{11.4}$$

where

$$G_0 \triangleq G_cG_pH. \tag{11.5}$$

$G_0(s)$ is called the *open loop transfer function*. $G_c(s)$ is defined so as to have a gain of 1.0.

It follows from Equation (11.4) that

$$\boxed{G_0(s_p) = -\frac{1}{K},} \tag{11.6}$$

where s_p is any closed loop pole. This equation between complex variables may be split into two equations, one for the magnitude and one for the angle:

$$\text{Angle criterion:} \quad \angle\, G_0(s_p) = 180°, \quad K > 0, \tag{11.7}$$

$$\text{Magnitude criterion:} \quad |G_0(s_p)| = \frac{1}{K}. \tag{11.8}$$

Equation (11.7) is a sufficient condition to find the locus of closed loop roots, in the s plane, of the characteristic equation, i.e., to find the locus of closed loop poles. This fact is illustrated by the following examples.

Example 11.1

Consider the velocity servo illustrated in Figure 11.2. The block diagram of this system is provided in Figure 11.3. Let

$$k' = \frac{K_c}{J}$$

and

$$\alpha = \frac{B}{J}.$$

Then

$$1 + \frac{k'}{s + \alpha} = 0. \tag{i}$$

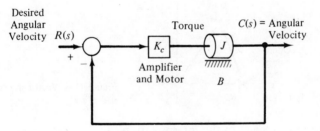

Figure 11.2. Velocity servo.

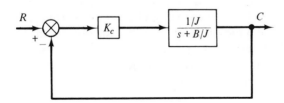

Figure 11.3. Block diagram of the velocity servo.

When the characteristic equation is in this form, k' is called the *sensitivity*. Thus, the characteristic equation may be written

$$\frac{1}{s_p + \alpha} = -\frac{1}{k'}.$$

If s_p is a closed loop pole, it follows from Equations (11.7) and (11.8) that the angle

$$\angle \frac{1}{s_p + \alpha} = 180° \tag{ii}$$

and that the magnitude

$$\frac{1}{k'} = \left| \frac{1}{s_p + \alpha} \right| = \frac{1}{r}. \tag{iii}$$

Recall that $(s + \alpha)^{-1}$ can be written in the polar form (see Appendix B):

$$\frac{1}{s + \alpha} = \frac{1}{re^{i\theta}} = \frac{1}{r} e^{-i\theta}. \tag{iv}$$

r and θ are defined in Figure 11.4.

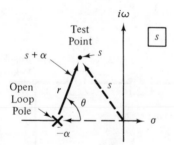

Figure 11.4. Vector interpretation of $s + \alpha$.

Equation (iv) leads to the interpretation of $|(s + \alpha)^{-1}|$ and $\angle (s + \alpha)^{-1}$ illustrated in Figures 11.5 and 11.6.

From Figure 11.6 and Equation (ii) it follows that the root locus is as shown in Figure 11.7. Notice that the locus of *closed loop* poles radiates in a radial fashion from the *open loop* pole. The locus in Figure 11.7 is calibrated, in terms of k', by using a graphical interpretation of Equation (iii).

Notice that as k' is increased, the closed loop time constant becomes smaller and smaller. Since high k' also means small steady state error, one might conclude that the optimum or best choice is $k' = \infty$. This point will be discussed in a later section.

* * *

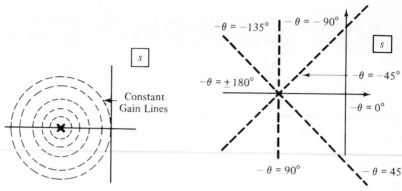

Figure 11.5. Lines of constant magnitude, $|1/(s + \alpha)|$.

Figure 11.6. Lines of constant angle, $< 1/(s + \alpha)$.

Figure 11.7. Locus of closed loop poles ($k' > 0$) for the velocity servo. The convention is to draw an arrow in the direction of increasing k'.

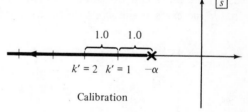

Example 11.2

The *open loop* transfer function is

$$G_0 = \frac{1}{(s + \alpha)^2}. \tag{i}$$

What is the locus of closed loop poles?

This question is again answered by interpreting Equation (11.7) using the polar form of G_0, i.e., by giving a graphical interpretation to

$$\frac{1}{(s + \alpha)^2} = \frac{1}{re^{i\theta}} \cdot \frac{1}{re^{i\theta}} = \frac{1}{r^2}e^{-2i\theta}, \tag{ii}$$

where r and θ are defined in Figure 11.8. It follows from Equation (ii) that the angle

$$-2\theta = \angle \frac{1}{(s_p + \alpha)^2} = 180° \tag{iii}$$

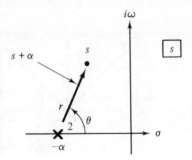

Figure 11.8. Double pole.

and that the magnitude

$$\frac{1}{k'} = \left| \frac{1}{(s_p + \alpha)^2} \right| = \frac{1}{r^2}. \tag{iv}$$

It follows from Equation (iii) that the lines of constant angle are as illustrated in Figure 11.9. Thus, the *root locus* is as shown in Figure 11.10. Again the locus can be calibrated using the magnitude criteria (iv). Notice that k' increases as the *square* of the distance from the open loop poles.

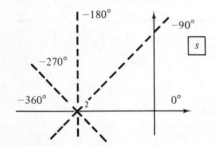

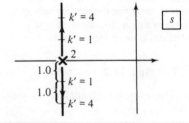

Figure 11.9. Lines of constant -2θ.

Figure 11.10. Root locus for $G_0 = 1/(s + \alpha)^2$.

For this example, the speed of response of the closed loop system does not change with k'. However, notice that the damping ratio of the closed loop poles does diminish with increasing k'. Thus, increasing k' degrades the damping in the closed loop system. A trade-off between damping and steady state error is in order.

* * *

Example 11.3

Consider the case where the open loop transfer function

$$G_0 = \frac{1}{(s + \alpha_1)(s + \alpha_2)} = \frac{1}{r_1 e^{i\theta_1} r_2 e^{i\theta_2}} \tag{i}$$

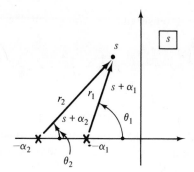

Figure 11.11. Case of two distinct open loop poles.

for which r_1, r_2, θ_1, and θ_2 are defined in Figure 11.11. In terms of these quantities the angle criterion is

$$\angle G_0 = -(\theta_1 + \theta_2) = 180° \tag{ii}$$

and the magnitude criterion is

$$\frac{1}{k'} = |G_0| = \frac{1}{r_1 r_2}. \tag{iii}$$

Equation (ii) indicates that those points in the s plane for which $\angle G_0 = 180°$ (i.e., the root locus) can be found by superimposing the angular contributions (θ_1 and θ_2) from the two open loop poles. The root locus so constructed is indicated in Figure 11.12. It follows from Equation (iii) that

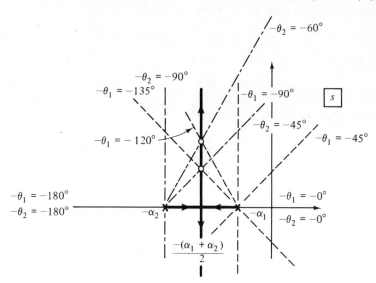

Figure 11.12. Root locus for Example 11.3. The arrows indicate the direction of increasing k'.

the value of sensitivity, k', for which the closed loop poles are *critically damped* is

$$\frac{1}{k'_c} = \frac{1}{r_{1c}r_{2c}} = \frac{4}{(\alpha_2 - \alpha_1)^2},\qquad\text{(iv)}$$

where r_{1c} is the distance from α_1 to the point where the root locus leaves the real axis and $r_{2c} = r_{1c}$ is the distance from α_2 to the same point.

If $k' > k'_c$, no decrease in response time is obtained since the real part of the closed loop pole, σ_p, remains constant; thus it would seem that $k' = k_c$ is an optimum choice of the value of k. This might not be the case if steady state error is an important consideration.

Consider Figure 11.13 and use Equation (11.3) to show that

$$\frac{E(s)}{R(s)} = \frac{1}{1 + \{k'/[(s + \alpha_1)(s + \alpha_2)]\}}.\qquad\text{(v)}$$

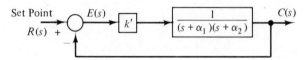

Figure 11.13. Block diagram of the control system of Example 11.3.

It follows from the final value theorem that

$$e(\infty) = \lim_{s\to 0} sE(s).$$

Thus, if

$$R(s) = \frac{1}{s},$$

it follows that the steady state error

$$e(\infty) = \frac{\alpha_1\alpha_2}{k' + \alpha_1\alpha_2} \longrightarrow 0,\quad\text{as } k' \to \infty.$$

In general, steady state error and transient response are competing criteria in design.

$$*\quad*\quad*$$

Example 11.4

Consider the case when

$$G_0(s) = \frac{1}{(1 + s)^3} = \frac{1}{r^3}e^{-i3\theta}.$$

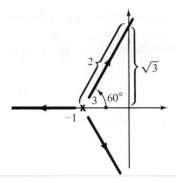

Figure 11.14. Three open loop poles.

It is left as an exercise to show that the root locus is as shown in Figure 11.14. Show that the value of gain at the stability limit is

$$k_u = r_u^3 = 8 \qquad (11.9)$$

and that the imaginary part of the closed loop pole at the stability limit is

$$\omega_u = \sqrt{3}. \qquad (11.10)$$

These same results were obtained in Examples 10.2 and 10.6.

* * *

11.2 Rules for the Rapid Construction of the 180° Root Locus Diagrams

The method demonstrated in the previous section for the construction of root locus diagrams is intuitively very satisfying; however, the method is rather cumbersome for problems only slightly more complicated than those discussed in Section 11.1. In this section, rules which aid the designer in the rapid construction of root locus diagrams will be presented. The rules allow the analyst to quickly *sketch* the root locus and, thereby, make qualitative judgments about the value of proposed compensation and feedback schemes. Of course, computer programs such as those presented in Appendix C can be used to obtain very accurate root locus diagrams.

It is suggested that the analyst begin by writing the characteristic equation in the alternative form

$$1 + k' \frac{(s_p + z_1)(s_p + z_2) \cdots (s_p + z_w)}{(s_p + p_1)(s_p + p_2) \cdots (s_p + p_v)} = 0. \qquad (11.11)$$

The z_k are called *open loop zeros* and the p_k are called *open loop poles*. The characteristic equation should be written in this form in order to simplify the calibration of the root locus. The multiplier, k', is called the *sensitivity* parameter. The above form should be contrasted with the so-called *frequency response* form of the characteristic equation

$$1 + \frac{k_0(T_1 s + 1)(T_2 s + 1) \cdots (T_w s + 1)}{(T_{p1} s + 1)(T_{p2} s + 1) \cdots (T_{pv} s + 1)} = 0 \qquad (11.12)$$

It follows from Equation (2.52) that k_0 is a gain. Notice the following fact about the sensitivity, k', and the gain, k_0:

$$k' = \frac{T_1 T_2 \cdots T_w}{T_{p1} T_{p2} \cdots T_{pv}} k_0; \quad \text{where } T_k = \frac{1}{z_k} \text{ and } T_{p_k} = \frac{1}{p_k}. \qquad (11.13)$$

Use the vector interpretation of the $s + \alpha$ term, indicated in Figure 11.4, to show that the open loop transfer function can be rewritten

$$\frac{(s + z_1)(s + z_2) \cdots (s + z_w)}{(s + p_1)(s + p_2) \cdots (s + p_v)} = \frac{r_{z1} r_{z2} \cdots r_{zw}}{r_{p1} r_{p2} \cdots r_{pv}} e^{i(\psi_1 + \psi_2 \cdots \psi_w - \phi_1 - \phi_2 \cdots \phi_v)}, \qquad (11.14)$$

where the angles ψ are associated with open loop zeros and the angles ϕ are associated with open loop poles, i.e.,

$$\psi_i = \angle (s + z_i) \qquad (11.15)$$

and

$$\phi_i = \angle (s + p_i). \qquad (11.16)$$

In terms of these quantities, the angle criterion, Equation (11.7), becomes

$$\psi_1 + \psi_2 + \cdots + \psi_w - \phi_1 - \phi_2 \cdots - \phi_v = \pm 180; \qquad (11.17)$$

if

$$k' > 0$$

and

$$p_i, z_i > 0 \quad \text{for all } i,$$

and the magnitude criterion, Equation (11.8), takes the form

$$\frac{r_{p1} r_{p2} \cdots r_{pv}}{r_{z1} r_{z2} \cdots r_{zw}} = k'. \qquad (11.18)$$

The rules for sketching root locus diagrams are easily remembered once they are learned. They are as follows:

Rule 1: The number of *branches* (i.e., the number of closed loop poles) is equal to the order of the characteristic equation (which usually equals the number of open loop poles).

Rule 2: Any portion of the real axis which lies to the left of an odd number of *critical points* (i.e., open loop poles or zeros) is on the root locus.

Comment: Notice that in Figure 11.15 the angles

$$\theta_2 + \theta_3 = 360°.$$

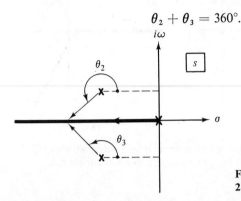

Figure 11.15. Demonstration of *Rule 2.*

Thus the open loop complex poles make no new contribution to the left-hand side of Equation (11.17) for *any* point, *s*, on the real axis. Rule 2 then follows from the fact that for a point, *s*, on the real axis, the left-hand side of Equation (11.17) is completely determined by the *real* open loop poles and zeros only.

Rule 3: A root locus branch *starts* at an open loop pole and *ends* at an open loop zero or, if $v > w$, at infinity (*start* means that $k' = 0$; *end* means that $k' = +\infty$).

This rule follows immediately from Equation (11.18) since $k' = 0$ implies that $r_p = 0$ and $k' = \infty$ implies that either $r_z = 0$ or (if $v > w$) $s_p = \infty$.

Rule 4: Straight line asymptotes of those branches which end at infinity exist and make angles

$$\gamma = \frac{(1 + 2m)180°}{(\text{Number of poles } G_0) - (\text{Number of zeros } G_0)}$$

$$= \frac{(1 + 2m)180°}{v - w}; \quad m = \text{any integer}, \tag{11.19}$$

with the real axis in the *s* plane.

Comment: The pole zero pattern for a $G_0(s)$, which has four poles and one zero, is shown in Figure 11.16a. At large *s*, the contribution from the

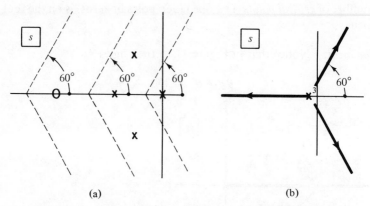

Figure 11.16. (a) Typical pole zero pattern for an open loop transfer function, (b) From large *s* the pattern in (a) appears as a triple pole. See Example 11.4.

zero to the left-hand side of Equation (11.17) is canceled by the contribution from one of the poles. Effectively, the zero "cancels" one of the poles; i.e., only three poles are seen from large *s*. This fact explains the root locus sketched in Figure 11.16b and is an illustration of the validity of Equation (11.17). Notice that this rule does not designate the point where the asymptotes intersect the real axis.

Proof of Rule 4. It follows from Equation (11.11) that

$$\frac{(s_p + z_1)(s_p + z_2) \cdots (s_p + z_w)}{(s_p + p_1)(s_p + p_2) \cdots (s_p + p_v)} = -\frac{1}{k'}.$$

For large *s*, this equation reduces to

$$s_p^{w-v} \simeq -\frac{1}{k'}.$$

Thus, the asymptote angles, γ, satisfy the relation

$$(r_p e^{i\gamma})^{w-v} = -\frac{1}{k'}.$$

Equating angles for these two complex numbers,

$$(w - v)\gamma = -n_0 \times 180°,$$

where n_0 is any odd integer. n_0 can be written $1 + 2m$, where *m* is any integer;

thus,

$$\gamma = \frac{(1 + 2m)180°}{v - w}. \quad \text{Q.E.D.}$$

* * *

Rule 5: The intersection of the real axis with the asymptotes, described in Rule 4, is called the *center of gravity* (c.g.) and is given by

$$\sigma_0 = \frac{\sum\limits_{c=1}^{v} Re(p_c) - \sum\limits_{c=1}^{w} Re(z_c)}{v - w}. \tag{11.20}$$

Proof of Rule 5. Notice that

$$(s_p + z_1)(s_p + z_2) \cdots (s_p + z_w) = s_p^w + \sum_{c=1}^{w} z_c s_p^{w-1} + \cdots + \prod_{c=1}^{w} z_c$$

and that

$$(s_p + p_1)(s_p + p_2) \cdots (s_p + p_v) = s_p^v + \sum_{c=1}^{v} p_c s_p^{v-1} + \cdots + \prod_{c=1}^{v} p_c.$$

Thus, division of the second polynomial by the first yields

$$\frac{s_p^v + \sum\limits_{c=1}^{v} p_c s_p^{v-1} + \cdots + \prod\limits_{c=1}^{v} p_c}{s_p^w + \sum\limits_{c=1}^{w} z_c s_p^{w-1} + \cdots + \prod\limits_{c=1}^{w} z_c} = s_p^{v-w} + \left[\sum_{c=1}^{v} p_c - \sum_{c=1}^{w} z_c\right] s_p^{v-w-1} + \cdots$$

If follows from Equation (11.11) that

$$s_p^{v-w} + \left[\sum_{c=1}^{v} p_c - \sum_{c=1}^{w} z_c\right] s_p^{v-w-1} + \cdots = -k'. \tag{i}$$

Transform the origin of the *s* plane to the c.g. σ_0; i.e., take

$$s' = s_p - \sigma_0;$$

then Equation (i) may be rewritten

$$(s')^{v-w} + \left[[v - w](-\sigma_0) + \sum_{c=1}^{v} p_c - \sum_{c=1}^{w} z_c\right](s')^{v-w-1} + \cdots = -k'.$$

But from Rule 4, if *s'* is on an asymptote,

$$(s')^{v-w} = -k'.$$

Thus, one must require that

$$[v - w](-\sigma_0) + \sum_{c=1}^{v} p_c - \sum_{c=1}^{w} z_c = 0$$

or

$$\sigma_0 = \frac{\sum_{c=1}^{v} p_c - \sum_{c=1}^{w} z_c}{v - w}. \quad \text{Q.E.D.}$$

* * *

Example 11.5

Consider the root locus when the open loop pole zero pattern for $G_0(s)$ is as shown in Figure 11.17. The real portion of the root locus is quickly constructed from Rule 2. Since there are two more open loop poles than open loop zeros, two branches must go to infinity.

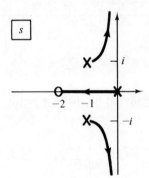

Figure 11.17. Root locus when G_0 has three poles and one zero.

From Rule 4, it follows that the straight line asymptotes to these branches make angles

$$\frac{180°}{3 - 1} = 90°$$

and

$$\frac{3(180°)}{3 - 1} = 270°$$

with the real axis.

From Rule 5, it follows that the asymptotes intersect the real axis at the point

$$\sigma_0 = \frac{-1 - 1 - (-2)}{3 - 1} = 0.$$

The asymptotes are, therefore, the positive and negative portions of the imaginary axis. The root locus is sketched in Figure 11.17. Notice that the above rules provide information only about asymptotes. The exact location of the locus for low gain, i.e., near the complex open loop pole, can be determined by a graphical trial and error procedure to find those points that satisfy Equation (11.17). The Routh array method, discussed in Chapter 10, is perhaps an easier way to achieve the same result.

Notice that in this example a favorable change occurs in the speed of response of the closed loop pole on the real axis branch as k' is increased. Unfortunately, a degradation in speed of response and damping occurs simultaneously for the closed loops on the complex branches as k' is increased.

$$* \quad * \quad *$$

Rule 6: The *breakaway point* on the real axis is the name given to that point where a locus leaves the real axis. This point is found using calculus or a numerical successive approximation scheme by noting that $k' = $ maximum (breakaway) or $k' = $ minimum (breakin) at this point.

Notice that closed loop poles at a breakaway or breakin point are said to be *critically damped*.

An example of the existence of a breakaway point on the root locus is illustrated in Figure 11.12 (refer to Example 11.3). As usual, the locus starts at the open loop poles for $k' = 0$. As k' increases, the poles move toward each other and coalesce at the breakaway point. For higher values of k', the locus leaves the real axis. Thus, if one considers *only the real axis portion* of the locus, k' is a maximum at the breakaway point.

It is convenient to have a method to calculate the location of a breakaway and breakin point and to calculate the associated value of sensitivity. Such a method will now be presented.

Restrict the closed loop, s_p, to real values, denoted σ_p, and rewrite Equation (11.11) in the form

$$k' = \frac{-\sum\limits_{\beta=0}^{v} b_\beta \sigma_p^\beta}{\sum\limits_{\beta=0}^{w} a_\beta \sigma_p^\beta}, \tag{11.21}$$

where the a_β and b_β are constants.

The location of the breakaway or breakin point will be denoted σ_c and satisfies the equation

$$f(\sigma_c) = \frac{dk'}{d\sigma_p}\bigg|_{\sigma_p=\sigma_c} = 0. \tag{11.22}$$

It follows from the description of Newton iteration, in Appendix B, that a numerical, iterative scheme for finding σ_c is

$$\sigma_{c(0)}, \quad \text{chosen arbitrarily,}$$

and

$$\sigma_{c(\alpha+1)} = \sigma_{c(\alpha)} - \frac{f(\sigma_{c(\alpha)})}{df/d\sigma_c}. \tag{11.23}$$

* * *

Example 11.6

Construct the root locus when the open loop transfer function has three poles, as illustrated in Figure 11.18. The real axis portion of the root locus is quickly sketched by using Rule 2. Notice that all three branches must end at infinity.

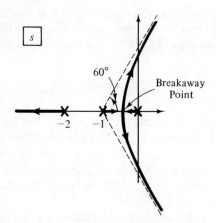

Figure 11.18. Root locus for an open system with three poles.

From Rule 4 the asymptotes are at

$$\gamma = \frac{180}{3} = 60°$$

and

$$\gamma = \frac{3(180)}{3} = 180°.$$

The c.g. is at

$$\sigma_0 = \frac{-1-2}{3} = -1.$$

The root locus is sketched in Figure 11.18. To find the breakaway point, use Equation (11.21), which has the form

$$k' = \sigma_p(\sigma_p + 1)(\sigma + 2) = \sigma_p^3 + 3\sigma_p^2 + 2\sigma_p. \tag{i}$$

From the figure, it is apparent that

$$-1 < \sigma_c < 0. \tag{ii}$$

It follows that

$$f(\sigma_c) = \frac{dk'}{d\sigma_p}\bigg|_{\sigma_p = \sigma_c} = 3\sigma_c^2 + 6\sigma_c + 2.$$

Thus, Equation (11.23) takes the form

$$\sigma_{c(\alpha+1)} = \sigma_{c(\alpha)} - \frac{3\sigma_{c(\alpha)}^2 + 6\sigma_{c(\alpha)} + 2}{6\sigma_{c(\alpha)} + 6}$$

or

$$\sigma_{c(\alpha+1)} = \frac{\frac{1}{2}\sigma_{c(\alpha)}^2 - \frac{1}{3}}{\sigma_{c(\alpha)} + 1}. \qquad \text{(iii)}$$

Observation (ii) is used to pick the initial guess, $\sigma_{c(0)}$, for the iteration scheme summarized in the following table:

$\sigma_{c(\alpha)}$	$\frac{1}{2}\sigma_{c(\alpha)}^2 - \frac{1}{3}$	$\sigma_{c(\alpha)} + 1$	$\sigma_{c(\alpha+1)}$
−0.50	−0.208	0.50	−0.416
−0.416	−0.246	0.584	−0.42
−0.42	−0.244	0.580	−0.42

Clearly, the scheme has converged so that

$$\sigma_c = -0.42. \qquad \text{(iv)}$$

It now follows from Equation (i) that

$$k_c' = |\sigma_c(\sigma_c + 1)(\sigma_c + 2)|$$
$$= 0.385.$$

An examination of Figure 11.18 indicates that if $k' > k_c'$, the poles on the complex branch will be dominant and the speed and damping will be inferior to a system with closed loop poles at the breakaway point. If $k' < k_c'$, the pole on the real branch emanating from the open loop pole at the origin will be dominant and the response will again be slower than that for the system with poles at the breakaway point. This system will never display a steady state error (why?). Thus, a setting of $k' = k_c'$ is the best choice.

* * *

Rule 7: The *angle of departure* of the root locus from complex open loop poles is a useful quantity for the purpose of sketching the root locus. To find this angle, determine the contribution to the left-hand side of Equation (11.17) from *all other* open loop poles and zeros assuming that the test point is on the complex pole in question. Call this contribution θ; then the angle of departure, θ_d, is given by

$$\theta_d = \theta + 180°.\tag{11.24}$$

The angle of approach to a complex zero is

$$\theta_a = -\theta - 180°.\tag{11.25}$$

This rule is illustrated by example.

Example 11.7

For this example, suppose that

$$G_0(s) = \frac{1}{s[(s+1)^2 + 1]}.$$

The asymptotes make angle

$$\gamma = \frac{(1 + 2m)180°}{3} = (1 + 2m)60°$$

with the real axis and the c.g. is at

$$\sigma_0 = \frac{-1 - 1 - 0}{3 - 0} = -\frac{2}{3}$$

The asymptotes are sketched in Figure 11.19. Concentrate on the upper open loop complex pole and show that

$$\theta = -135° - 90° = -225°.$$

Thus, by Equation (11.24),

$$\theta_d = -45°.$$

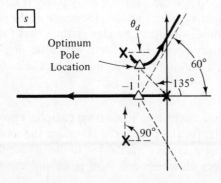

Figure 11.19. Root locus for Example 11.7.

The angle of departure is also shown in Figure 11.19. The root locus is also sketched in this figure, greatly aided by knowledge of γ and θ_d. Again, the exact location of the root locus might be determined by a trial and error graphical procedure.

As the sensitivity parameter k' is increased, the speed of response of the real pole decreases while that of the complex pole increases. Thus, the optimum closed loop pole location is that for which the complex and real poles lie on the same vertical line. This location is indicated by the triangles in Figure 11.19.

*   *   *

Example 11.8

Suppose that

$$G_0(s) = \frac{(s+1)^2 + \frac{1}{4}}{s^2 + \frac{1}{4}}.$$

The open loop pole zero pattern is shown in Figure 11.20.

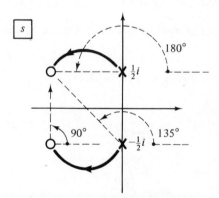

Figure 11.20. Root locus for Example 11.8.

Since no branches go to infinity, Rules 4 and 5 are not of much help here. Concentrate on the upper open loop pole for which

$$\theta = 0° + 45° - 90° = -45°,$$

so that

$$\theta_d = +135°.$$

This angle is sketched in Figure 11.20.

For the upper zero

$$\theta = +90° - 180° - 135° = -225°,$$

so that the angle of approach is

$$\theta_a = +225° - 180°$$

or

$$\theta_a = 45°.$$

The root locus is sketched in Figure 11.20.

*   *   *

Rule 8: If $w \leq v - 2$, then

$$\sum_{\alpha=1}^{v} s_{p\alpha} = \sum_{\alpha=1}^{v} p_{\alpha}, \tag{11.26}$$

where $s_{p\alpha}$ is a closed loop pole and p_{α} is an open loop pole.

This rule is used to simplify the calibration of root loci and to find optimum closed loop pole locations. Both of these assertions are demonstrated in succeeding examples.

Proof of Rule 8. The characteristic equation may be written

$$(s + p_1)(s + p_2)\cdots(s + p_v) + k'(s + z_1)(s + z_2)\cdots(s + z_w)$$
$$= (s + s_{p1})(s + s_{p2})\cdots(s + s_{pv})$$

or

$$s^v + \sum_{\alpha=1}^{v} p_{\alpha} s^{v-1} + \cdots + k'\left[s^w + \sum_{\alpha=1}^{w} z_{\alpha} s^{w-1} + \cdots\right]$$
$$= s^v + \sum_{\alpha=1}^{v} s_{p\alpha} s^{v-1} + \cdots.$$

Equation (11.21) follows immediately.

*   *   *

Example 11.9

Again consider the system discussed in Example 11.7. Denote the closed loop poles on the complex branch by $\sigma_p \pm i\omega$ and the closed loop pole on the real branch by σ_r. Equation (11.26) becomes

$$\sigma_r + \sigma_p - i\omega + \sigma_p + i\omega = 0 - 1 + i - 1 - i$$

or

$$\sigma_p = -1 - \frac{\sigma_r}{2}. \tag{i}$$

One need only calibrate the real portion of the root locus. When this task is performed, the calibration of the imaginary branch follows immediately from Equation (i).

The optimum closed loop pole location is that for which $\sigma_p = \sigma_r$. From Equation (i), this occurs when

$$\sigma_r = -1 - \frac{\sigma_r}{2}$$

or

$$\sigma_r = -\frac{2}{3}. \tag{ii}$$

From the magnitude criterion

$$k' = |s_p[(s_p + 1)^2 + 1]|.$$

The optimum sensitivity setting is found by using Equation (ii) to show that

$$k' = |-\tfrac{2}{3}[(\tfrac{1}{3})^2 + 1]| = \tfrac{20}{27}. \tag{iii}$$

* * *

The final rule deals with systems which have dead time as component parts:

Rule 9: If the open loop transfer function $G_0(s)$ contains an $e^{-\tau s}$ in the numerator, an infinite number of branches exist and the asymptotes of these branches are horizontal lines in the s plane. As $k' \rightarrow +0$, these asymptotes are in the right half-plane and are located at

$$\omega = \pm\frac{m\pi}{\tau}, \quad m = \text{any } odd \text{ integer} \tag{11.27}$$

as $k' \rightarrow 0$. The loci are in the left half-plane and are at

$$\omega = \begin{cases} \pm\dfrac{m\pi}{\tau}, & \text{if } v - w \text{ even,} \\[2mm] \pm 2\dfrac{m\pi}{\tau}, & \text{if } v - w \text{ odd,} \end{cases} \quad m = \text{any odd integer.} \tag{11.28}$$

To understand this rule, notice that the dead time term can be written in polar form as

$$e^{-s\tau} = e^{-(\sigma + i\omega)\tau} = re^{i\phi}, \tag{11.29}$$

where

$$r \triangleq e^{-\sigma\tau} \quad \text{and} \quad \phi \triangleq \omega\tau. \tag{11.30}$$

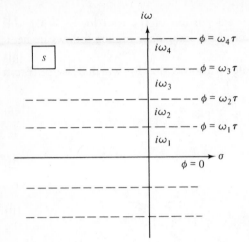

Figure 11.21. Lines of constant angle for the dead time element [Equation (11.29)].

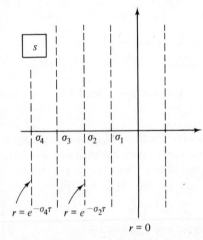

Figure 11.22. Lines of constant magnitude for the dead time element [Equation (11.29)].

The lines of constant angle and magnitude for the dead time element are, therefore, as shown in Figures 11.21 and 11.22, respectively. Figure 11.21 is the more important for the present discussion.

Rule 9 follows from the fact that as one takes a test point far to the right in the *s* plane, the poles and zeros of $G_0(s)$ contribute 0° to the left side of Equation (11.7). On the other hand, for a test point far to the left, the contribution to the left-hand side of Equation (11.7) from the poles and zeros of $G_0(s)$ is 0° if $v - w$ is even and 180° if $v - w$ is odd.

Example 11.10

Consider the feedback system illustrated in Figure 11.23.

Using Rule 9, two of the horizontal asymptotes are as shown in Figure 11.24. The root locus is also sketched in this figure. It is necessary to assume

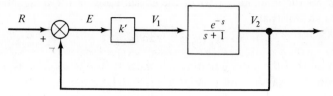

Figure 11.23. System of Example 11.10.

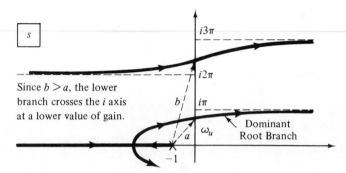

Figure 11.24. Root locus for feedback system shown in Figure 11.23.

that a breakaway point exists on the real axis. To find the breakaway point, use the characteristic equation to show that for real closed loop poles

$$k' = -(1 + \sigma_p)e^{\sigma_p}. \tag{i}$$

Thus, the location of the breakaway point, σ_c, is given by

$$\frac{dk'}{d\sigma_p}\bigg|_{\sigma_p=\sigma_c} = -e^{\sigma_c} - (1 + \sigma_c)e^{\sigma_c} = 0.$$

Thus,

$$\sigma_c = -2. \tag{ii}$$

It then follows from Equation (11.8) that

$$k'_c = |-e^{-2}| = 0.139.$$

To gain a better understanding of the operation of the system, some other interesting points on the root locus will be determined. As was shown in Example 10.3, the stability limit value of the sensitivity is given by

$$k'_u = -\frac{1}{\cos \omega_u}, \tag{iii}$$

where ω_u is the imaginary part of the closed loop on the imaginary axis. It was also shown that

$$\omega_u = -\tan(\omega_u). \qquad \text{(iv)}$$

The reason that the latter equation has more than one solution is now made clear from an examination of Figure 11.24. The different values of ω_u correspond to different branches crossing the imaginary axis. As is indicated in Figure 11.24, the magnitude criteria can be interpreted to mean that the dominant branch crosses the imaginary axis at the lowest value of k' and therefore is the crucial part of the root locus.

Take

$$f(\omega_u) = \omega_u + \tan(\omega_u).$$

Then, application of Newton iteration methods leads to the formula

$$\omega_{u,\alpha+1} = \omega_{u,\alpha} - \frac{(\omega_{u,\alpha} + \tan\omega_{u,\alpha})}{1 + \sec^2(\omega_{u,\alpha})}$$

$$= \frac{\omega_{u,\alpha} - \sin(\omega_{u,\alpha})\cos(\omega_{u,\alpha})}{1 + \cos^2(\omega_{u,\alpha})}.$$

The initial guess for the iteration, summarized in the following table, was obtained from an examination of Figure 11.24:

$\omega_{u,\alpha}$	$\cos^2(\omega_{u,\alpha})$	$\sin(\omega_{u,\alpha})\cos(\omega_{u,\alpha})$	$\omega_{u,\alpha+1}$
$3\pi/4 = 2.35$	0.50	0.50	1.90
1.90	0.10	0.31	2.01
2.01	0.18	0.39	2.04

Take

$$\omega_u = 2.04. \qquad \text{(v)}$$

It then follows from Equation (iii) that

$$k'_u = 2.35. \qquad \text{(vi)}$$

To study the steady state properties of the system, start with

$$\frac{E}{R} = \frac{1}{1 + [k'e^{-s}/(1 + s)]}.$$

It follows from the final value theorem that for $R(s) = 1/s$,

$$e(\infty) = \frac{1}{1 + k'}.$$

Thus, for the critical damping value of k',

$$e(\infty) = \frac{1}{1 + 0.136} = 0.88,$$

which represents a low-accuracy system indeed.

It follows that

$$\frac{r - v_2(\infty)}{r} = 0.88$$

or $v_2(\infty)$ is only 12% of what it should be in the steady state.
Near the stability limit value of k', the steady state error decreases to

$$e(\infty) = \frac{1}{1 + 2.35} = 0.30.$$

Thus, the feedback system is only 70% accurate even when a highly oscillatory transient response is used.

* * *

Example 11.11

For the feedback system of Figure 11.25,

$$\frac{E}{R} = \frac{1}{1 + [k'e^{-s}/s(s + 1)]}.$$

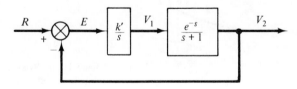

Figure 11.25. Feedback system with dead time. An integrator is added to eliminate steady state error.

Thus for $R = 1/s$,

$$e(\infty) \equiv 0$$

or the system is 100% accurate. The question is what has been lost in order to gain this steady state accuracy? The root locus for this system is shown in Figure 11.26. This locus was sketched using Rule 9 as a guide.

The characteristic equation can be used to show that for real closed loop poles, σ_p,

$$k' = (\sigma_p^2 + \sigma_p)e^{\sigma_p}.$$

To find the breakaway point, maximize k'; i.e., set

$$\frac{dk'}{d\sigma_p} = (2\sigma_p + 1 + \sigma_p^2 + \sigma_p)e^{\sigma_p} = 0.$$

The root of this equation which lies between 0 and -1 is

$$\sigma_p = -0.38,$$

which corresponds to a closed loop time constant of 2.6 sec. This compares to a time constant of

$$\frac{1}{\sigma_c} = 0.5 \text{ sec},$$

obtained in Example 11.10. Clearly, speed of response has been traded for steady state accuracy.

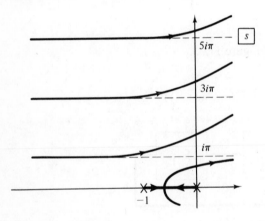

Figure 11.26. Root locus for the system illustrated in Figure 11.25.

* * *

Example 11.12

Consider the feedback system illustrated in Figure 2.31. A block diagram for this system is illustrated in Figure 11.27. It is assumed that the comparator is synthesized by an operational amplifier with gain k_c. The measurement device is taken to be a potentiometer. The characteristic equation is

$$1 + \frac{k'}{s}e^{-\tau s} = 0,$$

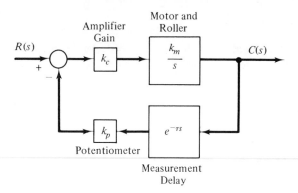

Figure 11.27. Block diagram of the thickness control system (refer to Problem 2.10).

where the sensitivity

$$k' = k_c k_m k_p.$$

The construction of the root locus diagram is found by summing contributions to the angle from the dead time element and from the pole at the origin of the s plane. The result of this method, which was discussed in Section 11.1, is presented in Figure 11.28.

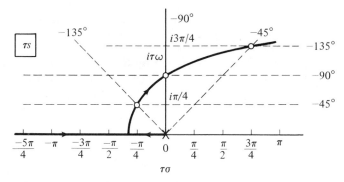

Figure 11.28. Dominant branch of the root locus for the thickness control system. Dashed horizontal lines are contribution to angle, $\angle G_0$, from dead time. Dashed radial lines are the contribution from the pole at the origin. Circles denote points in the s plane where the sum of the contributions is 180°.

* * *

11.3 Analytical Methods for Constructing the 180° Root Locus*

The rules presented in Section 11.2 provide guides or asymptotes for the root locus. These rules are most useful when some branches go to infinity. In other cases, the rules are not of much use for constructing the exact form of the imaginary portions of some branches.

In some cases, high school geometry or trigonometry can be used with the angle criteria [Equation (11.7)] to determine the root locus.

Example 11.13

Suppose that

$$G_0(s) = \frac{(s+z)^2}{(s+p)^2}.$$ (11.31)

Assertion: The root locus for this example is a circle and is illustrated in Figure 11.29.

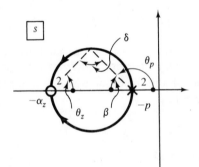

Figure 11.29. Root locus for Example 11.13.

Proof. Consider the circle with center on the real axis and which passes through the critical points. δ is an inscribed angle bounded by a diameter and, therefore, is a right angle. Thus, θ_z and the angle β are complementary so that

$$\theta_z + \beta = 90°,$$

and since θ_p and the angle β, defined in Figure 11.29, are supplementary,

$$\theta_p + \beta = 180°,$$

it follows that

$$\theta_p - \theta_z = 90°$$

* This section may be omitted during the first reading of the text.

or

$$2\theta_p - 2\theta_z = 180°. \quad \text{Q.E.D.}$$

* * *

Example 11.14

Suppose that

$$G_0(s) = \frac{s + z}{(s + p)^2}$$

Assertion: The complex portion of the root locus is a circle with center at the zero and which passes through the poles (see Figure 11.30).

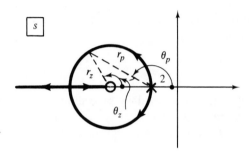

Figure 11.30. Root locus for Example 11.14.

Proof. Refer to Figure 11.14:

$$r_z \sin \theta_z = r_p \sin \theta_p$$

and

$$-r_z \cos \theta_z = -r_p \cos \theta_p - r_z.$$

Rewrite these equations as

$$(\sin \theta_z)r_z - (\sin \theta_p)r_p = 0$$

and

$$(-\cos \theta_z + 1)r_z + (\cos \theta_p)r_p = 0.$$

Since r_z and r_p are not zero, one must require that the determinant of the above system of equations is zero; this conclusion leads to the equation

$$-\tan \theta_p = \frac{\sin \theta_z}{-\cos \theta_z + 1} = \cot \frac{\theta_z}{2}.$$

Thus,

$$-\theta_p = \frac{\pi}{2} - \frac{\theta_z}{2}$$

or

$$2\theta_p - \theta_z = -\pi. \quad \text{Q.E.D.}$$

* * *

11.4 The Importance of the 0° Root Locus: Nonminimum Phase Compensators

Rewrite the characteristic equation (11.6) in the form

$$G_0(s_p) = -\frac{1}{k}. \tag{11.32}$$

For the case $k > 0$, this equation was interpreted as meaning the following: For a point, s, in the s plane to be a closed loop pole, *it had to satisfy the angle criterion:*

$$\sum_i^w \psi_i - \sum_i^v \phi_i = 180°. \tag{11.33}$$

This fact was used in preceding sections for the construction of loci of closed loop poles.

However, there are several cases when the correct angle criterion becomes

$$\sum_i^w \psi_i - \sum_i^v \phi_i = 0°. \tag{11.34}$$

Contrary to some popular opinion, there are important cases when Equation (11.34) is the appropriate criterion. The two *most* obvious and *least* important cases where the 0° root locus should be used are the cases of *negative gain* and of *positive feedback*, since if k is negative in Equation (11.32), then Equation (11.34) is the correct angle criterion for the case of negative gain. It is left as an exercise to show that if the actual output in a feedback system is *added* to the set point rather than subtracted from it, Equation (11.34) is the correct angle criterion.

Feedback systems with positive feedback or negative gain have poor stability and steady state error properties and will not be discussed further. Rather, the case when Equation (11.34) is the correct criterion, even though the feedback system in question employs *negative feedback and positive gain*, will be explored.

To understand this special case, the reader must understand some preliminary facts concerning two special compensators: the *negative gain compensator*, which has the transfer function

$$G_c(s) = \frac{k(T_1 s - 1)}{(T_2 s + 1)}, \tag{11.35}$$

and the *reverse action* or *nonminimum phase compensator*, which has the transfer function

$$G_c(s) = \frac{k(1 - T_1 s)}{(T_2 s + 1)}.$$ (11.36)

Both these compensators have the *same* pole zero pattern (shown in Figure 11.31). This fact is the reason that the wrong angle criterion is sometimes applied to feedback systems containing these compensators.

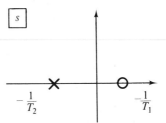

Figure 11.31. Pole zero diagram: negative gain *and* reverse action compensators.

Apply the final value theorem to show that the reverse action compensator has *positive gain* while the transfer function of Equation (11.35) has a negative gain. The reason for the term *reverse action* is illustrated in Figure 11.32 wherein is sketched the unit step response for the reverse action compensator. Notice that the output of the compensator changes, discontinuously, in the reverse or negative direction at $t = 0$.

It is important now to consider the contribution to $\angle G_0(s)$ made by a zero in the right half-plane. Consult the vector diagram in Figure (11.33a). Notice that the vector from a right half-plane zero to some test point is

$$s - \frac{1}{T} = -\left(\frac{1}{T} - s\right).$$

This fact leads to the following conclusions:

1. If $G_0(s)$ contains a $s - 1/T$ numerator term, the contribution to the angle, $\angle G_0$, is

$$\angle\left(s - \frac{1}{T}\right) = \psi.$$ (11.37)

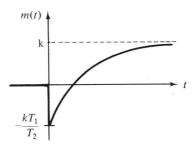

Figure 11.32. Unit step response of the reverse action compensator.

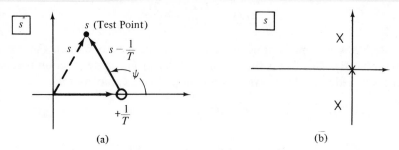

Figure 11.33. (a) Polar diagram of $s - 1/T$, (b) typical critical points for $G_p(s)$ of many servo systems.

2. If $G_0(s)$ contains a $1/T - s$ numerator term, the contribution to the angle, $\angle G_0$, is

$$\angle\left(\frac{1}{T} - s\right) = \psi + 180°. \tag{11.38}$$

Equations (11.38) and (11.17) lead to the following surprising conclusion: For *negative* feedback systems with *positive* sensitivity parameter, k', and which contain reverse action or nonminimum phase zeros, the locus of closed loop poles is the collection of all points, s, *which lie on the $0°$ root locus* [i.e., all points for which Equation (11.34) is satisfied].

The rules for construction of the $0°$ locus will not be stated here. The reader can deduce these rules for himself using the line of reasoning illustrated in Section 11.2. The importance of reverse action compensators (hence, of the $0°$ locus), however, will be demonstrated by example.

Example 11.15

Consider a feedback system which is a servo system for which

$c(t) \triangleq$ position of the controlled object,
$m(t) \triangleq$ voltage input to a solenoid and hydraulic valve-actuator system,
$e(t) \triangleq$ error voltage and compensator input.

Typically the critical points for the transfer function, $G_p(s)$, might be distributed as shown in Figure (11.33b). The imaginary poles near the imaginary axis tend to induce a highly oscillatory response.

The problem is to find a compensator that will induce a fast, well-damped closed loop response. In this example only first-order compensators will be considered. In particular, the following compensators are considered:

1. The *lag-lead* compensator:

$$G_c(s) = \frac{k_c(T_1 s + 1)}{T_2 s + 1}, \quad T_1 < T_2.$$

2. The *lead-lag* compensator:

$$G_c(s) = \frac{k_c(T_1 s + 1)}{T_2 s + 1}, \quad T_1 > T_2.$$

3. The negative gain compensator:

$$G_c(s) = \frac{k_c(T_1 s - 1)}{T_2 s + 1},$$

4. The reverse action compensator:

$$G_c(s) = \frac{k_c(1 - T_1 s)}{T_2 s + 1}.$$

To study the relative merits of these compensators, the root locus (for $k_c > 0$) diagrams were determined and are summarized in Figures 11.34–11.37. The 180° loci are appropriate for compensators 1, 2, and 3, while (because of the reverse action term) the 0° locus is appropriate for compensator 4. *Note:* In all sketches the critical points of the compensator are subscripted with c and Δ = closed loop poles selected in computer-aided design.

It is obvious from an examination of these diagrams that the reverse action compensator provides the best results. The optimum pole location is indicated by the triangles.

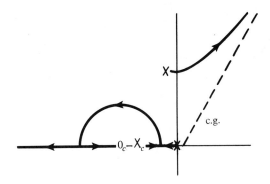

Figure 11.34. Sketch of positive gain locus for the lag-lead compensator.

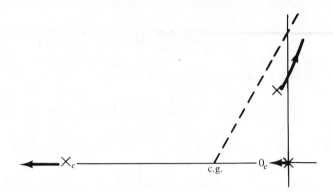

Figure 11.35. Positive gain locus for the lead-lag compensator (2).

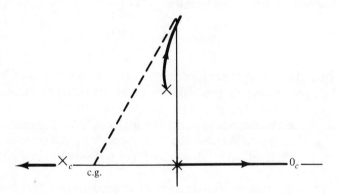

Figure 11.36. Positive gain locus for the negative gain (nonminimum phase) compensator.

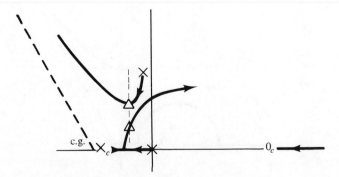

Figure 11.37. Positive gain locus for the reverse action compensator.

*   *   *

11.5 The Important Concept of *Parasitic* Poles in Control System Design

The term *parasitic* pole will be used to refer to those poles of an actual dynamic system which are not perceived by the designer because of an incomplete analysis. Time and resource constraints are such that *all* analyses are incomplete so that parasitic poles *always* exist. The effect of parasitic poles on the operation of a feedback system, once it is synthesized, will be illustrated by example.

Example 11.16

Consider the system illustrated in Figure 11.38. Use the methods presented in Chapter 5 to show that

$$\frac{H}{F} = \frac{R/\gamma}{1 + RCs}.$$

Also show that the displacement

$$E = \left(\frac{a + b}{b}\right)R - \frac{a}{b}H.$$

Finally, use the valve equation [Equation (7.34)] to show that

$$F = \frac{C_v K_v}{R_v}E,$$

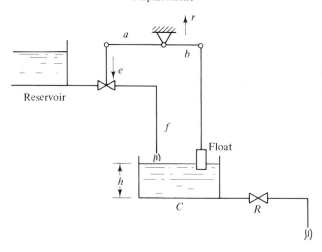

Figure 11.38. Liquid level control system.

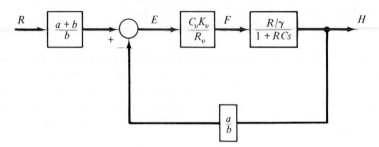

Figure 11.39. Block diagram of the system illustrated in Figure 11.38.

where K_v is the proportionality constant relating valve displacement to valve area. These equations are summarized in block diagram form in Figure 11.39.

It follows that the characteristic equation can be written

$$1 + \frac{k'}{s_p + \alpha} = 0, \tag{i}$$

where

$$\alpha = \frac{1}{RC} \tag{ii}$$

and the sensitivity

$$k' = \frac{aC_vK_v}{b\gamma R_vC}. \tag{iii}$$

This quantity is variable since the a to b ratio is variable. The root locus associated with Equation (i) is illustrated in Figure 11.7. If the analyst were to stop at this point, he might conclude that he should design the lever so that b is as small as possible, so that k' is as large as possible, in order to synthesize a feedback system with a very fast response.

At this point the analyst should attempt to identify parasitic poles and their effect on the response of a high-gain-feedback system.

A more complete analysis of the float will be an example of the existence and effect of parasitic poles. Refer to Figure 11.40 and use the notation

x = distance of water line of float below liquid surface,
$x_f = h - x$,
A_f = cross-sectional area of the float,
m = mass of the float,
b = viscous friction constant.

It follows that if the force of the lever on the float is negligible, then

$$m\ddot{x}_f = \gamma A_f x - b\dot{x}_f$$

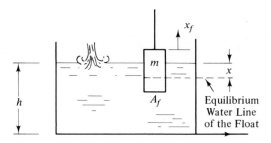

Figure 11.40. The inertia of the float leads to a delay in the response of the float to a change in liquid level.

since $\gamma A_f x$ is the buoyant force of the fluid on the float. Hence,

$$m\ddot{x}_f + b\dot{x}_f + \gamma A_f x_f = \gamma A_f h.$$

In the Laplace domain

$$\frac{X_f(s)}{H(s)} = \frac{\gamma A_f/m}{s^2 + (b/m)s + (\gamma A_f/m)}. \tag{11.39}$$

The block diagram must be modified in the manner indicated in Figure 11.41.

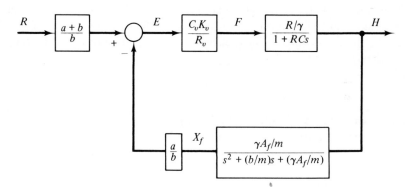

Figure 11.41. Modification of Figure 11.39. which illustrates the parasitic poles associated with the float.

The characteristic equation becomes

$$1 + \frac{k'}{(s + \alpha)[s^2 + (b/m)s + (\gamma A_f/m)]}. \tag{iv}$$

where

$$k' = \frac{aC_v K_v A_f}{bR_v Cm}. \tag{v}$$

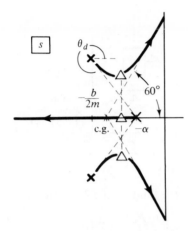

Figure 11.42. Root locus for the system described in Figure 11.41.

The root locus for Equation (iv) is sketched in Figure 11.42. The details of this sketch are left as an exercise. Also show that the real part of the optimum pole location is found from Equation (11.26) and that

$$\sigma_r = -\frac{1}{3}\left[\frac{1}{RC} + \frac{b}{m}\right]. \tag{11.40}$$

The important point is that a more complete analysis has indicated that a high value of k' (thus, a high value of a to b) could cause the closed loop system to be unstable. In this case, the effect of the parasitic poles associated with the float is dramatic. Such effects will always be present in high-gain (sensitivity) systems.

* * *

11.6 Examples of Preliminary Designs Obtained Using Root Locus Techniques

Examples of the use of the root locus method are provided in the following problems. The level of modeling used in these examples is that which might be used in preliminary design studies. Thus, the solutions to the example problems provide only an indication of the form of final design which would be developed after a more complete analysis followed by a testing program.

Example 11.17

Consider the feedback system illustrated in Figure 10.3. The block diagram for this system is shown in Figure 10.4. Assume that the armature has inductance, L, and that viscous friction, b, exists between the mass and its support. Then

$$\begin{bmatrix} M \\ I \end{bmatrix} = \begin{bmatrix} 1 & Ls \\ 0 & 1 \end{bmatrix} \begin{bmatrix} 0 & g \\ \frac{1}{g} & 0 \end{bmatrix} \begin{bmatrix} n & 0 \\ 0 & \frac{1}{n} \end{bmatrix} \begin{bmatrix} 1 & ms + b \\ 0 & 1 \end{bmatrix} \begin{bmatrix} 0 \\ sC \end{bmatrix}$$

$$= \begin{bmatrix} \cdots & \frac{nLs}{g}(ms + b) + \frac{g}{n} \\ \cdots & \cdots \end{bmatrix} \begin{bmatrix} 0 \\ sC \end{bmatrix}.$$

Thus,

$$G_p = \frac{C}{M} = \frac{g/(nmL)}{s[s^2 + (b/m)s + (g^2/n^2Lm)]}.$$

Assume that the voltage

$$M = K_c[R - C],$$

so that the characteristic equation is

$$1 + \frac{k'}{s_p[s_p^2 + 2\xi\omega_n s_p + \omega_n^2]} = 0, \tag{11.41}$$

where

$$k' = \frac{K_c g}{nmL}, \tag{11.42}$$

$$\omega_n = \frac{g}{n}\sqrt{\frac{1}{Lm}}, \tag{11.43}$$

$$\xi = \frac{bn}{2g}\sqrt{\frac{L}{m}}. \tag{11.44}$$

The root locus diagram for Equation (11.41) is illustrated in Figure 11.43. It is possible that breakin and breakaway points exist. To investigate this possibility, take s_p to be real and rewrite Equation (11.41) as

$$k' = -[\sigma_p^3 + 2\xi\omega_n\sigma_p^2 + \omega_n^2\sigma_p].$$

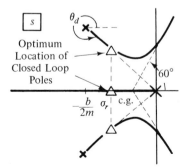

Optimum Location of Closed Loop Poles

Figure 11.43. Root locus for the system illustrated in Figure 10.3. if $(bn/2g)(L/m) < 0.87$.

Then

$$\left.\frac{dk'}{d\sigma_p}\right|_{\sigma_p=\sigma_c} = -3\sigma_c^2 - 4\xi\omega_n\sigma_c - \omega_n^2 = 0$$

has real roots if

$$(4\xi\omega_n)^2 - 4(3\omega_n^2) \geq 0,$$

i.e., if

$$\xi \geq \frac{\sqrt{3}}{2} = 0.87 \tag{11.45a}$$

or

$$\theta \leq 29.6°. \tag{11.45b}$$

The significance of the existence of breakin and breakaway points will be investigated in the problem sets.

It is left as an exercise to use Equation (11.26) to show that the real part of the optimum closed loop pole location is given by

$$\sigma_r = -\frac{b}{3m}. \tag{11.46}$$

Notice that no steady state error exists for a step input.

*   *   *

Example 11.18

The feedback system illustrated in Figure 11.44 is synthesized in order to control the angular velocity, f, to "follow" a recorded desired velocity, r. Samples placed on the centrifuge arms will then be subjected to prescribed programs of acceleration for testing purposes. Begin the analysis at the output of the amplifier and find

$$\begin{bmatrix} E_a \\ F_a \end{bmatrix} = \begin{bmatrix} 1 & Ls \\ 0 & 1 \end{bmatrix} \begin{bmatrix} 0 & g \\ \dfrac{1}{g} & 0 \end{bmatrix} \begin{bmatrix} 0 \\ sl\delta \end{bmatrix}$$

$$= \begin{bmatrix} \cdots & g \\ \cdots & \cdots \end{bmatrix} \begin{bmatrix} 0 \\ sl\delta \end{bmatrix}.$$

Thus,

$$\delta = \frac{1}{slg}E_a = \frac{K_c}{slg}(R - F). \tag{i}$$

Assume that

$$F_p = K_p\delta. \tag{ii}$$

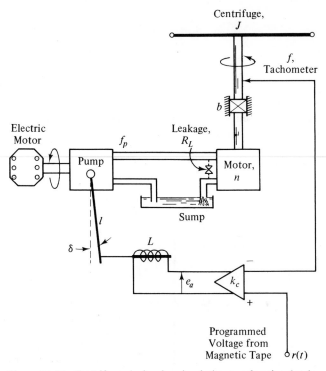

Figure 11.44. Centrifuge device for simulating acceleration loads.

Also show that

$$\begin{bmatrix} E_p \\ F_p \end{bmatrix} = \begin{bmatrix} 1 & 0 \\ \dfrac{1}{R_L} & 1 \end{bmatrix} \begin{bmatrix} n & 0 \\ 0 & \dfrac{1}{n} \end{bmatrix} \begin{bmatrix} 1 & b + Js \\ 0 & 1 \end{bmatrix} \begin{bmatrix} 0 \\ F \end{bmatrix}$$

$$= \begin{bmatrix} \cdots & \cdots \\ \cdots & \dfrac{n}{R_L}(b + Js) + \dfrac{1}{n} \end{bmatrix} \begin{bmatrix} 0 \\ F \end{bmatrix},$$

so that

$$G_p(s) = \frac{F(s)}{\delta(s)} = \frac{K_p}{(n/R_L)(b + Js) + (1/n)}. \tag{iii}$$

Combine Equations (i) and (ii) to show that the characteristic equation is

$$1 + \frac{k'}{s(s + \alpha)} = 0, \tag{iv}$$

where the sensitivity

$$k' = \frac{R_L K_p}{lgnJ} K_c \tag{v}$$

and

$$\alpha = \frac{R_L + n^2 b}{n^2 J}. \tag{vi}$$

The root locus diagram associated with Equation (iv) is illustrated in Figure 11.45. The closed loop pole location is chosen so as to maximize speed and damping. Notice that no steady state error exists for a step input. Combine Equation (v) of Example 11.3 and Equation (v) above to show that K_c should be chosen so that

$$K_c = \frac{lgnJ\alpha^2}{4R_L K_p} = \frac{(R_L + n^2 b)^2 lg}{4R_L K_p Jn^3}. \tag{11.47}$$

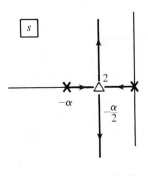

Figure 11.45. Root locus diagram for the system illustrated in Figure 11.44. The optimum location of closed loop pole is denoted by the triangles.

* * *

Example 11.19

As is indicated in Figure 11.45, the fastest uncompensated closed loop response that can be achieved is that associated with a time constant of $2/\alpha$. A preliminary selection of a compensator that induces a faster closed loop response will be made in this example.

In this example, the voltage e_m is the manipulated variable. Denote the error

$$e = r - f \tag{i}$$

and consider compensators of the form

$$\frac{E_m(s)}{E(s)} = \frac{(s + \alpha_z)}{(s + \alpha_p)}. \tag{ii}$$

Notice that

$$E_m(s) = E(s) + \frac{1}{s}[\alpha_z E(s) - \alpha_p E_m(s)]. \tag{iii}$$

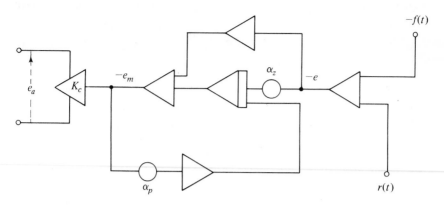

Figure 11.46. Synthesis of a compensator with one pole (α_p) and one zero (α_z) which is to be inserted into the feedback system illustrated in Figure 11.44.

This equation suggests the synthesis indicated in Figure 11.46 (refer to Chapter 8).

It will be left as an exercise to show that it is wise to select

$$\alpha_p < \alpha_z \tag{11.48}$$

and

$$\alpha_z \le \alpha, \tag{11.49}$$

where α is defined in Equation (v) of Example 11.18.

The root locus diagram will then take the form indicated in Figure 11.47. The complex branch will often contain the dominant poles of the closed loop system. For sufficiently high gain, poles will be associated with time constants less than $2/\alpha$, and improvement in the speed of response will be

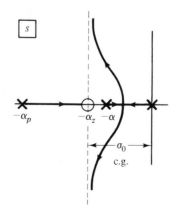

Figure 11.47. Modification of the diagram of Figure 11.45. when the compensator is added to the feedback loop.

achieved. In the limit as K_c becomes large, the closed loop time constant approaches $1/\sigma_0$.

The analyst should be warned against the choice of

$$\alpha_z = \alpha.$$

This choice is quite tempting because of the favorable closed loop damping that would, theoretically, be obtained. As was indicated in Section 10.10, pole zero cancellation is a tricky business and can lead to incorrect conclusions. The analyst is better advised to select the compensator sufficiently far to the left of the plant pole at α in order to ensure that inequality (11.49) is satisfied, even after a parameter variation.

* * *

References

1. D'Azzo, J., and Houpis, C., *Feedback Control System Analysis and Synthesis*, McGraw-Hill, New York, 1966.

2. Horowitz, I., *Synthesis of Feedback Systems*, Academic Press, New York, 1963.

3. Ogata, K., *Modern Control Engineering*, Prentice-Hall, Englewood Cliffs, N.J., 1970.

4. Takahashi, Y., Rabins, M., and Auslander, D., *Control*, Addison-Wesley, Reading, Mass., 1970.

Problems

11.1. Use the method of Section 11.1 to establish the validity of Figure 11.14 and Equations (11.9) and (11.10).

11.2. Prove the validity of Equation (11.13).

11.3. Establish the iteration scheme suggested in Equation (11.23).

11.4. Verify the validity of Figures 11.32 and 11.33.

11.5. Verify Equation (11.38).

11.6. Verify all of the transfer functions indicated in Figure 11.39.

11.7. Verify the root locus in Figure 11.42.

11.8. Sketch the root locus of the system of Example 11.17 when $\delta > 0.87$.

11.9. Verify Equation (11.46).

11.10. Establish inequalities (11.48) and (11.49).

11.11. Sketch the 180° root locus diagrams for the systems described by Figure 11.48. The only quantities that need be determined accurately are (a) asymptote angles, (b) c.g., (c) angles of departure and approach, and (d) breakaway and breakin points.

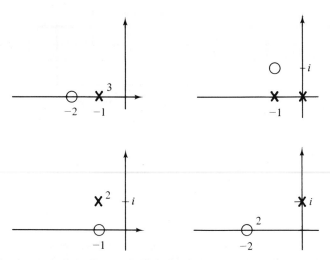

Figure 11.48.

11.12. Construct the root locus diagram when the open loop transfer function is

$$G_0(s) = \frac{1}{s([s+4]^2 + 4)}.$$

Hint: Calibrate the real axis, $-6 < \sigma < 0$, every half unit before you attempt to construct the complex portion of the root locus.

11.13. Consider the pitch control of a rocket ship (refer to Figure 11.49). Let $m(t)$ be the pilot's stick deflection and $\omega(t)$ the pitch rate. It can be shown that

$$\frac{\Omega(s)}{M(s)} = \frac{k_x(Ts+1)}{(s^2/\omega_n^2) + (2\zeta/\omega_n)s + 1}.$$

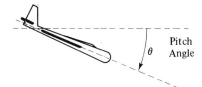

Figure 11.49.

It is desired to automate the control of pitch rate in the manner shown in Figure 11.50. *Sketch* the root locus as k_c is varied. At the normal operation condition,

Altitude = 35,000 ft,

Mach number = 0.3,

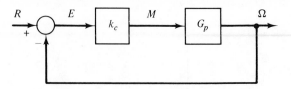

Figure 11.50. Feedback control of pitch rate.

and the parameters are

$$k_x = 0.1066,$$

$$\frac{1}{T} = 0.123,$$

$$\zeta\omega_n = 0.2604,$$

$$\omega_n = 1.988.$$

11.14. From the sketch obtained in the solution of Problem 11.13, what is the optimum location of the closed loop poles? Determine the value of k_c that will force the closed loop poles to this location. What is the steady state error for your design? What must you sacrifice in order to decrease steady state error? Is steady state error an important consideration in this design?

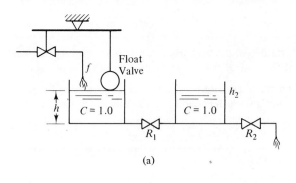

(a)

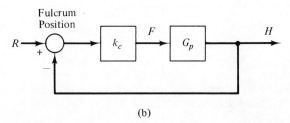

(b)

Figure 11.51. (a) Head control (side capacitance), (b) Block diagram.

11.15. Consider the head-control system shown in Figures 11.51a and 11.51b. Use the normal operating conditions:

$$F° = 10.0 \text{ cu ft/sec,}$$
$$H_2° = 1.0 \text{ ft,}$$
$$H° = 1.5 \text{ ft,}$$
$$\gamma = 100.0 \text{ lb/cu ft.}$$

Sketch the root locus and decide the optimum location of the closed loop poles. Determine the corresponding values of K_c and steady state error. Neglect inertances.

11.16. Repeat the procedure of Problem 11.15 for the system in Figure 11.52. Use the same parameter values as in Problem 11.15.

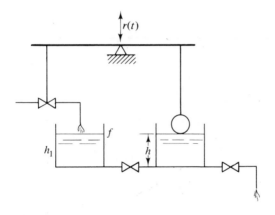

Figure 11.52. Head control (bilateral coupling).

11.17. *Sketch* the root loci for the feedback systems in Figures 11.53a–11.53c and 11.54. Indicate the optimum closed loop pole locations for each case. *Note:* Be sure to put the characteristic equation in standard root locus form.

11.18. P-D control is to be used with system (b) (Figure 11.53b). The block diagram is illustrated in Figure 11.55. *Sketch* the root locus for K_c varying for the following three cases:

Case a: $T_D = 0.50$ sec,
Case b: $T_D = 1.5$ sec,
Case c: $T_D = 4.0$ sec.

From your sketches choose the best value of T_D and then choose K_c to provide optimum closed loop pole location.

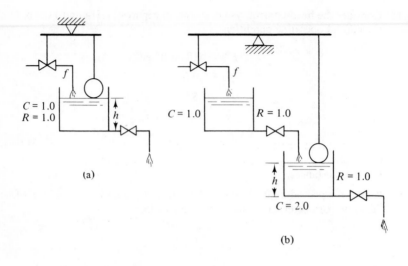

(a)

(b)

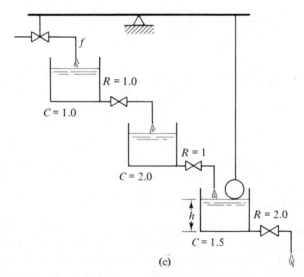

(c)

Figure 11.53.

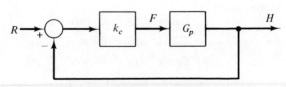

Figure 11.54. Block diagram.

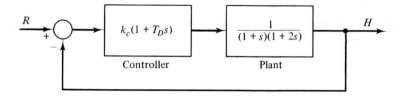

Figure 11.55.

11.19. Repeat the above problem using the P-I compensator

$$G_c = k_c\left(1 + \frac{1}{T_c s}\right).$$

What advantages does P-I control provide that P-D control does not? What is the trade-off?

11.20. Consider the thickness control system diagrammed in Figure 11.56. Sketch the root locus and determine the optimum location for the poles on the dominant branch. Find the corresponding value of k_c. Repeat this problem using the P-D compensator

$$k_c(0.5s + 1).$$

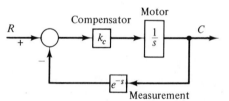

Figure 11.56. Block diagram of thickness control system.

11.21. Let $c(t)$ be the live population of a certain animal species that is to be controlled. Let $p(t)$ be the live population of a certain predator. Let $m(t)$ be the concentration of a pesticide that can be used against the predator. Then assume that

$$\frac{dc}{dt} = 2c - c^2 - pc - mc,$$

$$\frac{dp}{dt} = pc - \alpha mp,$$

$$\alpha = 1.0.$$

Linearize the above equations about the nominal steady state conditions $c° = 1.0, p° = 1.0$, and $m° = 1.0$. Take the Laplace transform and eliminate $P(s)$ to find the transfer function

$$G_p \triangleq \frac{C(s)}{M(s)}.$$

For the feedback system, what value of k_c should be used? What is the steady state error for this value of k_c?

11.22. For the system in Figure 11.57, show that the root locus is the unit circle in the s plane. Calibrate the root locus for k_c ranging from 0 to infinity and plot the following control system design parameters vs. k_c:
(a) Decay time (time for transient terms to fall to $1/e$ of initial value).
(b) Damping per cycle.
(c) Steady state error [for $R(s) = 1/s$].

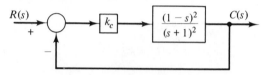

Figure 11.57.

11.23. What would be the effect of parasitic float poles on the root loci you constructed for Problems 11.15–11.17?

Design of Multivariable Control Systems by the Methods of Modal and Decoupling Control; The Use of Dynamic Observers* 12

The goals of multivariable (or state variable) control were introduced in Section 10.10. In the present chapter, some specific methods, which can be used to achieve these goals, are presented.

The basic problem is that given the open loop state variable model

$$\dot{x} = \mathcal{A}x + \mathcal{B}u \qquad (12.1)$$

and the output equation

$$c = \mathcal{D}x, \qquad (12.2)$$

determine the matrices $\mathcal{F}$ and $\mathcal{K}$, defined in Figure 10.22, such that the closed loop system meets the design objectives. It was demonstrated that the closed loop system is described by the closed loop state variable equation

$$\dot{x} = \{\mathcal{A} - \mathcal{B}\mathcal{K}\mathcal{F}\mathcal{D}\}x + \mathcal{B}\mathcal{K}r, \qquad (12.3)$$

$$c = \mathcal{D}x, \qquad (12.4)$$

where $r(t)$ is a vector of set points for the output $c(t)$.

Clearly, the closed loop transient response is quantified by the eigenvalues of

$$\mathcal{A}_c = \mathcal{A} - \mathcal{B}\mathcal{K}\mathcal{F}\mathcal{D}. \qquad (12.5)$$

* This chapter may be omitted during the first reading of the text.

Note that the scheme illustrated in Figure 10.22 is identical to that illustrated in Figure 12.1 if one makes the identifications

$$\mathcal{P} = \mathcal{K}, \tag{12.6}$$

$$\mathcal{F}_0 = \mathcal{K}\mathcal{F}\mathcal{D}. \tag{12.7}$$

In terms of $\mathcal{P}$ and $\mathcal{F}_0$, Equation (12.3) becomes

$$\dot{x} = \{\mathcal{C} - \mathcal{B}\mathcal{F}_0\}x + \mathcal{B}\mathcal{P}r, \tag{12.8}$$

$$c = \mathcal{D}x, \tag{12.9}$$

so that

$$\mathcal{C}_c = \mathcal{C} - \mathcal{B}\mathcal{F}_0. \tag{12.10}$$

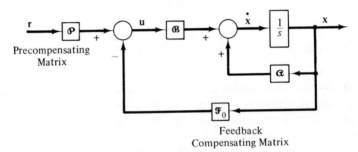

Figure 12.1. Vector-block diagram for a multivariable feedback system which is equivalent to that illustrated in Figure 10.22.

12.1 *Decoupling* Control for an Ideal Case

Assume first of all that

1. There are as many manipulated variables as there are states.
2. The inverse of the open loop input matrix, $\mathcal{B}^{-1}$, exists.
3. All of the states, $x_k(t)$, are directly measurable.

These assumptions are rarely valid in any real design, but this academic example is instructive and will be pursued.

Suppose that it is desired that the closed poles equal a prespecified set of values $\{-s_i\}$ and that the gains between set points and states equal a pre-specified set of values $\{k_i\}$.

One way of satisfying these design criteria is to choose $\mathcal{P}$ and $\mathcal{F}_0$ of Figure 12.1 such that

$$\mathbf{Q}_c = \begin{bmatrix} -s_1 & 0 & 0 & \cdots & 0 \\ 0 & -s_2 & 0 & \cdots & 0 \\ 0 & 0 & -s_3 & \cdots & 0 \\ \vdots & & & & \vdots \\ 0 & & & \cdots & -s_n \end{bmatrix} \tag{12.11}$$

and

$$\mathbf{\mathcal{B}P} = \begin{bmatrix} k_1 s_1 & 0 & 0 & \cdots & 0 \\ 0 & k_2 s_2 & 0 & \cdots & 0 \\ \vdots & & & & \vdots \\ 0 & & & \cdots & k_n s_n \end{bmatrix}. \tag{12.12}$$

It follows that Equation (12.12) is satisfied if

$$\mathbf{P} = \mathbf{\mathcal{B}}^{-1}\mathbf{\mathcal{B}}_c, \tag{12.13}$$

where $\mathbf{\mathcal{B}}_c$ is the matrix on the right-hand side of Equation (12.12). Similarly, Equation (12.11) is satisfied if

$$\mathbf{\mathcal{F}}_0 = \mathbf{\mathcal{B}}^{-1}(\mathbf{Q} - \mathbf{Q}_c). \tag{12.14}$$

When the compensators are chosen according to Equations (12.13) and (12.14), the resulting system is called a *decoupling feedback* system. The reason for this name will become apparent when the properties of the feedback system are explored.

It is left as an exercise for the reader to show that for decoupling feedback the steady state error can be set at any predetermined value (including **0**) by making a proper selection of the k_i in Equation (12.12). Also show that $e_i(\infty)$ depends *only* on k_i.

Consider now the closed loop transient response of a decoupling feedback system. The differential equation is

$$\dot{\mathbf{x}} = \mathbf{Q}_c\mathbf{x} + \mathbf{\mathcal{B}}_c\mathbf{r}. \tag{12.15}$$

First, the fundamental solution must be determined. From Equation (12.11)

$$s\mathbf{\mathcal{I}} - \mathbf{Q}_c = \begin{bmatrix} s + s_1 & 0 & 0 & \cdots & 0 \\ 0 & s + s_2 & 0 & \cdots & 0 \\ \vdots & & & & \vdots \\ 0 & & & \cdots & s + s_n \end{bmatrix}. \tag{12.16}$$

It is a simple matter to verify that

$$(s\boldsymbol{\mathcal{I}} - \boldsymbol{\mathcal{C}}_c)^{-1} = \begin{bmatrix} \dfrac{1}{s+s_1} & 0 & 0 & \cdots & 0 \\[2ex] 0 & \dfrac{1}{s+s_2} & 0 & \cdots & 0 \\ \cdot & \cdot & \cdot & & \cdot \\ \cdot & \cdot & \cdot & & \cdot \\ \cdot & \cdot & \cdot & & \cdot \\ 0 & 0 & 0 & \cdots & \dfrac{1}{s+s_n} \end{bmatrix}. \qquad (12.17)$$

Take the inverse Laplace transform of this equation to show that

$$\boldsymbol{\phi}_c(t) = \begin{bmatrix} e^{-s_1 t} & 0 & 0 & \cdots & 0 \\ 0 & e^{-s_2 t} & 0 & \cdots & 0 \\ \cdot & \cdot & \cdot & & \cdot \\ \cdot & \cdot & \cdot & & \cdot \\ \cdot & \cdot & \cdot & & \cdot \\ 0 & 0 & 0 & \cdots & e^{-s_n t} \end{bmatrix}. \qquad (12.18)$$

It follows from Equation (2.88) that

$$x_i(t) = e^{-s_i t} x_i(0^-) + k_i s_i \int_0^t e^{-s_i(t-\xi)} r_i(\xi)\, d\xi. \qquad (12.19)$$

This equation completely defines the transient response of the closed loop systems.

The benefits of decoupling control are immediately obvious from an examination of Equation (12.19):

1. The transients are composed entirely of decaying exponentials so that the closed loop damping is the best possible.

2. The speed of response of the ith variable is determined solely by s_i, which may be chosen arbitrarily in the specification of Equation (12.11).

3. $x_i(t)$ does not depend on $x_j(t), j \neq i$, for any t; i.e., the state variables are *decoupled*. Also, if a disturbance offsets only $x_i(t)$, none of the other states will be disturbed during the transient in x_i.

4. $x_i(t)$ does not depend on $r_j(t), j \neq i$, for any t; i.e., the inputs have been decoupled from each other and from the states. Also, if the desired value of x_i is to be changed by changing the value of r_i, none of the other state variables will be disturbed by the resulting transient in x_i.

It is seen that decoupling feedback causes the open loop state space variables, **x**, to become the *modal state space* for the closed loop system.

The difficulties in the practical realization of decoupling feedback, as outlined in this section, are many: The most obvious difficulty becomes apparent from an examination of Equations (12.13) and (12.14). Obviously, decoupling feedback is possible only if $\mathfrak{B}^{-1}$ exists.

Example 12.1

Apply decoupling control to the open loop system described by

$$\dot{\mathbf{x}} = \begin{bmatrix} 0 & -1 \\ 1 & 0 \end{bmatrix} \mathbf{x} + \begin{bmatrix} 1 & 0 \\ 0 & 1 \end{bmatrix} \mathbf{u}, \tag{12.20}$$

$$\mathfrak{D} = \mathfrak{I}.$$

Decoupling control will be compared to proportional scalar feedback control. Notice that for the open loop system

$$\mathbf{X}(s) = [s\mathfrak{I} - \mathfrak{C}]^{-1}\mathfrak{B}\mathbf{U}(s)$$

$$= \begin{bmatrix} s & 1 \\ -1 & s \end{bmatrix}^{-1} \begin{bmatrix} 1 & 0 \\ 0 & 1 \end{bmatrix} \mathbf{U}(s)$$

or

$$\mathbf{X}(s) = \begin{bmatrix} \dfrac{s}{s^2 + 1} & \dfrac{-1}{s^2 + 1} \\ \dfrac{1}{s^2 + 1} & \dfrac{s}{s^2 + 1} \end{bmatrix} \mathbf{U}(s). \tag{12.21}$$

Suppose that scalar feedback is to be achieved by feeding back $x_2(t)$ and using the single manipulated variable $u_1(t)$. It follows from Equation (12.21) that the plant transfer function

$$\frac{X_2(s)}{U_1(s)} = \frac{1}{s^2 + 1},$$

so that the block diagram illustrated in Figure 12.2a is appropriate. As is obvious, from an examination of the root locus plotted in Figure 12.2b, this type of feedback is completely ineffective in improving the performance of the system. Of course, if a compensator more complex than a simple gain were used, greater improvement could be expected.

For decoupling control, choose

$$\mathfrak{C}_c = \begin{bmatrix} -10 & 0 \\ 0 & -10 \end{bmatrix}$$

and

$$\mathfrak{B}_c = \begin{bmatrix} 10 & 0 \\ 0 & 10 \end{bmatrix}. \tag{12.22}$$

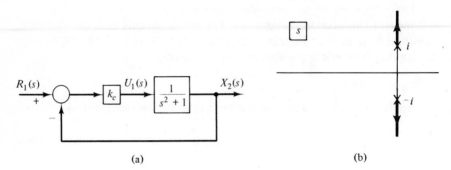

(a)   (b)

Figure 12.2. (a) Scalar control of the system of Example 12.1., (b) root locus for the system illustrated in (a).

It follows from Equations (12.13) and (12.14) that

$$\mathcal{F}_0 = \mathcal{J}^{-1}[\mathcal{Q} - \mathcal{Q}_c]$$

$$= \begin{bmatrix} 10 & -1 \\ 1 & 10 \end{bmatrix}$$

and that

$$\mathcal{P} = \mathcal{J}^{-1}\mathcal{B}_c$$

$$= \begin{bmatrix} 10 & 0 \\ 0 & 10 \end{bmatrix}. \tag{12.23}$$

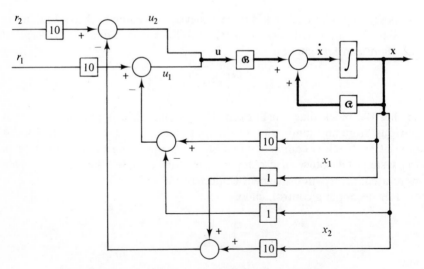

Figure 12.3. Decoupling control of the problem of Example 12.1. The feedback vectors are resolved into scalar signals in order to illustrate the complexity.

It follows from Equations (12.22) that the closed loop system has only real eigenvalues (corresponding to time constants of $\frac{1}{10}$ sec) and unity gains between set points and states. Obviously, considerable improvement in performance has been achieved.

One of the prices paid for this improvement is illustrated in Figure 12.3. Each of the gains and comparisons must be synthesized in the form of a piece of hardware. One cannot help but be concerned about the reliability of complex feedback systems. Part of this concern stems from the following question: If one of the gains or comparators fails, will the feedback system be stable?

$$* \quad * \quad *$$

12.2 Modal Control When the Number of Manipulated Variables and the Number of Outputs Are Equal to the Number of States[3]

Assume for a moment that the matrices $\mathfrak{B}$ and $\mathfrak{D}$ are square and have inverses. Assume further that the open loop eigenvalues are real and distinct. Consider the following choice of the $\mathfrak{K}$ and $\mathfrak{F}$ matrices (refer to Figure 10.22):

$$\mathfrak{F} = \mathfrak{J} \tag{12.24}$$

and

$$\mathfrak{K} = \mathfrak{B}^{-1}\mathfrak{J}\Sigma\mathfrak{J}^{-1}\mathfrak{D}^{-1}, \tag{12.25}$$

where $\mathfrak{J}$ is the *synthesizer* matrix for $\mathfrak{A}$ (refer to Section 8.7) and

$$\Sigma = \begin{bmatrix} \sigma_1 & 0 & 0 & \cdots & 0 \\ 0 & \sigma_2 & 0 & \cdots & 0 \\ \cdot & \cdot & \cdot & & \cdot \\ \cdot & \cdot & \cdot & & \cdot \\ \cdot & \cdot & \cdot & & \cdot \\ 0 & 0 & 0 & \cdots & \sigma_n \end{bmatrix}. \tag{12.26}$$

It follows that the closed loop equation (12.3) becomes

$$\dot{\mathbf{x}} = \{\mathfrak{A} - \mathfrak{J}\Sigma\mathfrak{J}^{-1}\}\mathbf{x} + \mathfrak{J}\Sigma\mathfrak{J}^{-1}\mathfrak{D}^{-1}\mathbf{r}. \tag{12.27}$$

By definition of the $\mathfrak{J}$ matrix

$$\mathfrak{A} = \mathfrak{J}\Lambda\mathfrak{J}^{-1}, \tag{12.28}$$

where Λ is the diagonal matrix of *open loop* eigenvalues. Combine Equations

(12.27) and (12.28) to show that in the closed loop

$$\dot{\mathbf{x}} = \mathfrak{T}\{\mathbf{\Lambda} - \mathbf{\Sigma}\}\mathfrak{T}^{-1}\mathbf{x} + \mathfrak{T}\mathbf{\Sigma}\mathfrak{T}^{-1}\mathfrak{D}^{-1}\mathbf{r}. \tag{12.29}$$

In other words, the closed loop coefficient matrix

$$\mathbf{\mathfrak{C}}_c = \mathfrak{T}\{\mathbf{\Lambda} - \mathbf{\Sigma}\}\mathfrak{T}^{-1}. \tag{12.30}$$

Two important and interesting facts are obtained from this equation:

1. The matrix, $\mathfrak{T}$, which diagonalizes the open loop matrix, $\mathbf{\mathfrak{C}}$, *also* diagonalizes the closed loop matrix, $\mathbf{\mathfrak{C}}_c$.
2. The closed loop eigenvalues

$$s_k = -(\lambda_k + \sigma_k). \tag{12.31}$$

Both of these conclusions apply only to the case when $\mathfrak{K}$ is selected as indicated by Equation (12.25).

It follows from Equation (12.31) that any open loop pole, λ_k, can be shifted to any desired closed location, s_k, by the proper choice of the parameter σ_k. Notice that each pole may be shifted independently.

The case when open loop poles are complex and/or repeated offers some complications which will not be discussed here but which have been dealt with in other places.

At this point, it is possible to consider more practical formulations of the modal control method.

12.3 Modal Control for the Case of Scalar Input and State Variable Feedback[3]

At this point the restrictive assumption that the number of manipulated variables equals the number of states (so that $\mathfrak{B}$ is square) is removed. In fact, it is assumed that only a single manipulated variable is available. It is also assumed that

$$\mathfrak{D} = \mathfrak{I}, \tag{12.32}$$

so that the open loop state variable equations become

$$\dot{\mathbf{x}} = \mathfrak{C}\mathbf{x} + \mathbf{b}u \tag{12.33}$$

and

$$\mathbf{y} = \mathbf{x}. \tag{12.34}$$

It is assumed that the system is controllable so that the lth element of the vector

$$(\mathfrak{T}^{-1}\mathbf{b})_l \neq 0 \tag{12.35}$$

for any l. Notice that the $\mathcal{K}$ matrix in Figure 10.22 is a row vector, which will be denoted $\mathbf{k}'$. Choose

$$\mathfrak{F} = \mathfrak{I}. \tag{12.36}$$

For this case the poles are shifted one at a time in the manner that will now be indicated. Choose the compensating vector

$$\mathbf{k}'_l = \boldsymbol{\sigma}'_l\mathfrak{T}^{-1}, \tag{12.37}$$

where $\mathfrak{T}^{-1}$ is the *analyzer* matrix for $\mathbf{C}$ and the vector $\boldsymbol{\sigma}_l$ has all zero components except the lth, which is denoted σ_l. It follows from Equation (12.3) that the closed loop system state equation becomes

$$\dot{\mathbf{x}} = \{\mathbf{C} - \mathbf{b}\boldsymbol{\sigma}'_l\mathfrak{T}^{-1}\}\mathbf{x} + \mathbf{b}\boldsymbol{\sigma}'_l\mathfrak{T}^{-1}\mathbf{r}, \tag{12.38}$$

so that the closed loop coefficient matrix

$$\begin{aligned} \mathbf{C}_c &= \mathbf{C} - \mathbf{b}\boldsymbol{\sigma}'_l\mathfrak{T}^{-1} \\ &= \mathfrak{T}\{\boldsymbol{\Lambda} - \mathfrak{T}^{-1}\mathbf{b}\boldsymbol{\sigma}'_l\}\mathfrak{T}^{-1}. \end{aligned} \tag{12.39}$$

Denote

$$\mathbf{h} = \mathfrak{T}^{-1}\mathbf{b}, \tag{12.40}$$

so that

$$\mathbf{C}_c = \mathfrak{T}\{\boldsymbol{\Lambda} - \mathbf{h}\boldsymbol{\sigma}'_l\}\mathfrak{T}^{-1}. \tag{12.41}$$

To understand the implications of the choice of the compensating matrix of Equation (12.37), it is necessary to determine the eigenvalues of $\mathbf{C}_c$, which are the closed loop poles s_i. These eigenvalues satisfy the characteristic equation

$$\begin{aligned} \det\{s\mathfrak{I} - \mathbf{C}_c\} &= \det\{s\mathfrak{I} - \mathfrak{T}[\boldsymbol{\Lambda} - \mathbf{h}\boldsymbol{\sigma}'_l]\mathfrak{T}^{-1}\} \\ &= \det\{\mathfrak{T}[s\mathfrak{I} - \boldsymbol{\Lambda} + \mathbf{h}\boldsymbol{\sigma}'_l]\mathfrak{T}^{-1}\} \\ &= \det\{\mathfrak{T}\mathfrak{T}^{-1}\}\det\{s\mathfrak{I} - \boldsymbol{\Lambda} + \mathbf{h}\boldsymbol{\sigma}'_l\} \\ &= 0. \end{aligned} \tag{12.42}$$

The next to last equality follows from the fact that $\det(\mathbf{C}\mathfrak{B}) = \det(\mathfrak{B}\mathbf{C})$ if $\mathbf{C}$ and $\mathfrak{B}$ are square. It follows that the closed loop eigenvalues satisfy the equation

$$\det\{s\mathfrak{I} - \boldsymbol{\Lambda} + \mathbf{h}\boldsymbol{\sigma}'_l\} = 0. \tag{12.43}$$

However,

$$s\mathcal{I} - \Lambda + \mathbf{h}\sigma'_l = \begin{bmatrix} s + \lambda_1 & 0 & \cdots & h_1\sigma_l & \cdots & 0 \\ 0 & s + \lambda_2 & \cdots & h_2\sigma_l & \cdots & 0 \\ \vdots & \vdots & & \vdots & & \vdots \\ 0 & 0 & \cdots & s + \lambda_l + h_l\sigma_l & \cdots & 0 \\ \vdots & \vdots & & \vdots & & \vdots \\ 0 & 0 & \cdots & h_n\sigma_l & \cdots & s + \lambda_n \end{bmatrix}$$

(12.44)

The determinant of this matrix is easily determined so that Equation (12.43) becomes

$$(s + \lambda_1)(s + \lambda_2) \cdots (s + \lambda_l + h_l\sigma_l) \cdots (s + \lambda_n) = 0. \qquad (12.45)$$

The conclusion is that the closed loop poles are the open loop poles except for the *l*th, which is given by

$$s_l = -(\lambda_l + h_l\sigma_l). \qquad (12.46)$$

Thus, the control law (12.37) has effected a shift in the *l*th eigenvalue only. It follows that this pole can be shifted to a desired location by the proper choice of σ_l *provided h_l is not zero*. However, the controllability condition (12.35) ensures that $h_l \neq 0$.

If it is desired to shift another pole, one merely repeats the above procedure except that $\mathbf{\alpha}_c$ of Equation (12.39) must now be used as the new open loop matrix. A new synthesizer matrix must be determined. This task is simplified somewhat by the fact that the new synthesizer differs from the previous synthesizer in only the *l*th column. The new *l*th column is found using Equation (12.46) and the eigenvector calculation method presented in Equation (8.92) or (8.93).

The method discussed in this section is much more practical than those discussed previously because one needs only a single manipulated variable in order to synthesize the feedback. Even more general methods are available. Of course, the requirement that all states be directly measurable is also somewhat restrictive. One of the applications of the theory presented in the next section is the removal of the requirement that all states be directly measurable.

12.4 *Dynamic Observers* for Linear Time-Invariant Dynamic Systems[4]

At this point the discussion of multivariable feedback design is put aside in order to introduce the seemingly unrelated concept of *dynamic observers*. As will be seen, however, observers play an important role in feedback design theory.

The goal of the user of an observer is illustrated in Figure 12.4. The observer is to provide an estimate, $z(t)$, of the state, $x(t)$, given only the input vector, $u(t)$, and the output vector, $c(t)$. For present purposes, only the case when $c(t)$ is a scalar need be considered. For this case, Equation (12.2) is

$$c(t) = d'x. \tag{12.47}$$

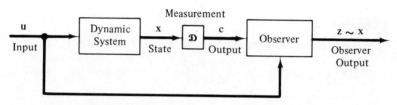

Figure 12.4. Schematic diagram of an observer which is to be designed so as to give an estimate of the state of a dynamic system.

It will be shown that an observer can be synthesized by applying the techniques of Chapter 8 to

$$\dot{z} = \mathcal{E}z + gc(t) + \mathcal{B}u, \tag{12.48}$$

where g is an arbitrary vector and

$$\mathcal{E} = \mathcal{C} - gd'. \tag{12.49}$$

To prove that Equation (12.48) is the proper starting point for the synthesis of an observer, notice that

$$\frac{d}{dt}\{x - z\} = \dot{x} - \dot{z}$$

$$= \mathcal{C}x + \mathcal{B}u - \{\mathcal{C} - gd'\}z - gd'x - \mathcal{B}u$$

or

$$\frac{d}{dt}\{x - z) = \{\mathcal{C} - gd'\}\{x - z\}. \tag{12.50}$$

Thus,

$$x(t) - z(t) = \phi(t)\{x(0) - z(0)\}$$

or

$$z(t) = x(t) + \phi(t)\{z(0) - x(0)\}, \tag{12.51}$$

where

$$\phi(t) = \mathcal{L}^{-1}\{(s\mathcal{I} - \mathbf{\alpha} - \mathbf{g}\mathbf{d}')^{-1}\}$$
$$= \mathcal{L}^{-1}\{(s\mathcal{I} - \mathbf{\epsilon})^{-1}\}. \tag{12.52}$$

The important conclusion to be obtained from Equations (12.51) and (12.52) is that if the eigenvalues of $\mathbf{\epsilon}$ are negative, $z(t) \rightarrow x(t)$; i.e., the dynamic system defined by Equation (12.48) is indeed an observer of the system defined by Equations (12.1) and (12.47). Good observer design is the selection of the vector $\mathbf{g}$ so that the eigenvalues of $\mathbf{\epsilon}$ [note Equation (12.49)] are such that $z(t)$ will approach $x(t)$ rapidly.

Example 12.2

Suppose that for the dynamic system defined by Equation (12.20),

$$\mathbf{d} = \begin{bmatrix} 0 \\ 1 \end{bmatrix}; \tag{12.53}$$

i.e., $x_2(t)$ is the output, $y(t)$. It follows from Equation (12.49) that the observer matrix

$$\mathbf{\epsilon} = \begin{bmatrix} 0 & -1 \\ 1 & 0 \end{bmatrix} - \begin{bmatrix} g_1 \\ g_2 \end{bmatrix} [0 \quad 1]$$
$$= \begin{bmatrix} 0 & -(1 + g_1) \\ 1 & -g_2 \end{bmatrix}, \tag{12.54}$$

so that the characteristic equation of the observer is

$$\lambda^2 + g_2\lambda + 1 + g_1 = 0, \tag{12.55}$$

where λ is an eigenvalue of the observer.
    Choose

$$g_2 = 2\sigma \tag{12.56}$$

and

$$g_1 = \sigma^2 - 1 \tag{12.57}$$

for some arbitrary σ. Then Equation (12.55) becomes

$$\lambda^2 + 2\sigma\lambda + \sigma^2 = (\lambda + \sigma)^2$$
$$= 0. \tag{12.58}$$

Thus, both eigenvalues are equal to the preselected value of $-\sigma$. Obviously, if σ is chosen to be large in magnitude, the observer output, $\mathbf{z}(t)$, will converge rapidly to the state, $\mathbf{x}(t)$.

*   *   *

It can be shown that the eigenvalues of $\boldsymbol{\varepsilon}$ can be preselected arbitrarily if the dynamic system is observable.[6]

12.5 The Application of Multivariable Techniques to the Design of Compensators for Scalar Variable Systems[4]

The design of scalar variable feedback systems was discussed in the previous chapter on root locus and will be dealt with, from a different point of view, in Chapter 16. When using these classical techniques, the design of compensators is a somewhat ambiguous trial and error procedure. In this section it is demonstrated that the design is a straightforward procedure when one combines the modern concepts of observer synthesis and modal control.

The scheme is illustrated in Figure 12.5. A dynamic observer is placed in the feedback loop to provide an estimate of a state variable representation

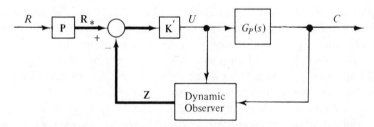

Figure 12.5. Solution of the scalar input-scalar output problem using observer synthesis and modal control.

associated with the plant transfer function, $G_p(s)$. The row vector, $\mathbf{k}'$, is then chosen in such a manner that some of the design objectives are met. For instance, Equation (12.37) could be used to fix $\mathbf{k}'$. The vector $\mathbf{p}$ is chosen so that $c(t)$ and its time derivatives will have the correct final values.

To illustrate some important and remarkable properties of the scheme illustrated in Figure 12.5, the transient properties of the feedback system will be investigated.

The state variable representation associated with $G_p(s)$ will have the form

$$\dot{\mathbf{x}} = \boldsymbol{\alpha}\mathbf{x} + \mathbf{b}u, \tag{12.59}$$

$$c = \mathbf{d}'\mathbf{x} \tag{12.60}$$

(refer to Chapter 8). Since

$$u = \mathbf{k}'(\mathbf{r}_* - \mathbf{z}),$$

it follows that for the feedback systems

$$\dot{\mathbf{x}} = \mathbf{\mathfrak{A}}\mathbf{x} + \mathbf{bk}'\mathbf{r}_* - \mathbf{bk}'\mathbf{z}. \tag{12.61}$$

Use Equation (12.51) to show that

$$\dot{\mathbf{x}} = \{\mathbf{\mathfrak{A}} - \mathbf{bk}'\}\mathbf{x} + \mathbf{bk}'\mathbf{r}_* + \mathbf{bk}'\boldsymbol{\phi}(t)\{\mathbf{x}(0) - \mathbf{z}(0)\}. \tag{12.62}$$

The matrix $\boldsymbol{\phi}(t)$ is defined by Equation (12.52) and, obviously, reflects the dynamic response of the observer. Notice that the final term in Equation (12.62) acts much like an additional input and is important only if an initial error exists between the state vector and the observer output.

It is important to compare Equation (12.62) with the feedback equation for the case where the state variables are directly measureable, i.e., for the case where a dynamic observer is not needed. It follows from Equation (12.3) that for this case

$$\dot{\mathbf{x}} = \{\mathbf{\mathfrak{A}} - \mathbf{bk}'\}\mathbf{x} + \mathbf{bk}'\mathbf{r}_*. \tag{12.63}$$

Comparison of Equations (12.62) and (12.63) leads to the remarkable conclusion that *the coefficient matrix for the system with an observer is identical to the coefficient matrix of the system where the state vector is directly measurable.* Since the closed loop poles are the eigenvalues of this matrix, *the closed loop poles of the system for which the state variables are directly measurable are also the closed loop poles of the scalar system with observer.* This fact greatly simplifies the task of calculating $\mathbf{k}'$. The presence of $\boldsymbol{\phi}$ in Equation (12.62) indicates that the *eigenvalues of the observer are also closed loop poles.*

To summarize, the designer chooses $\mathbf{k}'$ to meet objectives while assuming that the state variables are directly measurable. The observer is then designed so that the eigenvalues of the matrix $\mathbf{\boldsymbol{\varepsilon}}$ also satisfy constraints on the closed loop poles. The two design problems are conveniently decoupled.

Example 12.3

For the scalar system defined by

$$\frac{C(s)}{U(s)} = \frac{1}{s^2 + 1}, \tag{12.64}$$

design a feedback system so that all closed loop eigenvalues are less than $-10\,\text{sec}^{-1}$.

To use the scheme illustrated in Fig. 12.5, first assume that the state variables are directly measurable. Use the techniques of Chapter 8 to show that

$$\dot{\mathbf{x}} = \begin{bmatrix} 0 & -1 \\ 1 & 0 \end{bmatrix} \mathbf{x} + \begin{bmatrix} 1 \\ 0 \end{bmatrix} u \tag{12.65}$$

and

$$c = (0 \quad 1)\mathbf{x} \tag{12.66a}$$

is a state variable representation. It follows from Equation (12.63) that the closed loop poles satisfy

$$\det\{s_p\mathcal{I} - \mathcal{Q} + \mathbf{bk'}\} = \det\begin{bmatrix} s_p + k_1 & 1 + k_2 \\ -1 & s_p \end{bmatrix}$$

$$= 0.$$

Thus,

$$s_p^2 + k_1 s_p + 1 + k_2 = 0. \tag{12.66b}$$

If both poles satisfied the criterion, then

$$s_p^2 + 2(10)s_p + 10^2 = 0.$$

Compare these two equations and conclude that

$$\mathbf{k'} = (20 \quad 99). \tag{12.67}$$

The structure of the observer for this system was determined in Example 12.2. Since the eigenvalues of the observer will also be closed loop poles, it follows from Equation (12.58) that the analyst must specify

$$\sigma = 10$$

in order to meet the design specification so that

$$\mathbf{g} = \begin{bmatrix} 99 \\ 20 \end{bmatrix}. \tag{12.68}$$

The remaining task is to choose the **p** vector. It follows from Equation (12.62) that

$$\mathbf{x}(\infty) = -\{\mathcal{Q} - \mathbf{bk'}\}^{-1}\mathbf{bk'r_*}(\infty) \tag{12.69}$$

$$= \begin{bmatrix} -20 & -100 \\ 1 & 0 \end{bmatrix}^{-1} \begin{bmatrix} 20 & 99 \\ 0 & 0 \end{bmatrix} \mathbf{r_*}(\infty)$$

$$= \begin{bmatrix} 0 & 0 \\ 0.20 & 0.99 \end{bmatrix} \mathbf{r_*}(\infty).$$

It can then be shown that

$$c(\infty) = -\mathbf{d}'\{\mathfrak{R} - \mathbf{bk}'\}^{-1}\mathbf{bk}'\mathbf{p}r(\infty) \tag{12.70}$$

$$= (0 \quad 1)\begin{bmatrix} 0 & 0 \\ 0.20 & 0.99 \end{bmatrix}\begin{bmatrix} p_1 \\ p_2 \end{bmatrix}r(\infty)$$

$$= (0.20 \quad 0.99)\begin{bmatrix} p_1 \\ p_2 \end{bmatrix}r(\infty). \tag{12.71}$$

If the desired condition is that $c(\infty) = r(\infty)$, it follows from Equation (12.71) that one choice of the **p** vector is

$$\mathbf{p} = \begin{bmatrix} 0 \\ 1.01 \end{bmatrix}. \tag{12.72}$$

*   *   *

12.6 Comment on the Results of Linear Optimal Control Theory[1,2,5]

In previous sections of this chapter, the emphasis was placed on the synthesis of multivariable feedback systems which will have closed loop poles at desired locations. In this sense, the discussion has been an extension of Chapter 11 on root locus. For the sake of completeness, the results of another distinct branch of modern control theory will be stated and discussed in order to give the new student an opportunity to place the discussion of the previous sections in a certain perspective.

Once again, it is assumed that the open loop system is adequately described by Equations (12.1) and (12.2). It is now assumed that the feedback system is not to be designed to meet closed loop pole specifications but, rather, to minimize

$$I = \int_0^\infty \{\boldsymbol{\epsilon}'\boldsymbol{\mathcal{P}}\boldsymbol{\epsilon} + \boldsymbol{\xi}'\boldsymbol{\mathcal{W}}\boldsymbol{\xi}\} \, dt, \tag{12.73}$$

where $\boldsymbol{\mathcal{P}}$ and $\boldsymbol{\mathcal{W}}$ are *weighting matrices* and

$$\boldsymbol{\epsilon}(t) = \mathbf{c}(t) - \mathbf{c}(\infty), \tag{12.74}$$

$$\boldsymbol{\xi}(t) = \mathbf{u}(t) - \mathbf{u}(\infty). \tag{12.75}$$

This design criterion is discussed extensively in Section 10.10. The use of this type of criterion is common practice in *optimal control* theory.

The derivation of the synthesis of optimal control requires the use of advanced methods and will not be attempted here. The results will be stated and the computational methods described. The first result is that the structure of the optimal control is as indicated in Figure 10.22 and

$$\mathfrak{F} = \mathfrak{g}. \tag{12.76}$$

The remaining result that must be stated is the method for calculating the matrix $\mathfrak{K}$.

The calculation of the $\mathfrak{K}$ matrix is associated with the eigenvalue-eigenvector analysis of the matrix[1]

$$\mathfrak{M} = \left[\begin{array}{c:c} \mathfrak{C} & -\mathfrak{B}\mathfrak{W}^{-1}\mathfrak{B}' \\ \hdashline -\mathfrak{D}'\mathfrak{P}\mathfrak{D} & -\mathfrak{C}' \end{array}\right]. \tag{12.77}$$

Notice that $\mathfrak{M}$ is $2n \times 2n$ if the $\mathbf{x}$ vector is n-dimensional.

The eigenvalues of the $\mathfrak{M}$ matrix have some very interesting properties. First, if λ is an eigenvalue of $\mathfrak{M}$, then so is $-\lambda$. In other words, half of the eigenvalues of $\mathfrak{M}$ are the negatives of the other half. Second, and more importantly, *the eigenvalues of $\mathfrak{M}$ with negative real parts are the closed loop poles of the feedback system for which the choice of $\mathfrak{K}$ minimizes criterion* (12.73). In other words, the negative eigenvalues of $\mathfrak{M}$ are the closed loop poles of the optimally controlled feedback system.

It will now be shown that the matrix $\mathfrak{K}$ can be calculated by properly combining submatrices of the synthesizer matrix of $\mathfrak{M}$. As was indicated in Chapter 8, a fundamental result of the eigenvector analysis of a matrix is a diagonal matrix of the eigenvalues of the original matrix. For the $\mathfrak{M}$ matrix, partition this diagonal matrix as

$$\Lambda = \left[\begin{array}{c:c} \Lambda_+ & 0 \\ \hdashline 0 & \Lambda_- \end{array}\right], \tag{12.78}$$

where

$$0 = \text{matrix of zeros,} \tag{12.79}$$

$$\Lambda_+ = \text{diagonal matrix of the positive eigenvalues of } \mathfrak{M}, \tag{12.80}$$

$$\Lambda_- = \text{diagonal matrix of the negative eigenvalues of } \mathfrak{M}. \tag{12.81}$$

As stated in the previous paragraph,

$$\Lambda_- = -\Lambda_+. \tag{12.82}$$

Let $\mathfrak{T}$ denote the analyzer matrix of $\mathfrak{M}$ for Λ in the form indicated by Equation (12.78) and impose the partitioning

$$\mathfrak{T} = \left[\begin{array}{c:c} \phi_+ & \phi_- \\ \hdashline \psi_+ & \psi_- \end{array}\right]. \tag{12.83}$$

The fundamental result is that optimal control is achieved by selecting[1]

$$\mathcal{K} = \mathcal{W}^{-1}\mathcal{B}'\mathbf{\psi}_-\mathbf{\phi}_-^{-1}. \tag{12.84}$$

Thus, to calculate $\mathcal{K}$ one need only obtain the eigenvectors associated with the negative eigenvalues of $\mathfrak{M}$.

Example 12.4

Determine the optimal control and associated closed loop poles for the open loop system described by

$$\dot{\mathbf{x}} = \begin{bmatrix} 0 & -1 \\ 1 & 0 \end{bmatrix}\mathbf{x} + \begin{bmatrix} 1 \\ 0 \end{bmatrix}u, \tag{12.85}$$

$$c = (0 \quad 1)\mathbf{x}, \tag{12.86}$$

and for the choices

$$\mathcal{P} = \mathcal{J} \tag{12.87}$$

and

$$\mathcal{W} = 1.0. \tag{12.88}$$

For this case

$$\mathfrak{M} = \begin{bmatrix} 0 & -1 & -1 & 0 \\ 1 & 0 & 0 & 0 \\ 0 & 0 & 0 & -1 \\ 0 & -1 & 1 & 0 \end{bmatrix}. \tag{12.89}$$

The characteristic equation of $\mathfrak{M}$ is

$$(\lambda^2 + 1)^2 + 1 = 0,$$

so that

$$\lambda^2 + 1 = \pm i.$$

In other words, eigenvalues satisfy either

$$\lambda^2 = -1 + i$$
$$= \sqrt{2}\, e^{i(3\pi/4)} \tag{12.90}$$

or

$$\lambda^2 = -1 - i$$
$$= \sqrt{2}\, e^{-i(3\pi/4)}. \tag{12.91}$$

It follows from Equation (12.90) that

$$\lambda_1 = 0.455 + 1.08i,$$
$$\lambda_4 = -0.455 - 1.08i, \tag{12.92}$$

and it follows from Equation (12.91) that

$$\lambda_2 = 0.455 - 0.108i,$$
$$\lambda_3 = -0.455 + 1.08i. \tag{12.93}$$

Notice that Equation (12.82) is verified for this special case. As indicated in the text, λ_3 and λ_4 are the closed loop poles of the optimally controlled feedback systems. The complex closed loop poles are characteristic of optimally controlled systems.

Use Equation (8.92) to show that the last two eigenvectors of $\mathfrak{M}$ are

$$\mathbf{e}_3 = \begin{bmatrix} -i\lambda_3 \\ i \\ 1 \\ -\lambda_3 \end{bmatrix},$$

$$\mathbf{e}_4 = \begin{bmatrix} i\lambda_4 \\ i \\ 1 \\ -\lambda_4 \end{bmatrix}.$$

Since

$$\begin{bmatrix} \boldsymbol{\phi}_- \\ \hline \boldsymbol{\psi}_- \end{bmatrix} = [\mathbf{e}_3 \mid \mathbf{e}_4],$$

it follows that

$$\boldsymbol{\psi}_- \boldsymbol{\phi}_-^{-1} = \begin{bmatrix} 1 & 1 \\ -\lambda_3 & -\lambda_4 \end{bmatrix} \begin{bmatrix} -i\lambda_3 & i\lambda_4 \\ -i & i \end{bmatrix}^{-1}$$
$$= \begin{bmatrix} \dfrac{2i}{\lambda_3 - \lambda_4} & \dfrac{-i(\lambda_3 + \lambda_4)}{\lambda_3 - \lambda_4} \\ \dfrac{-i(\lambda_3 + \lambda_4)}{(\lambda_3 - \lambda_4)} & \dfrac{2i\lambda_3\lambda_4}{\lambda_3 - \lambda_4} \end{bmatrix}. \tag{12.94}$$

It follows from Equations (12.84), (12.88), and (12.94) that

$$\mathcal{K} = \begin{bmatrix} \dfrac{2i}{\lambda_3 - \lambda_4} & \dfrac{-i(\lambda_3 + \lambda_4)}{\lambda_3 - \lambda_4} \end{bmatrix}$$
$$= [0.91 \quad 0.42]. \tag{12.95}$$

* * *

References

1. BRYSON, A. E., and HALL, W. E., "Optimal Control and Filter Synthesis by Eigenvector Decomposition", *NAS 2–5143*, Dept. of Aeronautics and Astronautics, Stanford Univ., Dec. 1971.

2. "Special Issue on Linear-Quadratic-Gaussian Problem," *I.E.E.E. Transactions on Optimal Control, AC-16*, No. 6, Dec. 1971.

3. LOSCUTOFF, W. V., *Modal Analysis and Synthesis of Linear Lumped Parameter Systems*, Ph.D. Dissertation, Dept. of Mechanical Engineering, Univ. of California, Berkeley, 1968.

4. LUENBERGER, D. G., "An Introduction to Observers," *I.E.E.E. Transactions on Automatic Control, AC-16*, No. 6, Dec. 1971.

5. SCHULTZ, D. G., and MELSA, J. L., *State Functions and Linear Control Systems*, McGraw-Hill, New York, 1967.

6. WONHAM, W. M., "On Pole Assignment in Multi-Input Controllable Linear Systems," *I.E.E.E. Transactions on Automatic Control, AC-12*, No. 6., Dec. 1967, pp. 660–666.

Problems

12.1. Consider the multivariable servo system in Figure 12.6. The inputs are the *effort*, u_1, and the *velocity*, u_2. The outputs are the *stress*, σ_m, and the *velocity*, f_m. Determine the matrix-vector equation relating

$$\begin{bmatrix} \sigma_m \\ f_m \end{bmatrix} \quad \text{to} \quad \begin{bmatrix} u_1 \\ u_2 \end{bmatrix}$$

for this open loop system.

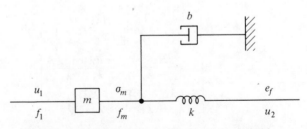

Figure 12.6. Multivariable control of stress and position.

12.2. Apply decoupling and then modal control feedback to the system in Figure 12.6. Construct schematic diagrams showing the synthesis of the two types of feedback (refer to Figure 12.3).

12.3. Consider the following biological control problem: Let

$$p_1 \triangleq \text{Insect population,}$$

$p_2 \triangleq$ Predator population,

$u_1 \triangleq$ Insecticide concentration,

$u_2 \triangleq$ Predator food supplied by human managers.

Assume that the growth of p_1 is equal to p_1 minus the predator population, p_2, minus the insecticide concentration, u_1, and that the rate of growth of p_2 is equal to p_2 plus p_1 plus the food supply, u_2, minus the insecticide level, u_1. Devise a feedback scheme for decoupling the insect-predator populations from one another and from desired population reference inputs. The desired growth time constants are T_1, T_2 for p_1, p_2, respectively.

12.4. Consider the simplified model of a hydraulic system illustrated in Figure 12.7. e and f are the outputs. Show that this system is described by the vector-matrix differential equation

$$\frac{d}{dt}\begin{bmatrix} e \\ f \end{bmatrix} = \begin{bmatrix} 0 & \dfrac{-1}{C_1} \\ \dfrac{1}{C_2} & -\left[\dfrac{1}{C_1} + \dfrac{2}{C_2}\right] \end{bmatrix}\begin{bmatrix} e \\ f \end{bmatrix} + \begin{bmatrix} \dfrac{1}{C_1} & 0 \\ \dfrac{1}{C_1} & \dfrac{1}{C_2} \end{bmatrix}\begin{bmatrix} u_1 \\ u_f \end{bmatrix}.$$

Devise a decoupling feedback control scheme for this system.

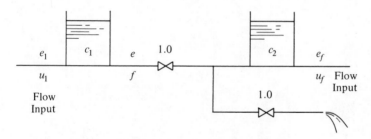

Figure 12.7. Open loop multivariable hydraulic system.

12.5. Show that if the lth and the mth poles are shifted using the method of Section 12.3, the net result is to set

$$\mathbf{k}' = \mathbf{k}'_l + \mathbf{k}'_m,$$

where $\mathbf{k}'_l$ and $\mathbf{k}'_m$ are defined by Equation (12.37).

12.6. Design an observer for

$$\dot{\mathbf{x}} = \begin{bmatrix} 0 & 1 \\ -1 & -2 \end{bmatrix}\mathbf{x} + \begin{bmatrix} 0 \\ 1 \end{bmatrix}u,$$

$$c = \begin{bmatrix} 1 & 0 \end{bmatrix}\mathbf{x}.$$

12.7. Apply the synthesis illustrated in Figure 12.5 to the system described in Problem 12.6. Assume that all closed loop poles must be less than -2 sec^{-1}.

12.8. Use Equations (12.66) and (12.94) to check Equations (12.92) and (12.93).

12.9. Determine the optimal feedback system for the first-order open loop system

$$\dot{x} = x + u.$$

Assume that

$$c = x$$

and that

$$\mathbf{\mathcal{P}} = 1.0$$

and that

$$\mathbf{\mathcal{W}} = w.$$

12.10. Indicate why the closed loop pole of the optimal system of Problem 12.9 varies with w in the manner that it does.

12.11. Show that for the ideal case for decoupling control, the steady state error can be set at any predetermined value by making the proper choice of the k_i in Equation (12.12).

Frequency Domain Analysis and Alternative Design Methods

13

Webster's Dictionary defines as follows:

Formalism—Structure, customs, outward forms of a discipline.

Chapter 13 is an introduction to a feedback design formalism which is an alternative to that discussed to this point. Table 13.1 contains a comparison of the pole zero (or root locus) formalism and the *frequency domain* formalism, which will be introduced.

Table 13.1
COMPARISON OF THE TWO ALTERNATIVE FORMALISMS

Attribute	Root locus	Frequency domain
Signals (inputs, outputs)	*Forcing* function described by pole zero diagrams in the Laplace domain or as functions of time	Superposition of sine wave (*spectral* or *harmonic*) components
Systems	Described by differential equations, transfer functions, or by vector-matrix (state variable) equations	Described by *Bode* or *Nyquist* diagrams (i.e., by *filter* properties); taken to be analogous to optical filters[4]
Chief limitation	Difficult to apply to distributed or delay systems	Only rough approximations of the parameters of transient response are obtained

The frequency domain formalism is also based on the use of an integral transform: the *Fourier* transform. The development of this subject will be

1. Start with the analysis of the Fourier series of periodic signals.
2. Extend the concepts to the more important set of aperiodic signals (Fourier transform theory).

These concepts can also be extended to an important subset of random signals, but such a step will not be taken here.

The order of topics in this text does not coincide with the historical order of development, which is

1. Frequency domain (circa 1925).
2. Root locus (1948).
3. State space (1958).

13.1 Fourier Series for Periodic Functions of Time[1, 2]

A periodic function of time with period T is a function which satisfies the equation

$$f(t) = f(t + nT) \tag{13.1}$$

for *all* positive *and* negative integers n (note Figure 13.1). Notice that periodic functions are not "causal" by definition.

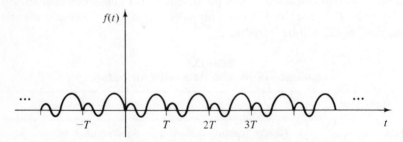

Figure 13.1. Graph of a periodic function.

One might ask, Is it possible to find constants a_0, $\{a_n\}$, $\{b_n\}$ such that

$$f(t) = \frac{a_0}{2} + \sum_{n=1}^{\infty} [a_n \cos (n\omega_0 t) + b_n \sin (n\omega_0 t)], \tag{13.2a}$$

$$\omega_0 \triangleq \frac{2\pi}{T}? \tag{13.2b}$$

The answer is yes if $f(t)$ satisfies the *Direchlet conditions*:

1. $f(t)$ has at most a finite number of discontinuities in one period.

$$(13.3)$$

2. $f(t)$ has at most a finite number of maxima and minima in one period.

$$(13.4)$$

3. The integral

$$\int_{-T/2}^{T/2} |f(t)| \, dt$$

exists. (13.5)

Moreover, it can be shown that

$$a_0 = \frac{2}{T} \int_{-T/2}^{T/2} f(t) \, dt, \tag{13.6}$$

$$a_n = \frac{2}{T} \int_{-T/2}^{T/2} f(t) \cos(n\omega_0 t) \, dt, \quad n = 0, 1, 2, \ldots, \tag{13.7}$$

$$b_n = \frac{2}{T} \int_{-T/2}^{T/2} f(t) \sin(n\omega_0) \, dt, \quad n = 1, 2, 3, \ldots. \tag{13.8}$$

Denote the right-hand side of Equation (13.2a) by $g(t)$; then $g(t) \equiv f(t)$ if $f(t)$ is *continuous*, but $g(t)$ is the *best approximation* of $f(t)$ if $f(t)$ is *discontinuous* in the sense that

$$\int_{-T/2}^{T/2} [f(t) - g(t)]^2 \, dt \tag{13.9}$$

is minimized when the a_n and b_n are defined by Equations (13.6)–(13.8). The minimum value of the above integral is not zero.

An example of the last statement is the square wave shown in Figure 13.2

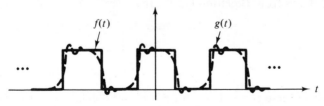

Figure 13.2. Square wave and its Fourier series approximation.

with a sketch of its Fourier series approximation. The fact that $g(t) \not\equiv f(t)$ is called the *Gibbs phenomenon*.

A slightly more complete introduction to Fourier series is provided in the next section.

13.2   Mathematical Aside: The Concepts of Orthogonality and Completeness for Fourier Series*

A set of *vectors* $\mathbf{x}_1, \mathbf{x}_2, \ldots, \mathbf{x}_k, \ldots$ is said to be *orthogonal* if

$$\mathbf{x}_i' \cdot \mathbf{x}_k = 0, \quad i \neq k. \tag{13.10}$$

A set of *vectors* $\mathbf{x}_1, \mathbf{x}_2, \ldots, \mathbf{x}_k, \ldots, \mathbf{x}_n$ is said to be *complete* if given any vector $\mathbf{g}$ there exists a set of numbers a_k such that

$$\mathbf{g} = a_1\mathbf{x}_1 + a_2\mathbf{x}_2 + \cdots + a_k\mathbf{x}_k + \cdots + a_n\mathbf{x}_n. \tag{13.11}$$

Notice that $\mathbf{e}_1, \mathbf{e}_2$ of Figure 13.3 do *not* form a set of complete vectors in three space; however, $\mathbf{e}_1, \mathbf{e}_2, \mathbf{e}_3$ do form such a set.

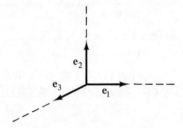

Figure 13.3. Unit vectors in three space.

Suppose that the set of vectors $\{\mathbf{x}_k\}$ is orthogonal *and* complete; then for any $\mathbf{g}$, Equation (13.11) is satisfied. To find the constant a_k, take the inner product of both sides of Equation (13.11) with $\mathbf{x}_k$ to find

$$\mathbf{x}_k' \cdot \mathbf{g} = a_1\mathbf{x}_k'\mathbf{x}_1 + a_2\mathbf{x}_k'\mathbf{x}_2 + \cdots + a_k\mathbf{x}_k'\mathbf{x}_k + \cdots + a_n\mathbf{x}_k'\mathbf{x}_n.$$

Thus, it follows from Equation (13.10) that

$$a_k = \frac{\mathbf{x}_k'\mathbf{q}}{\mathbf{x}_k'\mathbf{x}_k}. \tag{13.12}$$

To extend these concepts to sets of *functions*, define the *inner product* of two functions $\phi(t)$ and $\psi(t)$ with respect to a *weighting function*, $r(t)$, by

$$\langle \phi, \psi \rangle = \int_a^b r(t)\phi(t)\psi(t)\,dt. \tag{13.13}$$

---

* This section may be omitted during the first reading of the text.

A set of functions $\phi_0(t)$, $\phi_1(t)$, ..., $\phi_k(t)$, ... is said to be *orthogonal* if

$$\langle \phi_i(t), \phi_k(t) \rangle = 0, \quad i \neq k \tag{13.14}$$

[analogous to the definition of Equation (13.10)]. A set of functions $\phi_0(t) \cdots$ $\phi_k(t)$ is said to be *complete* in the interval (a, b) if given any function $g(t)$ there exists a set of numbers a_k such that

$$g(t) = a_0\phi_0(t) + a_1\phi_1(t) + \cdots + a_k\phi_k(t) + \cdots \tag{13.15}$$

in the interval (a, b).

If the set of functions $\{\phi_k(t)\}$ is orthogonal *and* complete, then the coefficients

$$a_k = \frac{\langle \phi_k, g \rangle}{\langle \phi_k, \phi_k \rangle}, \tag{13.16}$$

a fact which is proved in a manner similar to that used to establish Equation (13.12).

Example 13.1

In any table of integrals, find that

$$\int_{-T/2}^{T/2} \cos(n\omega_0 t) \cos(m\omega_0 t)\, dt = \begin{cases} 0, & \text{if } n \neq m, \\ \dfrac{T}{2}, & \text{if } n = m, \end{cases} \tag{13.17}$$

$$\omega_0 = \frac{2\pi}{T}, \quad n, m \text{ integer.}$$

Thus the set of functions

$$\phi_k(t) = \cos(k\omega_0 t), \quad k \text{ integer,} \tag{13.18}$$

is orthogonal. This set of functions is *not* complete in general.

* * *

Example 13.2

It can also be shown that

$$\int_{-T/2}^{T/2} \sin(n\omega_0 t) \sin(m\omega_0 t)\, dt = \begin{cases} 0, & \text{if } n \neq m, \\ \dfrac{T}{2}, & \text{if } n = m, \end{cases} \tag{13.19}$$

$$\omega_0 = \frac{2\pi}{T}, \quad n, m \text{ integer;}$$

thus,

$$\psi_k(t) = \sin(k\omega_0 t) \tag{13.20}$$

is orthogonal (but not complete in general).

* * *

Example 13.3

It is also true that

$$\int_{-T/2}^{T/2} \sin(n\omega_0 t) \cos(m\omega_0 t)\, dt = 0 \quad \text{for any } m, n$$

$$\omega_0 = \frac{2\pi}{T}. \tag{13.21}$$

Thus, the combined set of functions

$$\{\sin(k\omega_0 t),\ \cos(k\omega_0 t)\} \tag{13.22}$$

is orthogonal. This combined set of functions is complete if one restricts the membership of functions $g(t)$ as indicated by Equations (13.3)–(13.5).

It follows that Equations (13.16) and (13.17) combine to

$$a_n = \frac{\langle g(t), \cos(n\omega_0 t)\rangle}{\langle \cos(n\omega_0 t), \cos(n\omega_0 t)\rangle} = \frac{2}{T}\int_{-T/2}^{T/2} g(t)\cos(n\omega_0 t)\, dt.$$

Similarly,

$$b_n = \frac{2}{T}\int_{-T/2}^{T/2} g(t)\sin(n\omega_0 t)\, dt,$$

which are summaries of Equations (13.6) (13.7), and (13.9).

* * *

Bessel functions, *Legendre* polynomials, and *Hermite* polynomials are other examples of sets of functions which satisfy Equation (13.4) and which are complete for restricted but important classes of functions.

13.3   The Concept of a *Spectrum* of a Periodic Function

An elementary trigonometric identity is

$$a_n \cos(n\omega_0 t) + b_n \sin(n\omega_0 t) = r_n \sin(n\omega_0 t + \phi_n), \tag{13.23}$$

where

$$r_n = \sqrt{a_n^2 + b_n^2},\qquad(13.24)$$

$$\phi_n = \tan^{-1}\left\{\frac{a_n}{b_n}\right\},\qquad(13.25)$$

so that Equation (13.2a) may be rewritten

$$f(t) = \sum_{n=0}^{\infty} r_n \sin(n\omega_0 t + \phi_n),\qquad(13.26)$$

r_n being the *amplitude* of the nth harmonic and ϕ_n its *phase*.

Example 13.4

Consider the triangle wave illustrated in Figure 13.4a. Since $f(t)$ has no discontinuity, the Fourier series $g(t) \equiv f(t)$. Use Equations (13.6)–(13.8) to show that

$$b_n = 0\qquad(13.27)$$

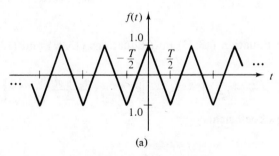

(a)

Figure 13.4. (a) Triangle wave.

for all n and that

$$a_n = 0\qquad(13.28)$$

if n is even; otherwise,

$$a_n = \frac{4}{T}\int_0^{T/2}\left[1 - \frac{4t}{T}\right]\cos\{n\omega_0 t\}\,dt$$

$$= \frac{-16}{T^2}\int_0^{T/2} t\cos\{n\omega_0 t\}\,dt$$

$$= \frac{-16}{T^2}\left[\frac{\cos\{n\omega_0 t\}}{n^2\omega_0^2} + \frac{t\sin\{n\omega_0 t\}}{n\omega_0}\right]_0^{T/2}.$$

Thus, if n is odd,

$$a_n = \frac{4}{n^2\pi^2}\{1 - \cos(n\pi)\}. \tag{13.29}$$

For this case $r_n = |a_n|$. The r_n are plotted against frequency in Figure 13.4b. Because of the similarity of this diagram to those encountered in the field of physical optics, the diagram will be referred to as the *spectrum* of the triangle wave.

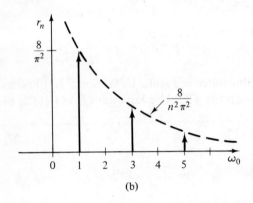

Figure 13.4. (b) Spectrum of the triangle wave.

(b)

*   *   *

Substitute Equation (13.23) into Equations (13.2a) and (13.6)–(13.8) to find

$$f(t) = \frac{a_0}{2} + \sum_{n=1}^{\infty}\left\{\left[\frac{a_n - ib_n}{2}\right]e^{in\omega_0 t} + \left[\frac{a_n + ib_n}{2}\right]e^{-in\omega_0 t}\right\}. \tag{13.30}$$

Define the new coefficients

$$\alpha_0 \triangleq \frac{a_0}{2},$$

$$\alpha_n \triangleq \frac{1}{2}\{a_n - ib_n\},$$

and

$$\alpha_{-n} \triangleq \frac{1}{2}\{a_n + ib_n\}. \tag{13.31}$$

Then Equation (13.24) becomes

$$f(t) = \alpha_0 + \sum_{n=1}^{\infty}[\alpha_n e^{in\omega_0 t} + \alpha_{-n}e^{-in\omega_0 t}]. \tag{13.32}$$

Compare Equations (13.24) and (13.25) with Equation (13.31) to show that the magnitude

$$|\alpha_n| = \sqrt{a_n^2 + b_n^2} = r_n \tag{13.33}$$

and the angle

$$<\alpha_n = \tan^{-1}\left\{-\frac{b_n}{a_n}\right\} = -\phi_n. \tag{13.34}$$

Use Equations (13.7) and (13.8) to demonstrate that

$$\begin{aligned} a_n &= a_{-n}, \\ b_n &= -b_{-n}. \end{aligned} \tag{13.35}$$

Thus, it follows from Equation (13.31) that

$$\alpha_n = \alpha_{-n}. \tag{13.36}$$

Change the sign of n in the second summation in Equation (13.32) and show that Equation (13.36) leads to the conclusion that

$$\sum_{n=1}^{\infty} \alpha_{-n} e^{-in\omega_0 t} = \sum_{n=-1}^{-\infty} \alpha_n e^{in\omega_0 t}.$$

Thus Equation (13.32) reduces to

$$f(t) = \sum_{n=-\infty}^{\infty} \alpha_n e^{in\omega_0 t}. \tag{13.37}$$

Notice that Equations (13.7), (13.8), and (13.31) lead to

$$\alpha_n = \frac{1}{2}\left[\frac{2}{T}\int_{-T/2}^{T/2} f(t)\cos(n\omega_0 t)\,dt - \frac{2i}{T}\int_{-T/2}^{T/2} f(t)\sin(n\omega_0 t)\,dt\right]$$

or

$$\alpha_n = \frac{1}{T}\int_{-T/2}^{T/2} f(t)e^{-in\omega_0 t}\,dt. \tag{13.38}$$

Equations (13.37) and (13.38) are a very compact form of the Fourier series for a periodic function. This form has the added advantage that the complex Fourier coefficients, α_n, may be interpreted in terms of the amplitude and phase of the nth harmonic.

13.4 Extension of the Concepts of Fourier Analysis to Aperiodic Functions of Time

A schematic diagram of the set of all signals is presented in Figure 13.5. To this point, only the Fourier analysis of signals in the *periodic* class has been considered.

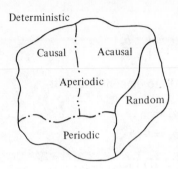

Figure 13.5. *Universe* of functions of time. The major division is between random and *deterministic* (i.e., non-random) signals.

The concepts of Fourier analysis can be extended to significant portions of all classes of signals represented in this figure *including* the class of random signals. The extension to random signals will be alluded to in an intuitive way, however; no formal introduction to these ideas will be made. The extension to the case of *aperiodic* signals [signals which do not satisfy Equation (13.1)] will, however, be discussed at some length. Notice that the set of causal functions, which is the domain of Laplace transformation, is a subset of the set of aperiodic functions.

To prepare the discussion, define

$$\omega(n) = n\omega_0. \tag{13.39}$$

Combine this definition with Equations (13.2b), (13.37), and (13.38) to show that

$$\alpha_n = \frac{\omega_0}{2\pi} \int_{-T/2}^{T/2} f(t)e^{-in\omega_0 t}\, dt \tag{13.40}$$

and

$$f(t) = \frac{1}{2\pi} \sum_{n=-\infty}^{\infty} e^{i\omega t} \int_{-T/2}^{T/2} f(t)e^{-in\omega_0 t}\, dt\omega_0. \tag{13.41}$$

The extension to aperiodic functions is accomplished by using the following trick: For the purposes of a Fourier analysis let

$$T \longrightarrow \infty. \tag{13.42}$$

Then one can consider *any* aperiodic $f(t)$ as periodic but with infinite period. The question to be considered now is, What form do the Fourier series formuli, Equations (13.40) and (13.41), take when $T \longrightarrow \infty$? This question will be answered in a *formal* (i.e., nonrigorous) fashion.

As $T \longrightarrow \infty$, one might suppose that

$$\omega_0 \longrightarrow d\omega$$

and that

$$\frac{1}{T} \longrightarrow \frac{d\omega}{2\pi}.$$

Hence Equation (13.41) becomes

$$f(t) = \sum_{\substack{\omega = -\infty \\ \omega_0}}^{\infty} e^{i\omega t} \frac{d\omega}{2\pi} \int_{-\infty}^{\infty} f(t)e^{-i\omega t}\, dt$$

$$= \int_{-\infty}^{\infty} e^{i\omega t} \frac{1}{2\pi} \int_{-\infty}^{\infty} f(t)e^{-i\omega t}\, dt\, d\omega. \tag{13.43}$$

Define the *Fourier transform* of $f(t)$:

$$F(\omega) \triangleq \mathfrak{F}[f(t)] = \int_{-\infty}^{\infty} f(t)e^{-i\omega t}\, dt. \tag{13.44}$$

It follows from Equation (13.43) that

$$f(t) \doteq \frac{1}{2\pi} \int_{-\infty}^{\infty} F(\omega)e^{i\omega t}\, d\omega. \tag{13.45}$$

Note:

$$\mathfrak{F}^{-1}[F(\omega)] \equiv f(t)$$

if $f(t)$ is continuous and is the best *approximation* if $f(t)$ is discontinuous [refer to the discussion preceding Equation (13.9)]. Notice the two common types of notation for Fourier transforms illustrated in Equation (13.44). Note also the similarity to Laplace transform notation.

It follows from Equations (13.40) and (13.44) that

$$F(n\omega_0)\omega_0 = \alpha_n,$$

which leads to the rough interpretation

$$F(\omega)\, d\omega = \alpha(\omega). \tag{13.46}$$

Thus, $|F(\omega)|$ is called the *relative frequency distribution* of the spectrum of $f(t)$ and is an amplitude *density* of the spectral content of $f(t)$.

Notice that for aperiodic functions, the spectrum is continuous as opposed to the discrete type of spectrum associated with periodic functions.

13.5 Relationship Between Laplace and Fourier Transforms

The basic definitions of the Fourier and Laplace transforms are

$$F(\omega) = \mathfrak{F}[f(t)] = \int_{-\infty}^{\infty} f(t)e^{-i\omega t}\, dt$$

and

$$F_L(s) = \mathcal{L}[f(t)] = \int_{0^-}^{\infty} f(t)e^{-(\sigma+i\omega)t}\,dt$$

for all σ for which the integral converges (i.e., in the region of convergence).
It follows that

$$F(\omega) = F_L(s)\Big|_{s=i\omega} \qquad (13.47)$$

if

1. $f(t) \equiv 0$, $t < 0$, and
2. The region of convergence for the Laplace transform includes $\sigma = 0$.

$$(13.48)$$

13.6 The *Uncertainty Principle* of Fourier Transform Theory

Suppose that $f(t)$ is Fourier transformable: $f(t)$ and its Fourier transform are sketched in Figures 13.6 and 13.7.

Assume that numbers Δt and $\Delta\omega$ exist such that

$$f(t) \simeq 0 \quad \text{for } |t| > \Delta t \qquad (13.49)$$

and

$$F(\omega) \simeq 0 \quad \text{for } |\omega| > \Delta\omega. \qquad (13.50)$$

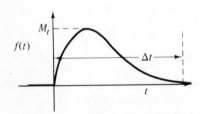

Figure 13.6. Aperiodic function.

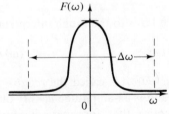

Figure 13.7. Fourier transform of $f(t)$ in Figure 13.6.

More specifically, $f(t)$, $F(\omega)$ may be taken as approximately zero when they reach some preselected, but arbitrary, percentage of their maximum values.

Since

$$e^{i\omega t} = \cos(\omega t) + i\sin(\omega t),$$

it follows that

$$|e^{i\omega t}| = 1.0 = |e^{-i\omega t}|. \qquad (13.51)$$

Thus, from Equations (13.44) and (13.51), it follows that

$$|F(\omega)| \leq M_t \cdot \Delta t \quad \text{for all } \omega, \tag{13.52}$$

where, by definition,

$$M_t = \max_t |f(t)|. \tag{13.53}$$

Similarly, it follows from Equations (13.45) and (13.51) that

$$|f(t)| \leq \frac{1}{2\pi} M_\omega \cdot \Delta\omega \quad \text{for all } t, \tag{13.54}$$

where, by definition,

$$M_\omega = \max_\omega |F(\omega)|. \tag{13.55}$$

In particular, Equation (13.52) indicates that

$$M_\omega \leq M_t \cdot \Delta t$$

and Equation (13.54) indicates that

$$M_t \leq \frac{1}{2\pi} M_\omega \, \Delta\omega.$$

Combining the last two inequalities leads to the conclusion that

$$M_t \leq \frac{1}{2\pi} \Delta\omega M_t \, \Delta t.$$

Finally,

$$\Delta\omega \cdot \Delta t \geq 2\pi \tag{13.56}$$

(ω in radians per second).

Equation (13.56) will be referred to as the *uncertainty principle* and will be useful in discussions of control system design.

13.7 The Convolution Theorem, Bandwidth, and Speed of Response

Consider a linear dynamic system described by the transfer function

$$C(s) = G(s)R(s). \tag{13.57}$$

As was shown in Chapter 2, if

$$g(t) = \mathcal{L}^{-1}G(s) \tag{13.58}$$

(interpreted as the *impulse response*), the output is given by

$$c(t) = \int_0^t g(\xi)r(t - \xi)\,d\xi. \tag{13.59}$$

One says that $g(t)$ is *convolved* in this way with the input to obtain the output.

The *convolution* integral, Equation (13.59), can be manipulated to yield a quantitative measure of that important dynamic system design criterion, *speed of response*. Begin this development with a careful consideration of each term in Equation (13.59).

The graph of $g(t)$, the impulse response of a dynamic system, will have the general appearance illustrated in Figure 13.8. This will often be true even if $G(s)$ is a transcendental function of s. Use physical reasoning to show that

$$g(t) \equiv 0, \quad t < 0. \tag{13.60}$$

A typical input to a dynamic system might be a step. The input term in Equation (13.59) has argument $t - \xi$ instead of t. The graph of $r(t - \xi)$ is shown in Figure 13.9. Combining Figures 13.8 and 13.9, one can sketch the integrand of the convolution integral as in Equation (13.59). This has been done for three cases in Figure 13.10 ($t < T_R$) and in Figures 13.11 and 13.12

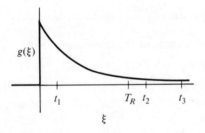

Figure 13.8. Graph of the impulse response of a typical dynamic system. T_R is that instant when $g(\xi)$ has essentially become zero.

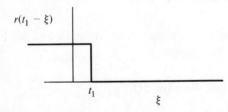

Figure 13.9. If the input, $r(t)$, to a system is a step, the graph of $r(t_1 - \xi)$ would be as illustrated.

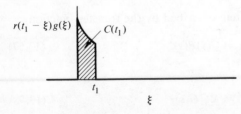

Figure 13.10. Graph of the integrand of Equation (13.59) when $t = t_1$.

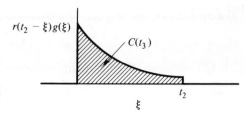

Figure 13.11. Graph of the integrand of Equation (13.59) when $t = t_2$.

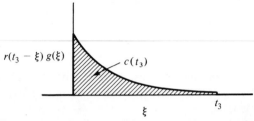

Figure 13.12. Graph of the integrand of Equation (13.59) when $t = t_3$.

$(t > T_R)$. The *area* under these graphs is the output response $c(t)$. The fact that the areas under the graphs in Figures 13.10 and 13.11 differ is interpreted as a difference in $c(t)$ at these two instants.

This interpretation of the area under the graph of $g(\xi)r(t - \xi)$ is of great practical significance. Suppose that t_2 and t_3 are two instants of time such that

$$t_3, t_2 > T_R. \tag{13.61}$$

It follows from an examination of Figures 13.11 and 13.12 that the area under the graph of the convolution integrand for $t = t_2$ is essentially the same as under the convolution integrand for $t = t_3$, i.e.,

$$c(t_2) \simeq c(t_3). \tag{13.62}$$

The important conclusion to draw from Equations (13.61) and (13.62) is that the output does not change much (thus has reached steady state) for $t > T_R$. In other words, the quantity T_R defined in Figure 13.8 illustrates a measure of the *duration* of the transient responses of a dynamic system. Conclude that T_R *is a measure of speed of response* of a dynamic system.

It follows from Equations (13.47), (13.58), and (13.60) that the Fourier transform of $g(t)$ is

$$\mathcal{F}[g(t)] = G(s)\Big|_{s=i\omega}, \tag{13.63}$$

where $G(s)$ is the familiar transfer function. Let $\Delta\omega$ be defined by

$$|G(i\omega)| \simeq 0, \quad |\omega| > \Delta\omega, \tag{13.64}$$

and define the *bandwidth*, ω_B, of a dynamic system by

$$\omega_B = \tfrac{1}{2}\Delta\omega. \tag{13.65}$$

It follows from the uncertainty principle, Equation (13.56), that

$$2\omega_B T_R \geq 2\pi,$$

so that the measure of speed of response

$$\boxed{T_R \geq \frac{\pi}{\omega_B}.} \tag{13.66}$$

This relationship between response time and bandwidth plays a central role in the chapter where frequency domain design of dynamic feedback systems is discussed.

13.8 Steady State Response of a Dynamic System to a Sinusoidal Input

The introduction to the structure of the frequency domain formalism continues in this section. As will be shown, it is useful to be able to rapidly determine the steady state response of a system when the input has the form

$$r(t) = A \sin(\omega t)\cdot u(t). \tag{13.67}$$

From a table of Laplace transforms, find

$$R(s) = \frac{A\omega}{s^2 + \omega^2}. \tag{13.68}$$

Thus, the transform of the output

$$C(s) = G(s)\frac{A\omega}{s^2 + \omega^2}. \tag{13.69}$$

Let $s_1, s_2, \ldots, s_n$ denote the poles of $G(s)$, and assume, for convenience, that they are distinct. After performing a heavyside expansion of Equation (13.69), find

$$c(t) = \underbrace{\sum_{\text{Poles of } G(s)} e^{+s_i t}\left|\frac{(s - s_i)G(s)A\omega}{s^2 + \omega^2}\right|_{s=s_i}}_{\text{Transient term}}$$

$$\underbrace{+ \left|\frac{e^{st}G(s)A\omega}{s + i\omega}\right|_{s=+i\omega} + \left|\frac{e^{st}G(s)A\omega}{s - i\omega}\right|_{s=-i\omega}}_{\text{Steady state term}} \tag{13.70}$$

Assume that the dynamic system is strictly stable (i.e., all of the s_i are in the left half of the s plane) so that in due time the transient term becomes negligibly small. Consider only the steady state term of Equation (13.70):

$$\frac{A\omega e^{i\omega t}G(i\omega)}{2i\omega} + \frac{A\omega e^{-i\omega t}G(-i\omega)}{-2i\omega} = \frac{A}{2i}[e^{i\omega t}G(i\omega) + e^{-i\omega t}G(-i\omega)]. \quad (13.71)$$

Notice that the complex conjugate (denoted with an overbar)

$$\overline{e^{-i\omega t}} = e^{i\omega t}. \quad (13.72)$$

It is left as an exercise for the reader to show that the complex conjugate

$$\overline{G(-i\omega)} = G(i\omega). \quad (13.73)$$

[*Hint*: Start with the fact that $\overline{(i\omega + a)} = (-i\omega + a)$.] Combine Equations (13.71)–(13.73) to show that

$$\underset{\substack{\text{Steady}\\\text{state}}}{c(t)} = \frac{A}{2i}[2i\,\text{Im}[e^{i\omega t}G(i\omega)]] = A\,\text{Im}[e^{i\omega t}G(i\omega)]. \quad (13.74)$$

Write $G(i\omega)$ in the polar form

$$G(i\omega) = |G(i\omega)|\,e^{i<G(i\omega)}. \quad (13.75)$$

Then, Equation (13.74) becomes

$$\boxed{\underset{\substack{\text{Steady}\\\text{state}}}{c(t)} = A|G(i\omega)|\sin[\omega t + <G(i\omega)].} \quad (13.76)$$

Thus, the output is also a sine wave. The system has amplified the amplitude by $|G(i\omega)|$ and added a phase lag $<G(i\omega)$ to the input wave.

Example 13.5

Consider the case of a triangle wave input to a system (Figure 13.13). The spectrum of this input was obtained in Example 13.4.

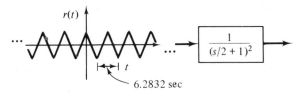

Figure 13.13. Dynamic system subjected to a triangle wave.

For the present case

$$\omega_0 = \frac{2\pi}{6.2832} = 1.0 \text{ radians/sec.}$$

One might ask the question, How is the spectrum of the output related to the spectrum of the input? The answer is supplied by Equation (13.76):

$$r'_n = |G(i\omega)| r_n, \tag{13.77}$$

where the r'_n are the amplitudes of the output spectral component and the r_n are the input spectral component amplitudes.

For the present example

$$G(i\omega) = \frac{1}{1 - (\omega^2/4) + i\omega} = \frac{1 - (\omega^2/4) - i\omega}{[1 + (\omega^2/4)]^2 + \omega^2};$$

thus,

$$|G(i\omega)| = \frac{1}{\{[1 - (\omega^2/4)]^2 + \omega^2\}^{1/2}}. \tag{i}$$

Also,

$$\omega = n\omega_0 = n$$

and, therefore, from Equation (i) and Equation (13.77),

$$r'_1 = 0.81 r_1,$$
$$r'_3 = 0.31 r_3,$$
$$r'_5 = 0.14 r_5,$$
$$\cdot$$
$$\cdot \quad \cdot \tag{ii}$$

The three elements of Equation (13.77) are plotted in Figures 13.14a, 13.14b, and 13.15. The sequence of figures illustrates the interpretation of linear systems as *filters*. These filters modify the input spectrum to produce the output spectrum in the same manner that an optical filter modifies a light spectrum passing through it. This viewpoint (formalism) for dynamic systems is exploited in Chapters 15 and 16, which contain a discussion of feedback system design. This spectral filter formalism should be contrasted with the transfer function formalism, in which a system is described by a transfer function, and the state space formalism, in which a dynamic system is described by a vector-matrix differential equation.

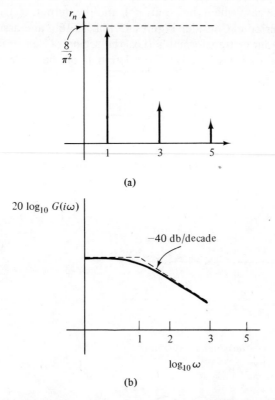

Figure 13.14. (a) Input spectrum, (b) system amplification factor.

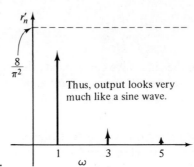

Figure 13.15. Output spectrum.

It is interesting to note that the term *transfer function* was first used in the field of physical optics.

* * *

The filter point of view introduced in the previous example is easily extended to the case where the input, $r(t)$, and the output, $c(t)$, of a dynamic system are aperiodic. Consider Figures 13.16a, 13.16b, and assume that $r(t)$ and $c(t)$ are Fourier transformable (i.e., the region of convergence of their Laplace transforms lies strictly in the left half of the s plane). Then, by Equation (13.76),

$$|\mathcal{F}[c(t)]| = |\mathcal{F}[r(t)]|\,|G(i\omega)| \qquad (13.78)$$

and the phase angle

$$<\mathcal{F}[c(t)] = <\mathcal{F}[r(t)] + <G(i\omega). \qquad (13.79)$$

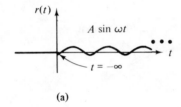

(a)

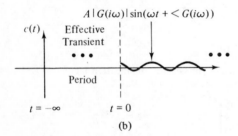

(b)

Figure 13.16. (a) Spectral component of an aperiodic input, (b) spectral component of the aperiodic output associated with the input of (a).

Because the sine wave components that make up the spectra for $r(t)$, $c(t)$ begin at $t = -\infty$, the *pseudo*-transients associated with these waves have decayed to negligible values for $t \geq 0$.

13.9 Bode Diagrams and Nyquist Diagrams of the Filtering Characteristics of Dynamic Systems

It is important to emphasize the close relationship between the transient response of a dynamic system and its *steady state* frequency response. The *filter* point of view presented at the end of the last section was one attempt to emphasize this fact.

Perhaps the best way to illustrate the intimate relationship between transient response and frequency response is to discuss Equation (13.76) in more detail. It is seen that the steady state frequency response of a dynamic system is completely determined by the quantities $|G(i\omega)|$ and $<G(i\omega)$, i.e., by the magnitude and angle of the complex number

$$G(i\omega) = G(s)\Big|_{s=i\omega}. \tag{13.80}$$

The last equation emphasizes the fact that $G(i\omega)$ is the transfer function, $G(s)$, evaluated at a point on the imaginary axis in the s plane. Since the transient response of a dynamic system is completely determined by the transfer function, $G(s)$, and steady state frequency response is completely determined by $G(i\omega)$, Equation (13.80) provides the desired demonstration of the relationship between the two types of response.

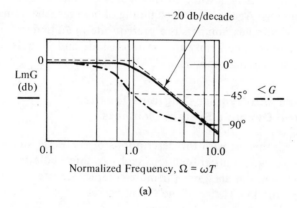

(a)

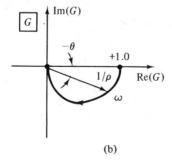

(b)

Figure 13.17. (a) Bode diagram for single time constant system, (b) Nyquist diagram for the single time constant system. The arrow points in the direction of increasing frequency.

As will be shown, frequency domain design techniques often involve graphical manipulations. The graphs are used to quantify the magnitude $|G(i\omega)|$ and the phase angle $<G(i\omega)$. Hopefully, the reader will, at this point, be disposed to believe that the manipulation of these two variables will influence the transient behavior of any system. There are several types of graphical representations of $|G(i\omega)|$ and $<G(i\omega)$. In this text, only two types of graphs will be discussed:

1. The *Bode diagram*, which is a graph of the gain term, $20 \log_{10} |G(i\omega)|$, and the phase term, $<G(i\omega)$, vs. $\log_{10} \omega$; and

2. The *Nyquist diagram*, which is a plot of the locus of points of $G(i\omega)$ (with ω as parameter) on the complex G plane.

An example of a Bode diagram is provided in Figure 13.17a. An example of a Nyquist diagram is shown in Figure 13.17b. The two diagrams contain essentially the same information, but in certain types of analysis one is more convenient to use than the other.

The next section is devoted to the details of construction of Bode and Nyquist diagrams. Again, an understanding of pole zero diagrams and of the polar form of complex numbers is a prerequisite to an understanding of the construction rules. The ability to construct Bode and Nyquist diagrams is exploited in succeeding chapters.

13.10 Bode and Nyquist Diagrams for the Basic Dynamic System Elements[3]

This section begins with the descriptions of the Bode and Nyquist diagrams of very simple transfer functions such as a single pole and a single zero. As will be shown, the diagrams for more complex transfer functions can be constructed from knowledge of the simpler cases.

First consider the *real pole* for which

$$G(s) = \frac{1}{Ts + 1}. \qquad . \qquad (13.81)$$

For this case

$$G(i\omega) = \frac{1}{Ti\omega + 1}. \qquad (13.82)$$

Using the symbols ρ, θ defined in Figure 13.18, it follows from Equation (13.82) that

$$|G(i\omega)| = \frac{1}{T\rho}, \qquad (13.83)$$

$$<G(i\omega) = -\theta. \qquad (13.84)$$

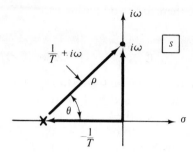

Figure 13.18. Vector interpretation of the $\alpha + i\omega$ term.

Equations (13.83) and (13.84) are suggestive of a rapid graphical method for determining the magnitude and phase as a function of frequency. One merely constructs the diagram in Figure 13.18 and carefully calibrates the ω axis. Then, for several values of ω, the distances, ρ, and the angles, θ, are measured and recorded. The conversion to gain and phase is accomplished with the aid of Equations (13.83) and (13.84). Note that the σ and ω axes must have the same scale.

For this simple case analytical expressions for the gain and phase are easily obtained. It is obvious from Figure 13.18 that

$$\rho = \sqrt{\frac{1}{T^2} + \omega^2},$$

$$\theta = \tan^{-1}(T\omega).$$

Thus, it follows from Equations (13.83) and (13.84) that

$$|G(i\omega)| = \frac{1}{\sqrt{1 + T^2\omega^2}} \tag{13.85}$$

and

$$<G(i\omega) = -\tan^{-1}(\omega T). \tag{13.86}$$

Exactly the same expressions could have been obtained by using the rules of complex algebra with Equation (13.82).

It follows from Equation (13.85) that

$$20\log_{10}|G(i\omega)| = -10\log_{10}(1 + T^2\omega^2).$$

It will be convenient to abbreviate the term $20\log_{10}|G(i\omega)|$ by $Lm(G)$ (read log-magnitude). Using this notation, the above equation becomes

$$Lm(G) = -10\log_{10}(1 + \omega^2 T^2). \tag{13.87}$$

Obviously, there are two limiting cases:

$$\mathrm{Lm(G)} \simeq \begin{cases} 0, & T\omega \ll 1, \\ -20 \log_{10}(\omega T), & T\omega \gg 1. \end{cases} \tag{13.88}$$

Thus if $\mathrm{Lm(G)}$ is plotted against $\log_{10}(\omega)$, two straight line asymptotes will bound the curve defined by Equation (13.87). This plot is shown in Figure 13.17a. Notice that the units of $\mathrm{Lm(G)}$ are called *decibels*. Notice that the high-frequency asymptote has a slope of -20 db/decade. This fact follows immediately from Equation (13.88).

The same information is displayed in the graphical form illustrated in Figure 13.17b, which is the Nyquist diagram.

The frequency at which the high-frequency and low-frequency asymptotes intersect in the Bode diagram is called the *corner frequency*. It follows from Equation (13.88) that the intersection occurs at that frequency for which

$$-20 \log_{10}(\omega T) = 0;$$

i.e., the corner frequency

$$\omega_c = \frac{1}{T}. \tag{13.89}$$

It is left as an exercise for the reader to show that, at the corner frequency,

$$\mathrm{Lm(G)} = -3 \text{ db}, \\ < G = -45° \tag{13.90}$$

for the first-order process.

The high- and low-frequency asymptotes for $<G(i\omega)$ are easily and quickly obtained from an examination of Figure 13.18. Obviously, $-\theta \rightarrow 0$ as $\omega \rightarrow 0$ and $-\theta \rightarrow +90°$ as $\omega \rightarrow \infty$.

One last point will be discussed before leaving this very simple example. The term *bandwidth*, introduced and discussed in Section 13.7, has not been precisely defined. It is conventional to use the corner frequency as the bandwidth. Using this convention, it follows from Equation (13.66) that for the first-order system

$$T_R \geq \frac{\pi}{\omega_c} = \pi T \simeq 3T. \tag{13.91}$$

A time equal to three time constants is a very good estimate of the speed of response of such systems.

Next consider the *real, left half-plane zero* for which the pole zero diagram is as illustrated in Figure 13.19 and for which

$$G(s) = 1 + Ts. \tag{13.92}$$

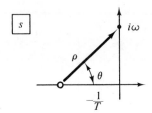

Figure 13.19. Left half-plane zero.

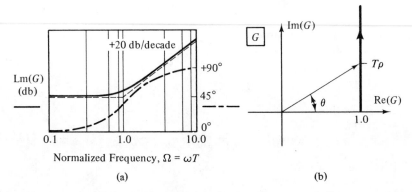

Normalized Frequency, $\Omega = \omega T$

(a)   (b)

Figure 13.20. (a) Bode diagram for a zero in the left half-plane, (b) Nyquist diagram for a zero in the left half-plane.

It is left as an exercise for the reader to show that the Bode and Nyquist diagrams are as shown in Figures 13.20a and 13.20b. For the Bode diagram, show LmG = +3 db at the corner frequency.

The Quadratic Factor. A transfer function with two complex poles can be written in one of two standard forms;

$$G(s) = \frac{\omega_n^2}{s^2 + 2\zeta\omega_n s + \omega_n^2} \tag{13.93}$$

or

$$G(s) = \frac{\alpha^2 + \beta^2}{(s + \alpha)^2 + \beta^2}. \tag{13.94}$$

The constants in these two standard forms are given the names:

$$\beta \triangleq \text{damped natural frequency}$$
$$\omega_n \triangleq \text{undamped natural frequency}$$
$$\alpha \triangleq \text{damping factor}$$
$$\zeta \triangleq \text{damping ratio}$$

and are interpreted in Fig. 13.21.

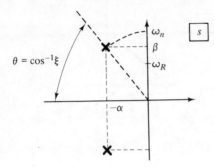

Figure 13.21. Pole zero pattern for quadratic factor showing geometrical interpretation of basic constants.

As shown in previous chapters

$$\beta = \omega_n\sqrt{1 - \xi^2} \tag{13.95}$$

$$\alpha = \xi\omega_n. \tag{13.96}$$

It follows that

$$G(i\omega) = \frac{\omega_n^2}{(i\omega)^2 + 2\xi\omega_n(i\omega) + \omega_n^2}. \tag{13.97}$$

It can be shown that

$$|G(i\omega)| = \frac{\omega_n^2}{\rho_1\rho_2} \tag{13.98}$$

and

$$<G(i\omega) = -\theta_1 - \theta_2 \tag{13.99}$$

where $\rho_1, \rho_2, \theta_1, \theta_2$ are defined in Figure 13.22. Equations (13.98) and (13.99) indicate the graphical method for finding the frequency response of a quadratic factor. The analytical method is pursued below.

Define a normalized frequency by

$$\Omega = \frac{\omega}{\omega_n}. \tag{13.100}$$

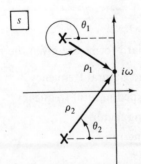

Figure 13.22. Vector interpretation of $G(i\omega)$ for the quadratic factor.

The use of this normalized variable reduces the number of parameters that must be carried through the calculations. In terms of Ω, Equation (13.97) becomes

$$G(i\Omega) = \frac{1}{1 - \Omega^2 + i2\xi\Omega}. \tag{13.101}$$

Multiply numerator and denominator by the conjugate of the denominator to find

$$G(i\Omega) = \frac{1 - \Omega^2 - i2\xi\Omega}{(1 - \Omega^2)^2 + 4\xi^2\Omega^2}. \tag{13.102}$$

It follows that

$$|G(i\Omega)| = \frac{1}{\sqrt{1 + \Omega^4 + 2\Omega^2(2\xi^2 - 1)}} \tag{13.103}$$

and that

$$<G(i\Omega) = -\tan^{-1}\left(\frac{2\xi\Omega}{1 - \Omega^2}\right). \tag{13.104}$$

The *resonant frequency* is that frequency which maximizes $|G(i\Omega)|$. To find the resonant frequency, one need only minimize the denominator of Equation (13.102) by setting

$$\frac{d}{d\Omega}[1 + \Omega^4 + 2\Omega^2(2\xi^2 - 1)]\Big|_{\Omega_R} = 4\Omega_R^3 + 4\Omega_R(2\xi^2 - 1) = 0.$$

Thus, the normalized resonant frequency

$$\Omega_R^2 = (1 - 2\xi^2),$$

or the nonnormalized resonant frequency is

$$\omega_R^2 = \omega_n^2(1 - 2\xi^2). \tag{13.105}$$

Comparing this equation with Equation (13.95), it follows that

$$\omega_R \le \beta \le \omega_n, \tag{13.106}$$

where the equalities apply only for the special case $\xi = 0$.

It is important to notice that ω_R exists *only if*

$$\xi < \frac{1}{\sqrt{2}} = 0.707. \tag{13.107}$$

This fact is a contradiction of the misconception sometimes held by undergraduates that any system with complex poles will have a resonant frequency.

Notice that Equation (13.107) demonstrates that resonance is displayed by second-order systems only if the angle, θ, in Figure 13.21 is greater than 45°.

Bode and Nyquist diagrams for the quadratic factor are shown in Figures 13.23a and 13.23b. It is not difficult to show that the Bode high-frequency asymptote for the quadratic factor is a straight line with a slope of -40 db/decade and that the corner frequency occurs at

$$\omega_c = \omega_n. \tag{13.108}$$

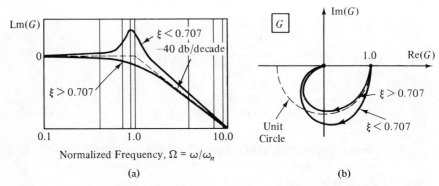

Figure 13.23. (a) Bode diagram for the quadratic factor, (b) Nyquist diagram for the quadratic factor.

It is conventional to use the corner frequency as the bandwidth, ω_B. For this convention it follows from Equation (13.66) that

$$T_R \geq \frac{\pi}{\omega_n}. \tag{13.109}$$

Use Equation (13.96) to show that

$$T_R \geq \frac{\xi\pi}{\alpha}. \tag{13.110}$$

The constant

$$T \triangleq \frac{1}{\alpha} \tag{13.111}$$

is the time constant associated with the envelope asymptotes shown in Figure 10.8a. Finally,

$$T_R \geq \xi\pi T. \tag{13.112}$$

$\xi\pi T$ is a good estimate of T_R *provided that ξ is not too small.* In other words, the corner frequency should be used as the bandwidth only if the damping factor, α, is of reasonable magnitude when compared to the natural frequency, ω_n.

The discussion turns now to the frequency response of the simple *dead time* element, i.e., the frequency response associated with the transfer function

$$G(s) = e^{-\tau s}. \tag{13.113}$$

This response is quickly obtained by noting that the term

$$G(i\omega) = e^{-\tau i\omega} \tag{13.114}$$

is in polar form. It follows that

$$|G(i\omega)| \equiv 1.0,$$
$$<G(i\omega) = -\tau\omega. \tag{13.115}$$

The Bode and Nyquist diagrams for dead time are shown in Figures 13.24 and 13.25. Notice that the phase lag grows indefinitely with increasing frequency.

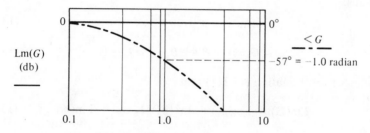

Figure 13.24. Bode diagram for dead time.

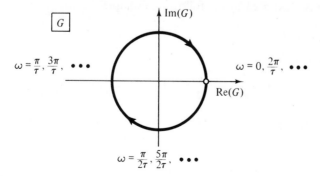

Figure 13.25. Nyquist diagram for dead time.

13.11   Bode and Nyquist Diagrams
for More Complex Transfer Functions

For transfer functions with several poles, zeros, quadratic factors, and/or
dead time, the Bode diagram is constructed by superimposing the Bode
diagrams of the basic elements of the previous section. To see why this is so,
notice that most transfer functions can be written in the form

$$G(s) = G_1(s) \cdot G_2(s) \cdot \ \cdots \ \cdot G_l(s), \tag{13.116}$$

where each of the $G_k(s)$ has one of the simple forms displayed in the previous
section. It follows that

$$G(i\omega) = G_1(i\omega)G_2(i\omega) \cdots G_l(i\omega). \tag{13.117}$$

Each of the complex numbers, $G_k(i\omega)$, has a gain, ρ_k, and a phase angle,
θ_k, so that

$$G(i\omega) = \rho_1 e^{i\theta_1} \rho_2 e^{i\theta_2} \cdots \rho_l e^{i\theta_l}$$
$$= \rho_1 \rho_2 \cdots \rho_l e^{i(\theta_1 + \theta_2 + \cdots + \theta_l)}. \tag{13.118}$$

Thus,

$$|G(i\omega)| = \rho_1 \rho_2 \cdots \rho_l, \tag{13.119}$$

$$<G(i\omega) = \theta_1 + \theta_2 + \cdots + \theta_l. \tag{13.120}$$

It follows from Equation (13.111) that

$$\text{Lm}(G) = \text{Lm}(\rho_1) + \text{Lm}(\rho_2) + \cdots + \text{Lm}(\rho_l). \tag{13.121}$$

Equations (13.120) and (13.121) establish the above assertion that one
need only superimpose the Bode diagrams of the basic elements of a transfer
function to construct the Bode diagram for that transfer function. The general
procedure is illustrated by the following example.

Example 13.6

Suppose that

$$G(s) = \frac{2(s+1)}{s(\frac{1}{2}s+1)} = 2 \cdot (s+1) \cdot \frac{1}{s} \cdot \frac{1}{(\frac{1}{2}s+1)}.$$

The Bode diagram for this transfer function is sketched in Figure 13.27 by
using superposition of the elementary diagrams (see Figure 13.26).

Actually, for the gain curves, only the asymptotes need be superimposed. The 3-db corrections are made at the corner frequencies [see Equation (13.90)].

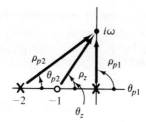

Figure 13.26. Pole zero pattern.

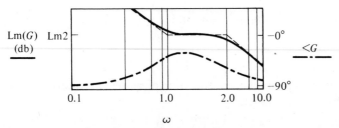

Figure 13.27. Bode diagram for Example 13.6.

Notice that, for this example,

$$|G(i\omega)| = |2| \cdot |(i\omega + 1)| \cdot \left|\frac{1}{i\omega}\right| \cdot \left|\frac{1}{i(\omega/2) + 1}\right|$$

$$= 2 \cdot \rho_z \cdot \frac{1}{\rho_{p_1}} \cdot \frac{2}{\rho_{p_2}} = \frac{4\rho_z}{\rho_{p_1}\rho_{p_2}}. \tag{i}$$

Similarly,

$$<G(i\omega) = 0 + \theta_z - \frac{\pi}{2} - \theta_{p_2} = \theta_z - \frac{\pi}{2} - \theta_{p_2}. \tag{ii}$$

Equations (i) and (ii) indicate the manner in which the entire frequency response can be obtained graphically with the help of the pole zero diagram. The latter method is an alternative to the method wherein the asymptotes of the basic elements are superimposed and corrections applied to the corner frequencies.

* * *

The Nyquist diagram can be constructed from the Bode diagram or directly from the pole zero diagram.

13.12 Right Half-Plane Zeros: The *Nonminimum Phase System*

A transfer function which has one or more zeros in the right half of the *s* plane is called a *nonminimum phase system*. Any other transfer function is called a *minimum phase system*. The reason for this terminology will become apparent in the discussion in this section. The peculiarities of nonminimum phase systems were illustrated in Section 11.4.

Concentrate on the particular examples of the negative gain and reverse action compensators discussed in Chapter 11. The transfer functions for these compensators were

$$G_1(s) = \frac{T_z s - 1}{T_p s + 1} \tag{13.122}$$

for the negative gain compensator and

$$G_2(s) = \frac{1 - T_z s}{T_p s + 1} \tag{13.123}$$

for the reverse action compensator. For purposes of comparison, the minimum phase transfer function

$$G_3(s) = \frac{T_z s + 1}{T_p s + 1} \tag{13.124}$$

will also be considered. The pole zero diagrams for these three transfer functions are shown in Figures 13.28a, 13.28b, and 13.28c.

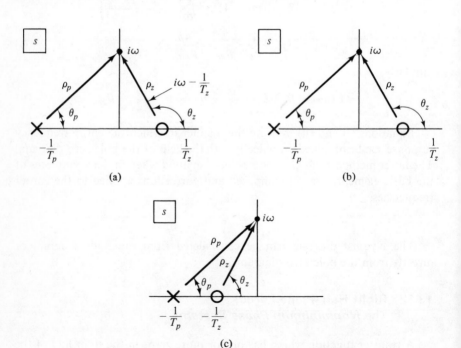

Figure 13.28. (a) Negative gain compensator, (b) reverse action compensator, (c) minimum phase compensator.

Notice that for *all three* transfer functions, the gain

$$|G(i\omega)| = \frac{T_p p_z}{T_z p_p}. \tag{13.125}$$

Thus, the Bode diagrams for the three systems differ only in the form of the phase curves.

For the transfer functions in Equations (13.122) and (13.124)

$$<G_2(i\omega) = \theta_z - \theta_p, \tag{13.126}$$

while for the transfer function of Equation (13.115) the phase

$$<G_2(i\omega) = \theta_z - \theta_p - 180°. \tag{13.127}$$

The 180° difference between $<G_1(i\omega)$ and $<G_2(i\omega)$ is due to the fact that the vector from a right half-plane zero to a point, $i\omega$, on the imaginary axis is

$$i\omega - \frac{1}{T_z} = -\left(\frac{1}{T_z} - i\omega\right). \tag{13.128}$$

This fact led to some complications in root locus analysis. The $-180°$ in Equation (13.127) is used in order to account for the first minus sign in the right-hand side of Equation (13.128).

The phase curves for the three transfer functions are sketched in Figures 13.29a, 13.29b, and 13.29c.

It is left as an exercise for the reader to sketch the Nyquist diagram for the three transfer functions discussed above. The reason for the use of the terms *minimum phase* and *nonminimum phase* is apparent from an examination of Figures 13.29a–13.29c.

There is an interesting relationship between passive networks and minimum phase transfer functions. Consider the passive system illustrated in Figure 13.30. Denote the output of this system by $C(s)$ and suppose that $C(s)$ is E_2, F_2, or X_2. Denote the input by $R(s)$ and suppose that $R(s)$ is E_1, F_1, or X_1.

Using the transfer matrix approach discussed in previous chapters, the input-output relation

$$[a_m s^m + a_{m-1} s^{m-1} + \cdots + a_0]R(s) = [b_n s^n + b_{n-1} s^{n-1} + \cdots + b_0]C(s) \tag{13.129}$$

is easily obtained. Suppose that, during the operation of this system, $r(t)$ is held at zero; from Equation (13.129), the resulting response, $c(t)$, is governed by the differential equation

$$[b_n D^n + b_{n-1} D^{n-1} + \cdots + b_0]c(t) = 0 \tag{13.130}$$

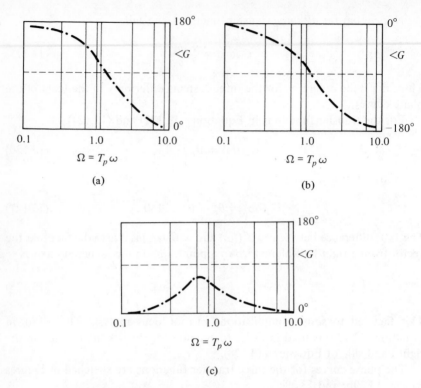

Figure 13.29. (a) Phase curve for the negative gain compensator, (b) phase curve for the reverse action compensator, (c) phase curve for the minimum phase system.

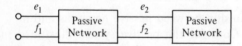

Figure 13.30. Dynamic system composed entirely of passive elements.

and the initial conditions $c(0^-)$, $Dc(0^-)$, ..., $D^{n-1}c(0^-)$. The system is passive; hence, $c(t)$ will remain bounded as $t \longrightarrow \infty$. This means that the roots of the polynomial in Equation (13.130) must lie in the left half of the complex plane. Similarly, if $c(t)$ is held equal to 0, $r(t)$ will vary as predicted by the differential equation:

$$[a_m D^m + a_{m-1} D^{m-1} + \cdots + a_0]r(t) = 0 \qquad (13.131)$$

and the initial conditions $r(0^-)$, $Dr(0^-)$, ..., $D^{n-1}r(0^-)$. Again, using physical arguments, it can be shown that the roots of the polynomial in Equation (13.131) must lie in the left half-plane.

The above argument proves the following assertion: *The transfer functions relating variables in a passive network must be minimum phase;* i.e., the poles *and* the zeros of such a transfer function must lie in the left half-plane. In other words, nonminimum phase systems can be synthesized only with *active* elements.

References

1. CHENG, D. K., *Analysis of Linear Systems*, Addison-Wesley, Reading, Mass., 1959.

2. LATHI, B. P., *Signals, Systems and Communication*, Wiley, New York, 1965.

3. OGATA, K., *Modern Control Engineering*, Prentice-Hall, Englewood Cliffs, N. J., 1970.

4. STONE, J. M., *Radiation and Optics*, McGraw-Hill, New York, 1963.

Problems

13.1. Determine the Fourier coefficients for the periodic function sketched in Figure 13.31 for one period.

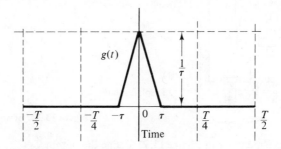

Figure 13.31. Periodic pulse function.

13.2. Sketch the envelope of the spectral lines of the periodic function of Problem 13.1 against frequency. As $\tau \rightarrow 0$, will more or fewer spectral components of the Fourier series have to be considered in order that a truncated version of the Fourier series approximate $g(t)$ to within a prescribed accuracy?

13.3. For the periodic function of Problem 13.1, set $\tau = T/4$ and develop a computer program which can be used to sum the first k terms of the Fourier series expansion of $g(t)$ for various values of t:

$$-\frac{T}{2} \leq t \leq \frac{T}{2}.$$

Evaluate the accuracy of the truncated series approximation for various values of k.

13.4. Justify Equation (13.60).

13.5. Establish Equation (13.73).

13.6. Start with Equation (13.82) and derive Equations (13.85) and (13.86).

13.7. Establish Equations (13.90).

13.8. Verify the validity of Figures 13.20 and 13.21.

13.9. Verify Equations (13.98) and (13.99).

13.10. Establish Equation (13.108).

13.11. Verify the uncertainty principle [i.e., show that relation (13.56) is valid] for
(a) $e^{-at}u(t)$.
(b) $u(t) - u(t - \tau), \tau > 0$.
(c) $e^{-t} \cos(t)u(t)$.
Equation (13.47) should prove useful.

13.12. Consider the *seismometer* in Figure 13.32. Determine the transfer function $X(s)/R(s)$. Plot the Bode diagram and pick m, b, k so that the spectrum of $r(t)$ is undistorted for all frequencies below 10 cycles/sec.

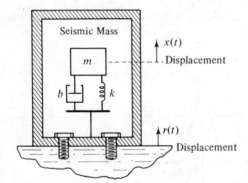

Figure 13.32. Seismometer.

13.13. For the seismometer in Figure 13.32, define the output

$$c(t) = r(t) - x(t).$$

Find $C(s)/R(s)$ and show that this device acts as an accelerometer. Select b, m, k so that the spectrum of $\ddot{r}(t)$ is undistorted for all frequencies below 10 cycles/sec.

13.14. Construct the Bode diagram for the accelerometer analyzed in Problem 3.9. For what range of frequencies does this apparatus act as a true accelerometer? What is the difference between the accelerometers of the Problem 13.13 and this problem?

13.15. Sketch the Bode and Nyquist diagrams for the lag compensator of Problem 3.14 and for the lead compensator of Example 3.4.

13.16. Sketch the Bode diagram for the filter of Problem 2.7.

13.17. Sketch the Bode and Nyquist diagrams for the thickness control system of Problem 2.10. Use the uncertainty principle to estimate the speed of response of this closed loop system.

13.18. Construct the Bode and Nyquist diagrams for a Padé filter for which the transfer function is

$$\frac{1 - Ts}{1 + Ts}. \qquad (13.132)$$

This filter is useful in simulation studies; can you guess why? (*Hint:* Compare your results with the diagrams of dead time.)

13.19. Plot the Nyquist locus for the linearized transfer function

$$\frac{H_1(s)}{F(s)}$$

for the two operating conditions shown in Figure 13.33.

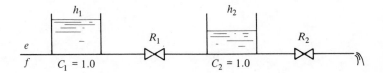

Condition	f	h_1	h_2
1	1.0	1.5	1.0
2	10.0	150.0	100.0

Figure 13.33. Two-tank system.

13.20. Sketch the Nyquist diagrams for both of the mechanical systems discussed in Example 2.9.

13.21. Construct the Bode and Nyquist diagrams for the transfer function of Equation (8.73).

13.22. Construct the Bode and Nyquist diagrams for the plant transfer function of the system discussed in Problem 8.22.

Frequency Domain
Methods for Empirical
Modeling, 14
Approximation, and
Simplification*

There are many problems of system theory which are so fundamental that systems analysts will always seek new and improved solutions even though the mathematical formalisms used will vary dramatically from generation to generation. Three such problems will be discussed in this chapter. The frequency domain formalism is used in this introductory discussion.

Empirical (or *functional*) *modeling* is the technique whereby the mathematical model of a dynamic system is determined by analyzing the actual recorded response of a dynamic system to certain signals. As will be shown, the analysis can be accomplished conveniently if the input signals are sine waves. The determination of the values of unknown model constants is an example of a simpler empirical modeling problem.

In this text, the terms *approximation* and *simplification* will refer to the technique of approximation of a complex transfer function, or state variable model, by a simpler transfer function or model. Such techniques might be used to replace a transcendental transfer function with a transfer function which is the ratio of two polynomials. Another application would be to replace a high-order model with a lower-order and more tractable model.

In this chapter, solutions to the above problems will be attempted by exploiting a single fact which follows from the discussion in Sections 13.8 and 13.9: If the filter properties of one system, as summarized in a Bode diagram, for example, closely approximate those of another system, the two systems will have nearly the same dynamic response. A complete discussion would

---

* This chapter may be omitted during the first reading of the text.

provide a precise definition of the terms *closely approximate* and *nearly the same* as used in the previous sentence. Such completeness will not be attempted here.

14.1 Functional Models for Dynamic Systems: The *Black Box* Problem[1]

In the first chapters of this book, the analysis of dynamic systems which lie in the shaded portion of Figure 14.1 was discussed. The term *physical system* was used to describe a system wherein the transients are due entirely

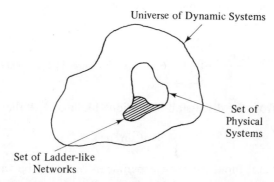

Figure 14.1. Venn diagram of the dynamic system universe.

to the transfer of the commonly recognized forms of physical energy (electrical, mechanical, thermal, magnetic). The concepts of power state and generalized impedance were used to derive transfer function and state space models for dynamic systems.

Transfer function and/or state space models can be developed for *any* linear, time-invariant system in the universe denoted°in Figure 14.1 and, of course, are not restricted to systems within the shaded portion of that figure. The significance of this fact is that the simulation and design techniques, the use of which require only the existence of a transfer function or state space model, are *not* restricted to the type of system discussed in the analysis portion of this book. Examples of nonphysical dynamic systems are economic, social, and population (*demographic*) systems.

It is clear that many important dynamic systems cannot be "broken up" (i.e., *reticulated*) into elements (generalized impedances, transformers, or gyrators) as can physical systems. Even for physical systems, this reticulation may prove unwieldy and inconvenient. Models which are developed from a reticulation of dynamic systems are called *structural models* and were the subject of the early chapters of this book.

Obviously, it would be convenient to have experimental techniques to determine transfer function or state space representations for dynamic systems. The type of experiment that will be discussed in this section is designed to determine a model for a dynamic system from an examination of the input record, $r(t)$, and the output record, $c(t)$. Models determined in this way are called *functional* or *phenomenological models.*

The value of functional models is that they serve as starting points for the simulation and design techniques discussed in this book. They might also be used in malfunction or pathology diagnosis.

The first question to ask in preparation of a functional model experiment is, What input, $r(t)$, should be used for the situation illustrated in Figure 14.2? (In this figure the term *black box* is used to emphasize the fact that the reticulation of the system is impossible or very difficult; hence, is in a "black," impenetrable "box.")

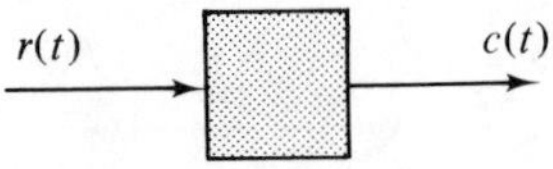

$r(t)$ → [box] → $c(t)$

Figure 14.2 Dynamic system in a black box.

It is clear from the discussion in Sections 13.8 and 13.9 that

$$r(t) = A \sin (\omega t) \tag{14.1}$$

would be a useful input for such experiments. ω could be varied and the steady state response

$$c(t) = A \, |G(i\omega)| \sin \{\omega t + <G(i\omega)\} \tag{14.2}$$
Steady
state

could be recorded. From these records one could determine the *amplitude ratio*

$$|G(i\omega)| = \frac{\text{Amplitude of } c(t) \text{ in steady state}}{\text{Amplitude of } r(t)} \tag{14.3}$$

and the *phase lag*

$$<G(i\omega) = \text{Phase difference between } c(t)$$

$$\text{in steady state and } r(t). \tag{14.4}$$

If the latter two quantities are plotted against ω, then the Bode diagram will have been determined experimentally. The problem of finding a functional model for a dynamic system then reduces to finding a transfer function, $G(s)$, whose theoretical Bode diagram fits the experimentally determined diagram.

It is left as an exercise for the reader to devise a scheme to estimate how long one should wait before he can safely call the output of a system the *steady state output*. (*Hint*: Consider the uncertainty principle.)

Of course, other types of inputs should be used if possible in order to provide checks on the results obtained using frequency response methods. Step responses are useful for finding accurate values of gain. An example of other types of information which can be obtained from step response data is provided in Example 2.9.

A more explicit description of the frequency response method is now provided. The output record of a typical frequency response test is as illustrated in Figure 14.3. The reader should be able to verify that

$$\omega = \frac{2\pi}{T}, \tag{14.5}$$

$$\text{Lm}[G(i\omega)] = 20 \log_{10} \frac{A_c}{A_r}, \tag{14.6}$$

and

$$<G(i\omega) = \frac{2\pi}{T}\tau \cdot (57°/\text{radian}). \tag{14.7}$$

One finds that when performing an interpretation of frequency data it is best to write quadratic factors in the standard form

$$G(s) = \frac{K\omega_n^2}{s^2 + 2\zeta\omega_n s + \omega_n^2}. \tag{14.8}$$

The following derivations are useful for finding the value of ζ from frequency

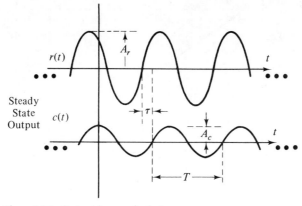

Figure 14.3. Output from a single frequency response measurement. For purposes of data reduction, the quantities A_r, A_c, T, and τ are measured and recorded.

response data. It was demonstrated in Chapter 13 that

$$|G(i\omega)| = \frac{K}{\sqrt{(1 - (\omega/\omega_n)^2)^2 + 4\xi^2(\omega/\omega_n)^2}}. \tag{14.9}$$

It was also demonstrated that the resonant frequency

$$\omega_R = \omega_n\sqrt{1 - 2\xi^2}, \quad \xi < \frac{1}{\sqrt{2}}. \tag{14.10}$$

It follows that

$$|G(i\omega_R)| = \frac{K}{2\sqrt{\xi^2 - \xi^4}} \tag{14.11}$$

and, in terms of the log-magnitude function,

$$\text{Lm}[|G(i\omega_R)|] - \text{Lm}(K) = -\tfrac{1}{2}\text{Lm}[\xi^2(1 - \xi^2)] - \text{Lm}(2). \tag{14.12}$$

The left-hand side of this equation is the *resonant peak* of the frequency response. This relationship is plotted in Figure 14.4 for the range

$$0.010 \le \xi \le 0.707.$$

An approximate formula derived from Equation (14.11) is

$$\xi \simeq \frac{K}{2|G(i\omega_R)|} \tag{14.13}$$

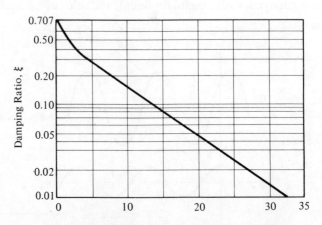

Resonant Peak (db), Lm $[G(i\omega_R)]$ − LmK

Figure 14.4. Relationship between the damping ratio and resonant peak of a quadratic factor. The graph is used to determine ξ from frequency response data.

and is valid for

$$\xi < 0.05.$$

For the reader's convenience, some other tips on the curve fitting of frequency response data are summarized below:

1. Low-frequency phase data must be asymptotic to an integer multiple of $\pm 90°$. The integer multiple is an indication of the powers of s in the numerator or denominator. The presence of nonminimum phase terms will modify this relationship. (14.14)

2. The high-frequency gain asymptote slope will be proportional to a multiple of 20 db/decade and will indicate the difference between the number of poles and the number of zeros. (14.15)

3. A rapidly decreasing phase at high frequency indicates the presence of dead time. The value of dead time can be obtained by plotting phase vs. ω (not $\log_{10} \omega$) and measuring the negative slope of the resulting straight line. [See Equation 13.15).] (14.16)

4. Sketch the high- and low-frequency gain asymptotes and verify whether or not the system is nonminimum phase. Remember that asymptotes must have slopes which are some integer multiple of ± 20 db/decade.

(14.17)

5. Add poles or zeros to a proposed $G(s)$ to explain deviation of middle-frequency data from the asymptotes drawn in step 4. Remember that a zero causes increasing phase. A quadratic factor can be confused with a zero if gain data alone are consulted. (14.18)

6. Remember that deviations from asymptotes are expected at corner frequencies; i.e., not *all* data must lie on the asymptote for the $G(s)$ which is finally used. (14.19)

7. If one or more poles exist at the origin, it is left as an exercise for the reader to show that the gain

$$K = 10^{\mathrm{LmG_1}}, \qquad (14.20)$$

where G_1 is that value of gain where the *extension of the low-frequency asymptote* intersects the vertical line at $\omega = 1.0$ radians/sec.

Example 14.1

The frequency response data for a dynamic system are presented in Figure 14.5a. To find an empirical transfer function, notice the following facts: (1) The phase data approach $-90°$, and the gain grows indefinitely as ω approaches 0. These facts indicate the presence of a pole at the origin. (2) The phase approaches $-270°$ as ω approaches ∞, indicating that the system has three more poles than zeros *if it is a minimum phase system*.

Once one has made a decision about the high- and low-frequency portions of the empirical transfer function, it is a good idea to draw in the corresponding asymptotes before any further speculations are made. This task was performed in Figure 14.5b.

Since the −60-db/decade line fits the high-frequency data rather well, one may conclude that the system is indeed a minimum phase system.

The remaining task is to explain the deviation of the middle-frequency data from the gain asymptotes. This deviation might be due to the presence of a zero in $G(s)$, but this hypothesis is inconsistent with the phase data, which

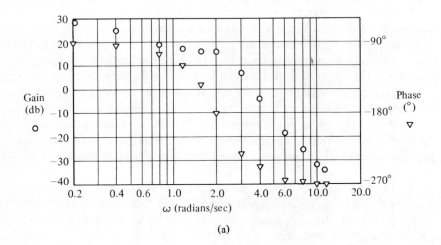

(a)

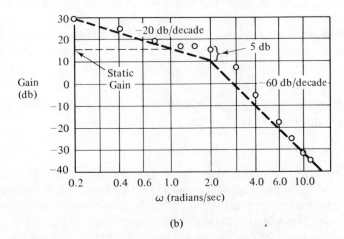

(b)

Figure 14.5. (a) Experimentally determined frequency response data, (b) high- and low-frequency asymptotes for the experimental data presented in (a).

does not increase in this frequency range. The remaining possibility is the presence of a quadratic factor.

The form of an empirical transfer function which is consistent with the above conclusions is

$$G(s) = \frac{K}{s((s^2/\omega_n^2) + (2\xi/\omega_n)s + 1)}. \tag{14.21}$$

Note Figure 14.5b. It follows immediately from Equation (14.20) that

$$K = 10^{15/20} = 5.6 \tag{14.22}$$

and

$$\omega_n = 2.0 \text{ radians/sec.} \tag{14.23}$$

The value of ξ which will give a 5-db deviation from the asymptote at the break frequency is found using Figure 14.4 so that

$$\xi = 0.3.$$

* * *

14.2 Comment on *Physical Parameter* Identification as Opposed to *Model Parameter* Identification

Using the techniques discussed in the previous section, one can also determine, experimentally, the coefficients of a transfer function for a dynamic system. This procedure is called *model parameter identification*. For many purposes, including simulation and design, this type of identification is sufficient.

Another type of identification is the *physical parameter identification*. This type of identification is the experimental determination of the value of resistances (R), inertances (L), gyrator constants (g), etc. Obviously, physical parameter identification is used in conjunction with a mathematical model.

To illustrate the differences (and some of the similarities) between *model* parameter and *physical* parameter identification, consider the simple system illustrated in Figure 14.6. If f_0 is selected as the output and f_i is selected as the input, then the system transfer function

$$\frac{F_0(s)}{F_i(s)} = \frac{1}{1 + RCs}. \tag{14.24}$$

This transfer function has a Bode diagram similar to that illustrated in Figure 13.17a. The model parameter, RC, can be found experimentally by

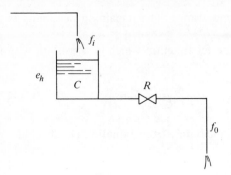

Figure 14.6. Single tank system.

finding the corner frequency (i.e., the frequency at which the gain falls 3 db below zero frequency gain or at which the phase $= -45°$). Notice, however, that no experiment can be devised which will allow one to determine the physical parameters R and C individually.

If, for some reason, the analyst has decided that physical parameter identification is important, a simple redefinition of the output variable will yield the desired result. If the effort, e_h, is used as an output, then the system transfer function

$$\frac{E_h(s)}{F_i(s)} = \frac{R}{1 + RCs},\tag{14.25}$$

so that the static gain

$$K = R\tag{14.26}$$

and the corner frequency

$$\omega_b = \frac{1}{RC}.\tag{14.27}$$

K and ω_b can be determined experimentally and R and C can then be determined from Equations (14.26) and (14.27).

The important conclusion is that one must be very careful in the selection of output variables if he is to perform physical parameter identification.

14.3 The Approximation Problem

The frequency domain method for substituting a simpler mathematical description for a complex description will be illustrated by example. The particular problem will be to find an approximation for

$$G(s) = e^{-\sqrt{2s}}.\tag{14.28}$$

The adequacy of an approximation depends on the ultimate application. It will be assumed that a proportional feedback system is to be designed for the plant represented by Equation (13.28).

Notice that

$$G(i\omega) = e^{-\sqrt{\omega}\sqrt{2i}}.$$

But

$$\sqrt{2i} = \sqrt{2}\,(e^{i\pi/2})^{1/2}$$
$$= \sqrt{2}\,e^{i(\pi/4)}$$
$$= 1 + i; \tag{14.29a}$$

thus,

$$G(i\omega) = e^{-\sqrt{\omega}}e^{-i\sqrt{\omega}}. \tag{14.29b}$$

This equation is in polar form so that

$$\mathrm{Lm}[G(i\omega)] = 20\log_{10} e^{-\sqrt{\omega}}$$
$$= -8.7\sqrt{\omega}, \tag{14.30}$$

and

$$\angle G(i\omega) = -\sqrt{\omega}. \tag{14.31}$$

Equation (14.30) is plotted in the form of a Bode diagram in Figure 14.7. Notice that the high-frequency portion is well approximated by an asymptote at -20 db/decade. This fact suggests that

$$G_1(s) = \frac{0.5}{s + 0.5} \tag{14.32}$$

might be a satisfactory approximation for $G(s)$. It is easily shown that

$$\angle G_1(i\omega) = -\tan^{-1}[2\omega]. \tag{14.33}$$

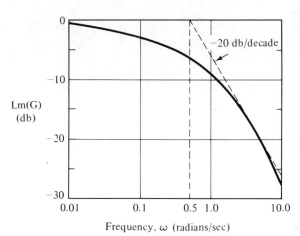

Figure 14.7. Log-magnitude curve for $e^{-\sqrt{2s}}$.

Notice that

$$\lim_{\omega \to \infty} <G(i\omega) = \infty, \tag{14.34}$$

while

$$\lim_{\omega \to \infty} <G_1(i\omega) = -90°. \tag{14.35}$$

These equations lead to the conclusion that the approximation could be improved by the addition of a dead time element; i.e., take

$$G_*(s) = e^{-\tau s}G_1(s)$$

$$= \frac{e^{-\tau s}}{2s + 1} \tag{14.36}$$

as the approximate transfer function.

Notice that

$$<G_*(i\omega) = -\tau\omega - \tan^{-1}[2\omega]. \tag{14.37}$$

Thus, the limit of $<G_*(i\omega)$ is the same as that indicated in Equation (14.34) for the original transfer function.

Choose a value for τ such that the difference between $<G(i\omega)$ and $<G_*(i\omega)$ is reduced for a wide range of ω; i.e., choose τ such that

$$-\tau\omega - \tan^{-1}[2\omega] + \sqrt{\omega} \simeq 0$$

or

$$\tau \simeq \frac{\sqrt{\omega} - \tan^{-1}[2\omega]}{\omega}. \tag{14.38}$$

One cannot choose τ to satisfy this equation for all ω. Choose the characteristic value of $\omega = 10$ radians/sec and find

$$\tau \simeq \frac{3.14 - 1.52}{10}$$

$$= 0.16. \tag{14.39}$$

Finally, choose the approximation

$$G_*(s) = \frac{e^{-0.16s}}{2s + 1} \tag{14.40}$$

for the transfer function in Equation (13.28).

The phase angles of $G(i\omega)$ and $G_*(i\omega)$ are compared in Figure 14.8. One means of checking this approximation would be to compare the step

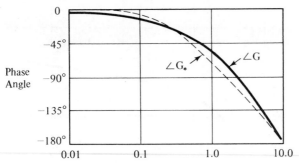

Figure 14.8. Angle $<G(i\omega)$ compared to the angle of the approximating transfer function, $G_*(i\omega)$.

responses. The approximate step response

$$c_*(t) = \mathcal{L}^{-1}\left\{\frac{e^{0.16s}}{s(2s+1)}\right\}, \qquad (14.41)$$

while the actual step response

$$c(t) = \mathcal{L}^{-1}\left\{\frac{e^{-\sqrt{2s}}}{s}\right\}. \qquad (14.42)$$

Apply the methods of Chapter 2 to show that

$$c_*(t) = \{1 - e^{-0.5(t-0.16)}\}u(t - 0.16). \qquad (14.43)$$

The use of advanced techniques[2] leads to

$$c(t) = \left\{1 - \text{erf}\left[\frac{1}{\sqrt{2t}}\right]\right\}u(t), \qquad (14.44)$$

where erf( ) is the *error function* or *probability integral* tabulated in many handbooks.

Both $c(t)$ and $c_*(t)$ are plotted in Figure 14.9. Both responses have the same final value of 1.0. The divergence about $t = 4.0$ sec is disappointing and can be associated with the divergence in the frequency responses at 1.0 radians/sec (notice, for instance, the $\text{Lm}G_*$ is 3 db below the break point illustrated in Figure 14.7, while $\text{Lm}G$ is 6 db below this same point). The agreement in the middle time range could be improved by introducing more complexity into $G_*(s)$, say, another pole and a zero.

On the other hand, the agreement between $c(t)$ and $c_*(t)$ for $t < 1.0$ sec is impressive and is due, in large part, to the presence of the dead time in $G_*(s)$.

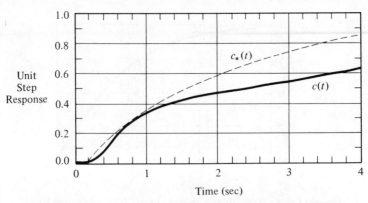

Figure 14.9. Exact unit step response, $c(t)$, for $G(s) = e^{-\sqrt{2s}}$ and the step response, $c_*(t)$, for the approximate transfer function.

The significance of the dead time for the stated application of the approximation, the design of a proportional feedback system, is demonstrated by a comparison of Figures 11.7 and 11.24.

The approach taken to approximation in this section is very similar to the approach to functional modeling presented in earlier sections. The chief difference is that in the former case, the frequency response of a rational function of s is fit to an actual system response obtained analytically instead of experimentally.

14.4 Low-Order State Variable Models for Distributed Parameter Systems

The discussion in this section will draw on the material presented in Sections 6.5 and 6.7. It will be demonstrated that if the approximation problem is associated with a physical process, a more straightforward procedure than that indicated in steps (14.14)–(14.20) can be employed. A method for obtaining a low-order state variable model of a distributed parameter system will also be demonstrated.

The particular system to be studied is illustrated in Figure 14.10. It follows from Equation (6.31) that

$$G(s) = \frac{E(L, s)}{E(0, s)}$$

$$= \frac{1}{\cosh\{\sqrt{\mathfrak{R}\mathfrak{C}sL}\}}, \tag{14.45}$$

where $\mathfrak{R}$ and $\mathfrak{C}$ are, respectively, the resistance and capacitance per unit length. An approximation for this transfer function will be obtained.

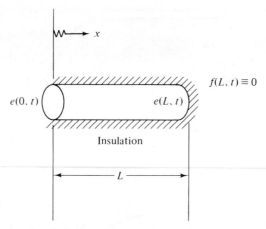

Figure 14.10. Flow of heat in a one-dimensional rod. The temperature, $e(0, t)$, is the output.

It will be assumed that

$$\Re\mathfrak{C} = \frac{2}{L^2},\tag{14.46}$$

so that

$$RC = 2.0 \text{ sec.}\tag{14.47}$$

R and C are the total resistance and total capacitance of the rod; thus,

$$G(s) = \frac{1}{\cosh\{\sqrt{2s}\}}$$

$$= \frac{2}{e^{\sqrt{2s}} + e^{-\sqrt{2s}}}.\tag{14.48}$$

Use Equation (14.29a) to show that

$$G(i\omega) = \frac{2}{e^{\sqrt{\omega}}e^{i\sqrt{\omega}} + e^{-\sqrt{\omega}}e^{-i\sqrt{\omega}}}.\tag{14.49}$$

Then use Equation (B.78) of Appendix B and a great deal of algebra to show that

$$|G(i\omega)| = \frac{\sqrt{2}}{\sqrt{\cosh(2\sqrt{\omega}) + \cos(2\sqrt{\omega})}}\tag{14.50}$$

and that

$$\angle G(i\omega) = -\tan^{-1}\{\tan(\sqrt{\omega})\tanh(\sqrt{\omega})\}.\tag{14.51}$$

These exact frequency response formulas are plotted in Figure 14.11.

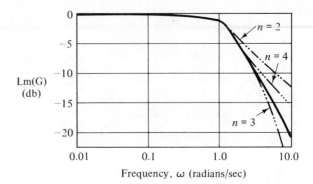

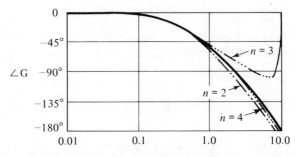

Figure 14.11. Exact frequency response of the rod illustrated in Figure 14.10 and the approximate responses of various nth-order distributed mode representations. For the approximate transfer functions, $K_2 = 1.18$, $K_3 = 0.906$, and $K_4 = 1.09$.

An approximation technique will now be developed based on the modal state space representation of the rod. Refer to Section 6.7 and verify that

$$e(x, t) = \sum_{k=1}^{\infty} z_k(t) \sin(\omega_k x), \qquad (14.52)$$

where

$$\omega_k = \frac{(2k - 1)\pi}{2L} \qquad (14.53)$$

bnd the *distributed modes*, $z_k(t)$, are solutions of the differential equations

$$\dot{z}_k = -\frac{\omega_k^2}{\Re\mathbb{C}} z_k + \frac{2\omega_k}{\Re\mathbb{C}L} e(0, t), \quad k = 1, 2, \ldots, \infty. \qquad (14.54)$$

Transform to the Laplace domain, and use Equation (14.46) and the fact that

$$\sin(\omega_k L) = (-1)^{k-1} \qquad (14.55)$$

to show that Equations (14.52)–(14.54) lead to the conclusion that

$$E(L, s) = \left(\sum_{k=1}^{\infty} \frac{(-1)^{k-1}(2k-1)\pi}{2\{s + [(2k-1)^2\pi^2/8]\}} \right) E(0, s). \tag{14.56}$$

To avoid the infinite sum, choose for an approximate transfer function the truncated version

$$G_*(s) = K_n \sum_{k=1}^{n} \frac{(-1)^{k-1}(2k-1)\pi}{2\{s + [(2k-1)^2\pi^2/8]\}}. \tag{14.57}$$

The factor K_n is added in order to force $G_*(s)$ to have the same gain as $G(s)$ [see Equation (14.45)]. The need for this factor is brought about, of course, by the truncation. It is left as an exercise to show that for a given n

$$\frac{1}{K_n} = \sum_{k=1}^{n} \frac{(-1)^{k-1}4}{(2k-1)\pi}. \tag{14.58}$$

The frequency responses $G_*(i\omega)$ for $n = 2$, 3, and 4 are plotted in Figure 14.11. Note the convergence to $G(i\omega)$ as n is increased. Note that the convergence to the phase curve is more rapid for the even values of n. The convergence to the magnitude curve is the more rapid for odd values of n.

Examine Equation (14.57) and notice that $G_*(s)$ is the sum of single pole transfer functions. The poles of $G_*(s)$ will be such that the break frequencies of $G_*(i\omega)$ will equal

$$\frac{(2k-1)^2\pi^2}{8} \tag{14.59}$$

for $1 \leq k \leq n$. These break frequencies are indicated in Table 14.1.

Notice that all break frequencies are beyond the range of frequencies used

Table 14.1

BREAK FREQUENCIES FOR THE DISTRIBUTED MODE
APPROXIMATION, $G_*(s)$, FOR THE CONDUCTION OF HEAT IN A ROD

k	Break frequency, $(2k-1)^2\pi^2/8$ (radians/sec)
2	1.23
3	11.1
4	30.8
5	60.5
10	445

in Figure 14.11 for $k \geq 3$. At first it is, therefore, surprising that the frequency responses for various n, as indicated in Figure 14.11, differ from one another as much as they do. This perplexing fact is resolved only after a study of the numerator dynamics of $G_*(s)$ for various n. The reader will be asked to make this very interesting study in the problem assignments.

To increase the reader's interest in the numerator dynamics of $G_*(s)$, some other facts are now pointed out. As indicated by Equation (14.57), $G_*(s)$ will have one zero and two poles when $n = 2$. Normally, one would expect that, for this case, $<G_*(i\omega) \rightarrow -90°$ for high values of frequency; yet, as indicated in Figure 14.11, $<G_*(i\omega)$ attains values of $-180°$ and less. How can this be?

References

1. CALDWELL, W. I., COON, G. A., and ZOSS, L. M., *Frequency Response for Process Control*, McGraw-Hill, New York, 1959.

2. CHURCHILL R. V., *Operational Mathematics*, McGraw-Hill, New York, 1958.

Problems

14.1. Try to approximate the closed loop transfer function for the thickness control system (Problem 13.17) with a transfer function of the form

$$\frac{K(Ts + 1)}{(s^2/\omega_n^2) + (2\xi/\omega_n)s + 1}.$$

14.2. Determine a functional transfer function model for a system for which the following data were taken:

ω (radians/sec)	Gain, $|G|$	Phase, $<G$	ω (radians/sec)	Gain, $|G|$	Phase, $<G$
0.08	50	$-90°$	4.0	1.2	$-135°$
0.10	80	$-90°$	6.0	0.31	$-140°$
0.2	40	$-91°$	8.0	0.38	$-145°$
0.4	20	$-99°$	12.0	0.092	$-153°$
1.0	7.4	$-110°$	16.0	0.112	$-157°$
2.0	2.6	$-123°$	32.0	0.052	$-169°$
3.0	0.88	$-130°$	40.0	0.0196	$-163°$

14.3. Determine a functional transfer function model for a system for which the following data were taken:

ω (cycles/sec)	$|G|$	$<G$	ω (cycles/sec)	$|G|$	$<G$
			1.42	1.45	$-205°$
0.77	2.4	$-97°$	1.85	0.78	$-222°$
0.39	1.25	$-27°$	2.25	0.59	$-235°$
0.23	1.05	$-19°$	2.68	0.45	$-241°$
0.54	1.5	$-50°$	3.0	0.38	$-245°$
0.83	2.5	$-95°$	4.5	0.24	$-256°$
0.94	2.8	$-118°$	6.28	0.17	$-252°$
1.0	2.8	$-132°$			

14.4. The closed loop transfer function for the feedback system illustrated in Figure 14.12 is

$$G(s) = \frac{C(s)}{R(s)}$$

$$= \frac{1}{2e^s + 1}.$$ (14.60)

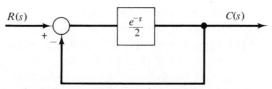

Figure 14.12. Feedback about a dead time element.

Determine a rational transfer function approximation for this closed loop signature.

14.5. For the system illustrated in Figure 14.12, determine $c(t)$ when $r(t)$ is the unit step. An intuitive approach will be more expedient than a Laplace transform analysis. Compare this response with the unit step response of the approximate transfer function obtained in the solution of Problem 14.4.

14.6. Take $n = 1$ in Equations (14.57) and (14.58) and compare the resulting $G_*(s)$ with $G(i\omega)$ of the one-dimensional rod.

14.7. Investigate the numerator dynamics of $G_*(s)$ for $n = 2$ in Equations (14.57) and (14.58) and answer the question posed at the very end of the chapter. The material in Section 13.7 should prove useful.

14.8. Use the state variable representation of Equation (6.65) to find a $G_*(s)$ for the rod system. Take $n = 1, 2$, and then 3 and deduce whether the finite difference or the distributed mode representations will lead to the minimum-order approximate state variable representation.

The Analysis of Closed Loop Stability by Frequency Domain Methods

15

Obviously, one of the important tasks of a feedback designer is to ensure stability of the feedback system. A stability analysis procedure,[1,2] based on the material presented in Chapter 13, is introduced in this chapter. The analysis technique is more general than most of the time domain techniques summarized in Chapter 10 because open loop transfer functions with transcendental terms present no special difficulties. As a matter of fact, it will be demonstrated that the analysis technique can be extended to the case when *hard* nonlinearities (such as relays) are used in the synthesis of feedback systems.

The discussion begins with an interpretation of the Nyquist diagram. This interpretation will be useful in succeeding sections of the present chapter and in Chapter 16, where feedback design by frequency domain methods is explored more thoroughly.

15.1 An Interpretation of the Nyquist Diagram

Consider the standard feedback control system illustrated in Figure 15.1. Denote the open loop transfer function

$$G(s) \triangleq G_c G_p H. \tag{15.1a}$$

Then the characteristic equation is

$$1 + G(s_p) = 0. \tag{15.1b}$$

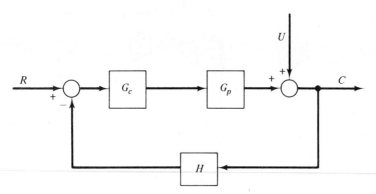

Figure 15.1. Scalar control system.

To work with a concrete example, consider the special case

$$G(s) = \frac{1}{(1 + T_1 s)(1 + T_2 s)}. \tag{15.2}$$

The root locus associated with this open loop transfer function is shown in Figure 15.2, as is the Nyquist diagram.

The Nyquist diagram is a plot of

$$G(i\omega) = G(s)|_{s = +i\omega} \tag{15.3}$$

in the G plane. Thus, the Nyquist diagram may be thought of as a *mapping* of the positive $i\omega$ axis in the s plane to the G plane. This type of mapping is illustrated in Figure 15.2. It is useful to consider that the entire imaginary axis is mapped to the G plane. The image of the $-i\omega$ axis is the mirror image of the Nyquist diagram. The image of the entire imaginary axis will be called the *Nyquist locus*. It is important to notice that Equation (15.1b) leads to the interpretation that *all* closed loop poles map to the -1 point in the G plane. This fact is illustrated in Figure 15.2.

For the particular transfer function in Equation (15.2), the Nyquist locus is shown in the G-plane plot of Figure 15.3. The question to be investigated now is, Where are the images in the G plane of vertical lines *near* the $i\omega$ axis in the s plane?

If the reader will imagine himself standing on the $i\omega$ axis in the s plane and facing the $\omega = +\infty$ direction, the line in the right half-plane will lie to his right. If he now stands on the image in the G plane of the $i\omega$ axis (i.e., on the Nyquist locus) and again faces the direction of increasing ω, he would, correctly, expect the *image* of the right half-plane line to lie to his right. Continuing this line of thinking, one would expect the entire right half of the s plane to map to the right of the Nyquist locus. Similarly, the left half of the s plane will map to the left of the Nyquist locus.

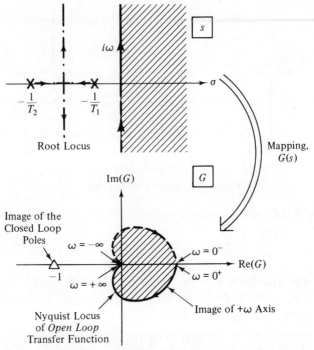

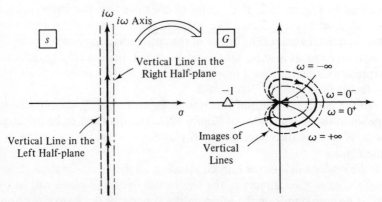

Figure 15.2. Interpretation of the Nyquist diagram as a mapping from the s plane to the G plane.

Figure 15.3. Mapping of vertical lines in the s plane to the G plane for the simple second-order transfer function of Equation (15.2).

Continue this intuitive line of thought, and notice that the Nyquist locus in Figure 15.3 is a closed curve in the G plane; this is often the case. If one accepts, for the moment, the tentative conclusions of the previous paragraph, he is led to the conclusion that the right half of the s plane maps to the right of the Nyquist locus and that the left half of this plane maps to the left of the Nyquist locus.

To summarize,

1. The imaginary axis of the s plane maps to the Nyquist locus in the G plane.
2. All *closed loop* poles in the s plane map to the -1 point in the G plane.
3. The right half of the s plane maps to the right of the Nyquist locus.

These interpretations lead to the following stability criterion for closed loop systems:

> *Imagine yourself moving along the Nyquist locus from $\omega = -\infty$ to $\omega = +\infty$. If the -1 point always lies in the region which lies to the left of your path, then the closed loop system is stable.* (15.4)

Example 15.1

Suppose that

$$G(s) = \frac{1}{(1 + Ts)^3}.$$

The Nyquist locus is shown in Figure 15.5. For this configuration, criterion (15.4) indicates that the closed loop system is stable. Increased gain forces

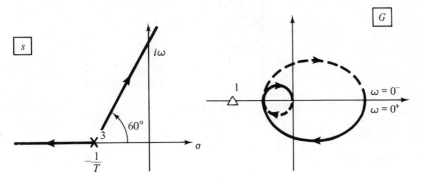

Figure 15.4. Root locus for a triple pole at $-1/T$.

Figure 15.5. Low-gain locus for the system of Example 15.1.

the locus from the origin as is illustrated in Figure 15.6. (Why?) It follows from criterion (15.4) that at high gain the closed loop system will become unstable (since the -1 point is then to the right of the Nyquist locus). The above conclusions are similar to those that would have been obtained from an examination of the root locus (see Figure 15.4).

* * *

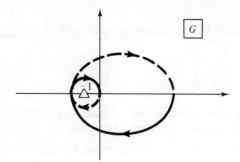

Figure 15.6. High-gain locus for the system of Example 15.1.

Example 15.2

Suppose that

$$G(s) = \frac{1}{1 + Ts}.$$

The Nyquist locus is shown in Figure 15.7. Obviously, the system is stable for all gains. The reader should verify this result with a root locus plot.

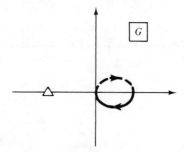

Figure 15.7. Nyquist locus for open loop transfer function.

* * *

Example 15.3

Suppose that

$$G(s) = \frac{e^{-\tau s}}{1 + Ts}.$$

The low- and high-gain loci are indicated in Figures 15.8 and 15.9.

The reader is invited to apply criterion (15.4) and to reconcile the results with those obtained using a root locus diagram.

* * *

Notice that criterion (15.4) need not be modified if $G(s)$ contains transcendental terms.

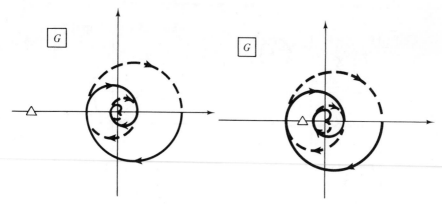

Figure 15.8. Low-gain locus for the first-order system with dead time.

Figure 15.9. High-gain locus for the first-order system with dead time.

The above presentation is an effort to establish an intuitive relationship between stability and the Nyquist locus. The Nyquist locus stability criterion can be put on a more rigorous basis for a special class of open loop transfer functions.

15.2 A Slightly More Rigorous Approach to Nyquist Stability for Nontranscendental Transfer Functions

The characteristic equation for a closed loop system can be rewritten in the form

$$1 + G(s) = 1 + \frac{k'(s - \alpha_1)(s - \alpha_2) \cdots (s - \alpha_m)}{(s - p_1)(s - p_2) \cdots (s - p_n)} = 0, \quad m < n, \quad (15.5)$$

or

$$1 + G(s) = \frac{K(s - z_1)(s - z_2) \cdots (s - z_n)}{(s - p_1)(s - p_2) \cdots (s - p_n)} = 0 \quad (15.6)$$

if the open loop transfer function, $G(s)$, is a ratio of polynomials in s and, as is usually the case, if the order of the numerator, m, is not greater than the order of the denominator, n. The z_i denote zeros of the characteristic equation; hence, they are *closed loop* poles. The p_i denote poles of $G(s)$; hence, the p_i are *open loop* poles.

Consider Figure 15.10 wherein the z_i and the p_i for a typical system are plotted. Imagine that a test point, s, is moved in a clockwise fashion around the closed circular contour indicated in this figure. The vector $s - z_1$ will undergo one clockwise rotation during this traverse. On the other hand, the

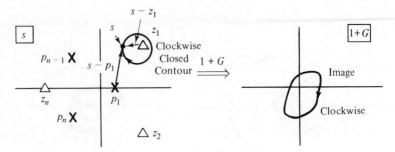

Figure 15.10. Image of a closed contour about a closed loop pole for the mapping $1 + G$.

vector $s - p_1$ will *not* undergo any *net* change in angle as the point is moved around the closed contour. In fact the vectors $s - p_i$ and $s - z_i$ will not undergo a *net* change in angle for any z_i or p_i not *inside* the contour. From Equation (15.6) it follows that the complex quantity $1 + G(s)$ will undergo one net clockwise rotation as s is moved about the contour. This fact leads one to conclude that the image of the closed contour will have the form indicated in Figure 15.10. Notice that the image encloses the origin. If the contour had enclosed an open loop pole, the image would be a counter-clockwise closed contour since the term $s - p_1$ appears in the denominator of Equation (15.6) (see Figure 15.11).

Consider the closed contour illustrated in Figure 15.12. The vector $s - z_1$ undergoes one clockwise rotation as does $s - p_1$. No other vectors perform a *net* rotation. In Equation (15.6), $s - z_1$ is in the numerator and $s - p_1$ is the denominator so that the quantity $1 + G$ undergoes *no* net rotation; i.e., the rotations of the two vectors cancel one another. Thus, the image must be as indicated in Figure 15.12, where no rotations about the origin of the $1 + G$ plane are indicated.

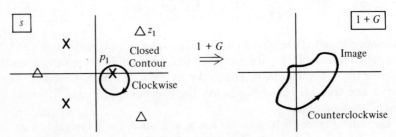

Figure 15.11. Image of a closed contour about an open loop pole for the mapping $1 + G$.

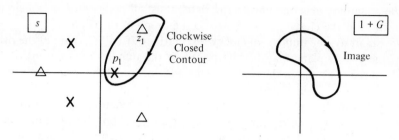

Figure 15.12. Image of a closed contour which encircles a single open loop pole and a single closed loop pole.

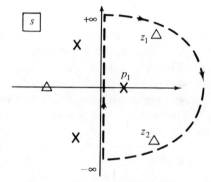

Figure 15.13. Contour enclosing the entire right half-plane. The semi-circle is at infinity.

Finally, consider the closed contour, indicated in Figure 15.13, which encloses the entire right half-plane. Let

N_0 = number of *open loop* poles in the right half s plane,
N_c = number of *closed loop* poles in the right half s plane,
N = number of net clockwise rotations of $1 + G(s)$ when s is moved around the contour indicated in Figure 15.13. (15.7)

It follows from the above discussion and Equation (15.6) that

$$N = N_c - N_0. \tag{15.8}$$

This equation is the basis for the *Nyquist stability criterion*, which will be discussed below. Clearly,

$$N_c = N + N_0. \tag{15.9}$$

It is left as an exercise for the reader to argue that *all points on the semi-circular portion of the contour illustrated in Figure 15.13 map to the +1 point in the 1 + G plane.* The G plane is obtained from the 1 + G plane by shifting the imaginary axis one unit to the right; thus, all points on the semicircular portion of the closed contour map to the *origin of the G plane.*

Similarly, the number of clockwise rotations of $1 + G(s)$ about the origin of the $1 + G$ plane may be interpreted as the number of rotations of $G(s)$ about the -1 point in the G plane. The only portion of the contour illustrated in Figure 15.13 which does not map to the origin of the G plane is the imaginary axis of the s plane. Thus, N may be interpreted as the number of clockwise rotations of $G(i\omega)$ (i.e., the Nyquist locus) about the -1 point in the G plane. This fact demonstrates again the intuitive relationship between closed loop stability and the Nyquist locus introduced in the previous section.

Equation (15.9) is used to determine closed loop stability in the following way: (1) Construct the Nyquist diagram of the open loop transfer function, $G(i\omega)$, and determine N; (2) the number of open loop poles in the right half-plane, N_0, is determined by examining the denominator of $G(s)$ (N_0 is usually zero); (3) N_c is determined from Equation (15.9). For closed loop stability, N_c must, of course, be zero. Used in this way, Equation (15.7) is called the *Nyquist criterion.*

Difficulties occur with the above procedure when $G(s)$ has poles at the origin. The manner in which this case is handled is illustrated in the following examples.

Example 15.4

The s-plane portrait for

$$G(s) = \frac{1 + Ts}{s^2}$$

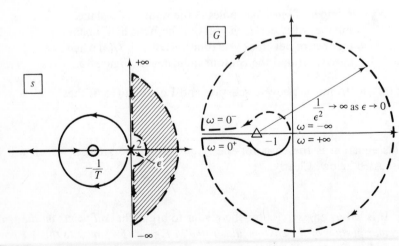

Figure 15.14. Root locus and half-plane contour for Example 15.4.

Figure 15.15. Nyquist locus for Example 15.4.

is illustrated in Figure 15.14. The Nyquist locus is shown in Figure 15.15. A small semicircular detour of radius ϵ is added to the contour in Figure 15.14. This detour maps as shown in Figure 15.15. Notice that since a double open loop pole exists at the origin, the image of the ϵ contour must rotate 360° clockwise as the test point, s, moves 180° counterclockwise about the ϵ contour. Since $N = 0$ and $N_0 = 0$, it follows from Equation (15.9) that $N_c = 0$; thus, no closed loop poles exist in the shaded portion of Figure 15.14. To cover the entire right half-plane, allow $\epsilon \rightarrow 0$. In Figure 15.15, the radius of the detours image, $1/\epsilon$ approaches ∞. In this limit $N = 0$ so that $N_c = 0$. Conclude that closed loop poles do not exist in any portion of the right half-plane. This conclusion is consistent with the root locus diagram constructed in Figure 15.14 in order to provide the reader with a check of results obtained using Equation (15.9).

* * *

Example 15.5

Plots similar to those in Example 15.4 are shown in Figures 15.16 and 15.17 for

$$G(s) = \frac{1 + Ts}{s^3}.$$

Notice that a detour of radius ϵ is included in order to avoid the singularities at the origin. Compare the G-plane image of this detour with the image of the detour of Example 15.4, illustrated in Figure 15.15.

It is clear that $N = 2$ as $\epsilon \rightarrow 0$. From Equation (15.9), it follows that two closed loop poles are in the right half s plane and that the system is unstable.

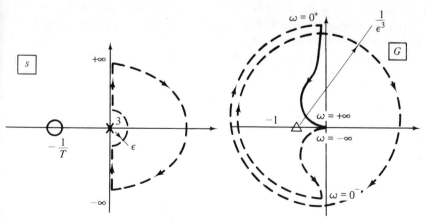

<table>
<tr><td>**Figure 15.16.** s-plane contour.</td><td>**Figure 15.17.** G-plane contour.</td></tr>
</table>

Increasing or decreasing the gain does not change the value of N, and, hence, does not change the value of N_c.

It is left as an exercise for the reader to confirm the conclusions of the previous paragraph by constructing the root locus diagram associated with the open loop transfer function of the previous example.

* * *

It is probably apparent that when $G(s)$ is the ratio of polynomials, the Nyquist stability criterion is a bit awkward to apply when compared to Routh array or root locus methods; however, as will become apparent, a designer may find certain feedback system designs easier to execute with the help of G-plane plots than with any other aid. If this be the case, then the Nyquist stability criterion becomes convenient since the designer will have constructed the Nyquist locus for other purposes anyway.

Another reason for introducing frequency domain stability criteria is that these notions may be extended to a class of important *nonlinear* feedback systems. A few of the elementary extensions are introduced in the next section.

15.3 Nonlinear *Regulators*: Nonlinear Elements and Their Block Diagram Representations*

Frequency domain stability analysis will presently be extended to nonlinear feedback systems. Before these extensions are discussed, it is useful to distinguish between two types of feedback systems: (1) control systems and (2) regulators. *Control systems* may be represented schematically as in Figure 15.1. The role of control systems is to force the output, $c(t)$, to *follow* the input, $r(t)$, as closely as possible.

Regulators are feedback systems designed to maintain the output as closely as possible to some *fixed* (i.e., time-invariant) level in the face of unavoidable noise and/or disturbance signals. It is clear that many control systems must also serve as regulators.

The distinction between the two types of feedback systems is not necessary in the discussion of the stability of *linear* feedback systems but must be made in the nonlinear case.

A series of nonlinear elements is described before a method for the stability analysis of feedback systems containing such devices is provided.

For the electric heater in Figure 15.18,

$$q = \frac{e^2}{R} = Rf^2, \qquad (15.10)$$

which is a nonlinear relationship.

---

* This section may be omitted during the first reading of the text.

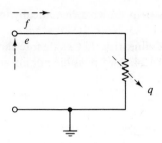

Figure 15.18. Electric heater.

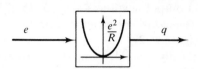

Figure 15.19. Block diagram symbol for a heater.

Nonlinearities will be given block representations much like transfer function relations. Instead of a transfer function being placed in the block, the graph relating the input to the output of the component is placed in the block. For instance, for the heater the block diagram notation is as shown in Figure 15.19.

As another example of a nonlinearity, consider the gear set sketched in Figure 15.20. The block diagram symbol is as shown in Figure 15.21. The graph is meant to denote the gear *backlash* or *hysteresis* associated with clearance.

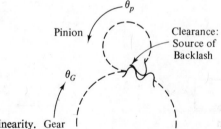

Figure 15.20. Gear set nonlinearity.

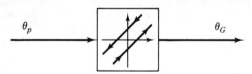

Figure 15.21. Symbol for backlash (hysteretic) behavior in a gear set.

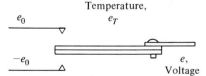

Figure 15.22. Bimetallic strip. The voltages e_0 and $-e_0$ are applied to the contacts.

The input-output relation for the bimetallic strip shown schematically in Figure 15.22 is represented as shown in Figure 15.23. Note the *dead zone*.

Coulomb friction is another example of a nonlinearity. Let e_f denote the friction force on the mass illustrated in Figure 15.24. Two possible nonlinear input-output relations are symbolized in Figure 15.25.

The electric or pneumatic or fluidic relay (Figure 15.26) is represented as in Figure 15.27.

Finally, consider amplifiers, hydraulic transmissions, servo-valves, and

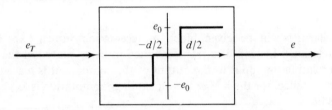

Figure 15.23. Dead zone in a bimetallic strip.

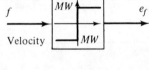

or

M_s is the coefficient of static friction.

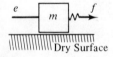

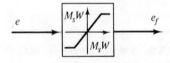

Figure 15.24. Coulomb friction.

Figure 15.25. Block diagram symbols for coulomb friction. W is the weight of the mass in Figure 15.24.

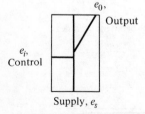

Figure 15.26. Fluidic relay.

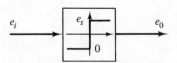

Figure 15.27. Ideal relay representation.

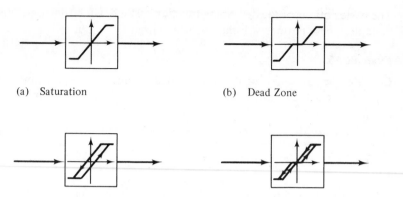

(a) Saturation   (b) Dead Zone

(c) Hysteresis or Backlash   (d) A device with all of the above characteristics.

Figure 15.28. Commonly encountered nonideal characteristics.

flapper nozzles. Commonly encountered nonideal characteristics are as illustrated in Figure 15.28.

15.4 Describing Functions and the Stability of Nonlinear Regulators*

Consider the nonlinear regulator illustrated in Figure 15.29. N is the graph of a non-linearity. The non-linearity may be the unfortunate property of a component or may be added intentionally. For instance, relays are cheap and reliable and are often included in the synthesis of a feedback system.

For given G_c, G_p, and N, the analyst will want to determine conditions for continuous cycling of the system. This nonlinear oscillation is called a *limit cycle* and often represents a stability limit condition. The *describing function* method is a frequency domain method which will allow the analyst to predict the possible existence of limit cycles and to determine approximate values of the amplitude and frequency of these oscillations.

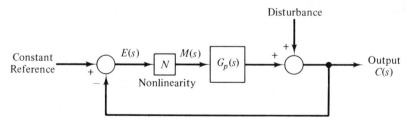

Figure 15.29. Nonlinear regulator. N denotes the input-output graph of a nonlinearity.

* This section may be omitted during the first reading of the text.

The basic notions of the describing function method will be demonstrated by example. These notions will then be generalized.

Example 15.6

Consider the particular regulator illustrated in Figure 15.30. Assume that the regulator is at the stability limit so that the error signal is approximately given by

$$e(t) = A \cos (\omega_0 t), \tag{15.11}$$

where the constants A and ω_0 are unknown. Notice that

$$
\begin{aligned}
m(t) &= A^3 \cos^3 (\omega_0 t) = A^2[1 - \sin^2 (\omega_0 t)] \cos (\omega_0 t) \\
&= A^2[\cos (\omega_0 t) - \sin^2 (\omega_0 t) \cos (\omega_0 t)].
\end{aligned} \tag{15.12}
$$

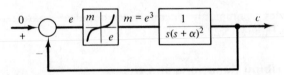

Figure 15.30. Nonlinear regulator with a cubic compensator.

Use the elementary trigonometric identity

$$\sin (A) \cos (B) = \tfrac{1}{2}\{\sin (A + B) + \sin (A - B)\} \tag{15.13}$$

to show that

$$\sin^2 (\omega_0 t) \cos (\omega_0 t) = \sin (\omega_0 t)\frac{\sin (2\omega_0 t)}{2}. \tag{15.14}$$

Now use the identity

$$\sin (A) \sin (B) = \tfrac{1}{2}\{\cos (A - B) - \cos (A + B)\} \tag{15.15}$$

to show that

$$\sin^2 (\omega_0 t)\cos (\omega_0 t) = \tfrac{1}{4}\{\cos (\omega_0 t) - \cos (3\omega_0 t)\}. \tag{15.16}$$

Combine Equations (15.12) and (15.16) to find

$$m(t) = A^3[\tfrac{3}{4} \cos (\omega_0 t) + \tfrac{1}{4} \cos (3\omega_0 t)]. \tag{15.17}$$

Thus, the spectrum of $m(t)$ is as illustrated in Figure 15.31.

The *harmonic approximation* is to neglect the high-frequency component at $3\omega_0$. One of the reasons for making this approximation is that the higher

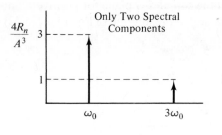

Figure 15.31. Spectrum (r_n) of the error signal when the regulator is in a limit cycle.

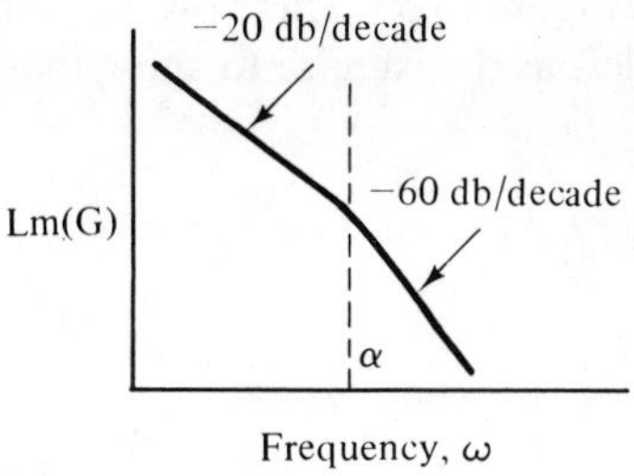

Figure 15.32. Gain curve of the Bode diagram for the plant in Figure 15.30.

frequency is attenuated more than the lower-frequency term by the plant (see Figure 15.32). The approximation is

$$c(t) \simeq \frac{3A^3}{4} |G(i\omega_0)| \cos(\omega_0 t + <G(i\omega_0)). \tag{15.18}$$

An examination of Figure 15.30 leads one to impose the compatibility condition

$$\frac{3A^3}{4} |G(i\omega_0)| \cos[\omega_0 t + <G(i\omega_0)] = -A \cos \omega_0 t, \tag{15.19}$$

which can be satisfied only if

$$|G(i\omega_0)| = \frac{4}{3A^2} \tag{15.20}$$

and

$$<G(i\omega_0) = \pi. \tag{15.21}$$

Equations (15.20) and (15.21) are two equations in the two unknowns ω_0 and A. The most convenient way to solve these equations is graphically. Define the *describing function*

$$N(A) = \frac{3A^2}{4}. \tag{15.22}$$

Notice that the right-hand side of Equation (15.20) is the magnitude of $-[1/N(A)]$ and that the right-hand side of Equation (15.21) is its angle. Thus, Equations (15.20) and (15.21) can be written in the single complex variable equation

$$G(i\omega_0) = -\frac{1}{N(A)}. \tag{15.23}$$

$G(i\omega_0)$ and $-[1/N(A)]$ are plotted on the complex G plane for positive A and ω_0 in Figure 15.33. These two loci intersect at the solution of Equation (15.23). It is left as an exercise to show that

$$\omega_0 = \alpha \tag{15.24}$$

and that

$$A^2 = \tfrac{8}{3}\alpha^3$$

or

$$A = \sqrt{\tfrac{8}{3}\alpha^3}. \tag{15.25}$$

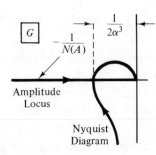

Figure 15.33. Graphical solution of Equation (15.23). The arrow on the graph of $-1/N(A)$ indicates the direction of increasing amplitude, A.

The result of Equation (15.24) gives the analyst confidence in the assumption of Equation (15.11). (Why?)

*   *   *

The generalization of the technique illustrated above is

1. Assume that the regulator is in a limit cycle so that

$$-c(t) = e(t) \simeq A\cos(\omega_0 t), \quad A, \omega_0 \text{ unknown}. \tag{15.26}$$

2. Use trigonometric identities, if possible, or a Fourier series expansion to determine the spectrum of the manipulated variable, $m(t)$.

3. Make the *harmonic approximation*; i.e., neglect higher-frequency harmonics in $m(t)$ (good approximation if the plant is a *low-pass* filter) so that for some F and ϕ,

$$m(t) \simeq F(A)\cos\{\omega_0 t + \phi(A)\}. \tag{15.27}$$

4. Impose the compatibility condition

$$c(t) \simeq F(A)|G(i\omega_0)|\cos\{\omega_0 t + \phi(A) + <G(i\omega_0)\}$$
$$= -A\cos\{\omega_0 t\}, \tag{15.28}$$

so that

$$F(A)|G(i\omega_0)| = A \tag{15.29}$$

and

$$<G(i\omega_0) + \phi(A) = -\pi. \tag{15.30}$$

These equations are usually solved graphically, in the G plane, to determine the unknowns A and ω_0.

Define the *describing function* of the nonlinearity as the complex-valued function of amplitude, A, given by

$$N(A) = \frac{F(A)}{A} e^{i\phi(A)}. \tag{15.31}$$

Then Equations (15.29) and (15.30) can be written in the compact form

$$G(i\omega_0) = -\frac{1}{N(A)}. \tag{15.32}$$

The graphical solution of this equation is obtained by plotting the Nyquist diagram, $G(i\omega)$, and the *amplitude locus*, $-[1/N(A)]$, in the same complex plane. The intersection of these loci provides values of the amplitude and frequency of the limit cycle.

Most often, Fourier series, rather than trigonometry, will be the tool used for harmonic approximation. For this case,

$$m(t) \simeq A_1 \cos(\omega_0 t) + B_1 \sin(\omega_0 t), \tag{15.33}$$

where the Fourier coefficients

$$A_1 = \frac{2}{T} \int_{-T/2}^{T/2} m(t) \cos(\omega_0 t)\, dt \tag{15.34}$$

and

$$B_1 = \frac{2}{T} \int_{-T/2}^{T/2} m(t) \sin(\omega_0 t)\, dt. \tag{15.35}$$

The use of the generalized procedure will now be illustrated with an example.

Example 15.7

Suppose that the nonlinearity is an ideal relay (see Figure 15.27). A graphical method for finding $m(t)$ when $e(t)$ is given by Equation (15.26) is illustrated in Figure 15.34. Notice that the graph of $e(t)$ is rotated 90° in order to determine $m(t)$ by projection.

For this nonlinearity, Equation (15.35) becomes

$$B_1 = 0. \tag{15.36}$$

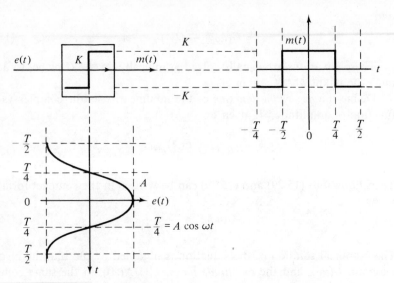

Figure 15.34. Graphical determination of the output, $m(t)$, of a relay given the input, $e(t)$.

Thus, no phase lag between $e(t)$ and $m(t)$ exists; i.e., the describing function has zero imaginary part. Equation (15.34) becomes

$$A_1 = \frac{8}{T} \int_0^{T/4} K \cos(\omega t)\, dt$$

$$= \frac{8K}{T\omega} \sin(\omega t) \Big]_0^{T/4} = \frac{8K}{2\pi} = \frac{4K}{\pi}. \tag{15.37}$$

Thus, the describing function is found from Equation (15.31) to be

$$N(A) = \frac{4K/\pi}{A} e^{i \cdot 0} = \frac{4K}{\pi A}. \tag{15.38}$$

It follows that the amplitude locus for an ideal relay has the form

$$\boxed{-\frac{1}{N(A)} = -\frac{\pi A}{4K}.} \tag{15.39}$$

The last quantity can be plotted on the G plane along with the Nyquist diagram to determine stability limit conditions.

Define a normalized amplitude by

$$A' = \frac{\pi A}{4K}. \tag{15.40}$$

Then

$$-\frac{1}{N(A)} = -A'.$$ (15.41)

The plot of $-1/N$ on the G plane is shown in Figure 15.35. A limit cycle occurs for the system represented in Figure 15.35 when the frequency

$$\omega = \omega_3$$

and

$$A' = \frac{1}{2}$$

or the amplitude

$$A = \frac{4K}{\pi} \cdot \frac{1}{2} = \frac{2K}{\pi}.$$ (15.42)

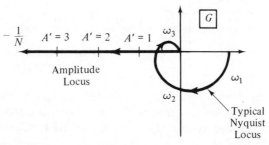

Figure 15.35. Plot of the amplitude locus for an ideal relay on the G plane.

Notice that $<G \neq -180°$ as $\omega \to \infty$. Thus, an infinite frequency limit cycle (*chatter*) will not occur since the origin is not a point where Equation (15.30) is satisfied.

* * *

15.5 *Stability* of Limit Cycles: A Heuristic Approach*

Often, Equation (15.32) will be satisfied for certain values of A and ω, but experimentally one finds that the corresponding limit cycle never occurs or is not sustained in the actual system. Such limit cycles are said to be *unstable*. To see how one might predict an *instability* of a limit cycle, consider the regulator with a relay (see Example 15.7). Suppose that the plant transfer

* This section may be omitted during the first reading of the text.

function has three poles and two zeros so that the amplitude and Nyquist diagram would appear as in Figure 15.36.

Notice that two possible limit cycles exist. It will now be shown that only one of these limit cycles is stable.

The correct analysis of the stability phenomenon is beyond the scope of this presentation. The analysis presented here starts with the crude approximation

$$M(s) \simeq N(A^*)E(s) \tag{15.43}$$

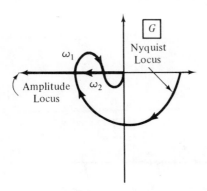

Figure 15.36. Relay and a three-pole, two-zero plant.

where $N(A^*)$ is the describing function evaluated at a specific value of amplitude. Equation (15.43) will be thought of as representing a transfer function relation for a *linearized* system. For this linearized regulator system the characteristic equation is

$$1 + N(A^*)G(s_p) = 0$$

or

$$G(s_p) = -\frac{1}{N(A^*)} = \text{Constant}, \tag{15.44}$$

where s_p is a closed loop pole for the linearized system (see Figure 15.29). That portion of the Nyquist diagram near the amplitude locus intersections is shown in Figure 15.37.

By applying the Nyquist criterion, arrows which indicate what will happen to the amplitude, A, for initial increases or decreases in A from limit cycle conditions, can be drawn. For example, consider the limit cycle at ω_2. A perturbation which reduces the amplitude to A^* from the limit cycle condition will cause the image of the closed loop poles of the linearized system to move to the position marked by the triangle just to the right of the ω_2 intersection [see Equation (15.44) and note the direction of increasing amplitude on the amplitude locus]. The Nyquist criterion indicates that the linearized system is stable so that the amplitude will *continue* to decrease. Similar

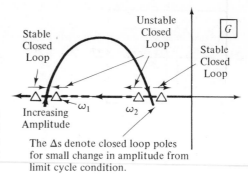

Figure 15.37. Enlarged view of a portion of Figure 15.36. The triangles denote $-1/N(A^*)$, where A^* is slightly less than or slightly greater than the limit cycle condition.

The Δs denote closed loop poles for small change in amplitude from limit cycle condition.

reasoning will lead to the conclusion that an increased amplitude will lead to further increases.

It is seen that the limit cycle at ω_2 is unstable since any change in amplitude will cause continued change in amplitude. By the same reasoning, the limit cycle at ω_1 is stable. The output record of this regulator might look like that sketched in Figure 15.38 if the regulator is started in such a way that the limit cycle at ω_2 is initially excited.

Note the convention in Figure 15.37 where arrows are placed over the triangles to denote what changes in amplitude are excited by small perturbations.

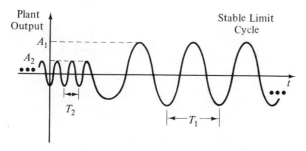

Figure 15.38. Transition from the unstable to the stable cycle.

15.6   Relay with Dead Zone (Bimetallic Strip)*

The output of a relay with dead zone for a cosine input is as illustrated in Figure 15.39. The harmonic approximation is defined by

$$m(t) = A_1 \cos(\omega t) + 0 \sin(\omega t),$$

* This section may be omitted during the first reading of the text.

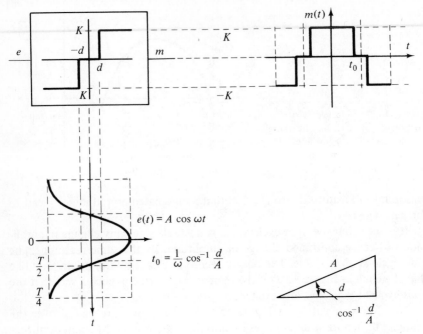

Figure 15.39. Output of a relay with dead zone.

Figure 15.40. Angle whose cosine is d/A.

where

$$A_1 = \frac{8}{T} \int_0^{T/4} y[A \cos (\omega t)] \cos (\omega t) \, dt$$

$$= \frac{8}{T} \int_0^{(1/\omega) \cos^{-1} (d/A)} K \cos (\omega t) \, dt$$

$$= \frac{8}{T\omega} \sin (\omega t) \Big]_0^{(1/\omega) \cos^{-1} (d/A)}$$

or

$$A_1 = \frac{4K}{\pi} \sin \left\{ \cos^{-1} \left(\frac{d}{A} \right) \right\}. \tag{15.45}$$

From the sketch in Figure 15.40, it follows that

$$\sin \left\{ \cos^{-1} \left(\frac{d}{A} \right) \right\} = \frac{\sqrt{A^2 - d^2}}{A}. \tag{15.46}$$

Notice that the harmonic approximation is not defined for $A < d$.

To summarize, the harmonic approximation is

$$m(t) = \begin{cases} 0, & \text{for } A \le d, \\ \dfrac{4K}{\pi} \dfrac{\sqrt{A^2 - d^2}}{A} \cos(\omega t), & \text{for } A > d. \end{cases} \qquad (15.47)$$

Thus, the describing function is defined by

$$N(A) = \begin{cases} 0, & \text{if } A \le d, \\ \dfrac{4K}{\pi} \dfrac{\sqrt{A^2 - d^2}}{A^2}, & \text{if } A > d. \end{cases} \qquad (15.48)$$

This relationship is plotted in Figure 15.41. Let

$$N_0 \triangleq \frac{N \cdot \pi \cdot d}{4K} \qquad (15.49)$$

and

$$A_0 \triangleq \frac{A}{d}. \qquad (15.50)$$

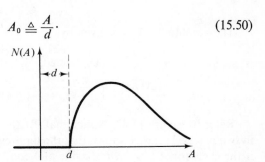

Figure 15.41. Describing function for a relay with dead zone.

Then, the normalized describing function

$$N_0 = \frac{\sqrt{A_0^2 - 1}}{A_0^2}. \qquad (15.51)$$

To find the maximum value of N_0, set

$$\frac{dN_0}{dA_0} = \frac{1}{2} \frac{(2A_0)}{\sqrt{A_0^2 - 1} A_0^2} - 2A_0^{-3}\sqrt{A_0^2 - 1} = 0.$$

This equation implies that

$$A_0^2 - 2(A_0^2 - 1) = 0$$

or, at the maximum of N_0, that

$$A_0^2 = 2$$

and the maximum value

$$N_0 = \tfrac{1}{2}.$$

Using nonnormalized variables, the above relations indicate that N is maximized when

$$A^2 = 2d^2 \qquad (15.52)$$

and the maximum value

$$N = \frac{2K}{\pi d}. \qquad (15.53)$$

In terms of $-1/N$, the amplitude locus is minimized when

$$A = \sqrt{2}\,d \qquad (15.54)$$

and the minimum value is

$$-\frac{1}{N} = -\frac{\pi d}{2K}. \qquad (15.55)$$

The amplitude locus is given by

$$-\frac{1}{N(A)} = \begin{cases} \text{Undefined,} & \text{if } A \le d, \\[2mm] -\dfrac{\pi A}{4K}\dfrac{1}{\sqrt{1 - (d/A)^2}}, & \text{if } A > d. \end{cases} \qquad (15.56)$$

Refer to Figure 15.42. Notice that if $G(i\omega)$ intersects $-1/N$, it intersects in two places. It can be shown that the limit cycle for $A < \sqrt{2}\,d$ is unstable if the cycle for $A > \sqrt{2}\,d$ is stable and conversely.

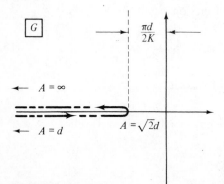

Figure 15.42. Amplitude locus for a relay with dead zone.

Example 15.8

Consider the regulator diagrammed in Figure 15.43. For the plant, several points on the Nyquist diagram follow.

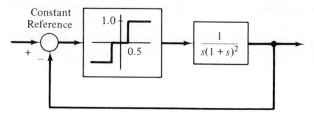

Figure 15.43. Regulator containing a relay with dead zone.

ω	$|G|$	$<G$
0.5	1.60	$-144°$
1.0	0.50	$-180°$
1.5	0.226	$-206°$
2.0	0.108	$-130°$

For this regulator,

$$\frac{\pi d}{2K} = 0.785.$$

The amplitude locus and Nyquist diagram are illustrated in Figure 15.44. Because the loci do not intersect, Equation (15.32) is not satisfied; thus, the system will not cycle. What about accuracy? Notice that the system could be at equilibrium for the output range

$$-0.5 < c(\infty) < 0.5$$

or in general,

$$-d < c(\infty) < d. \tag{15.57}$$

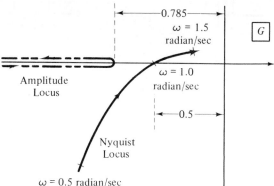

Figure 15.44. *G*-plane loci for regulator containing relay with dead zone.

Thus, the accuracy can be improved by decreasing the width of the dead zone (move the contacts closer to the bimetallic strip). The smallest value of d for a regulator system which does not limit-cycle is given by the condition

$$\frac{\pi d}{2K} = |G(i\omega)|_{\omega = 1.0}.$$

For this example,

$$d = \frac{2(1)}{\pi}.$$

*   *   *

15.7   A Relay with Hysteresis*

If the nonlinearity in the regulator of Figure 15.29 is a relay with hysteresis, the output of the relay during limit-cycle operation would be as illustrated in Figure 15.45.

Notice that A must be less than d to activate the relay.

One could use Fourier transformation to find the fundamental, but it is

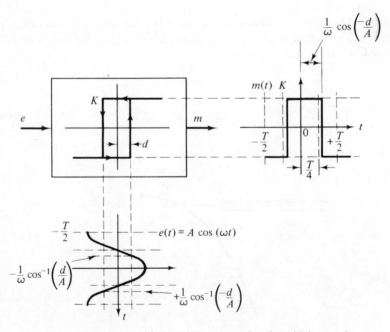

Figure 15.45. Determination of the output of a relay with hysteresis.

* This section may be omitted during the first reading of the text.

easier to use Equation (15.37). Noting the similarity between the outputs in Figures 15.34 and 15.45, it follows that the harmonic approximation

$$m(t) = \frac{4K}{\pi} \cos(\omega t - \phi), \tag{15.58}$$

where

$$\phi = \left[-\frac{T}{4} + \frac{1}{\omega} \cos^{-1}\left(-\frac{d}{A}\right) \right] \cdot \frac{2\pi}{T} \tag{15.59}$$

or

$$\phi = -\frac{\pi}{2} + \cos^{-1}\left(-\frac{d}{A}\right). \tag{15.60}$$

Thus,

$$\left(\phi + \frac{\pi}{2}\right) = \cos^{-1}\left(-\frac{d}{A}\right)$$

or

$$\cos\left(\frac{\pi}{2} + \phi\right) = -\frac{d}{A}. \tag{15.61}$$

But

$$\cos\left(\frac{\pi}{2} + \phi\right) = -\sin(\phi);$$

thus,

$$\sin(\phi) = \frac{d}{A} \tag{15.62}$$

and

$$\cos(\phi) = \left[1 - \frac{d^2}{A^2}\right]^{1/2}. \tag{15.63}$$

From Equation (15.58), it follows that the describing function

$$N(A) = \frac{4K}{\pi A} e^{-i\phi}. \tag{15.64}$$

The amplitude locus is therefore given by

$$-\frac{1}{N(A)} = \frac{\pi A}{4K} e^{i(\phi - \pi)} = -\frac{\pi A}{4K} \cos(\phi) - i\frac{\pi A}{4K} \sin(\phi). \tag{15.65}$$

Finally, combine Equations (15.62), (15.63), and (15.65) to show that

$$-\frac{1}{N(A)} = -\frac{\pi[A^2 - d^2]^{1/2}}{4K} - i\frac{\pi d}{4K}. \tag{15.66}$$

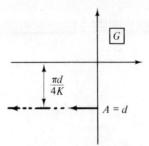

Figure 15.46. Amplitude locus for a relay with dead zone.

The amplitude locus for this type of relay is sketched in Figure 15.46. This is the first example of a complex-valued describing function. Notice that the imaginary part of the amplitude locus is independent of amplitude.

Example 15.9

For the regulator illustrated in Figure 15.47,

$$\frac{\pi d}{4K} = \frac{\pi(0.5)}{4 \cdot 1} = 0.392.$$

The amplitude and Nyquist diagram are sketched in Figure 15.48. The intersection implies that Equation (15.32) is satisfied and that a limit cycle is

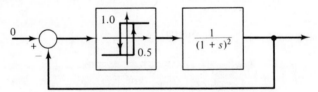

Figure 15.47. Regulator containing a hysteretic relay.

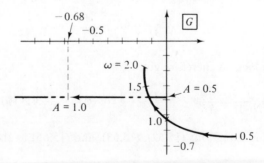

Figure 15.48. *G*-plane loci for Example 15.9.

possible ($\omega \simeq 1.4$ radians/sec and $A \simeq 0.6$). Notice that the system could be stabilized by using a compensating gain

$$= \frac{0.392}{0.5}$$

$$= 0.785.$$

* * *

15.8 Dependency of the Imaginary Part of the Describing Function on Hysteresis*

Consider the general type of nonlinearity illustrated in Figure 15.49. Denote the real and imaginary parts of the describing function by R and I, respectively, so that

$$N(A) = R(A) + iI(A). \tag{15.67}$$

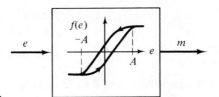

Figure 15.49. Nonlinear device.

Suppose that the regulator is in a limit cycle so that the input to the nonlinearity is given by

$$e(t) = A \cos \omega t = \text{Re} \{Ae^{i\omega t}\}. \tag{15.68}$$

From a Fourier series expansion of the output, $m_1, m_2, \ldots$ exist such that

$$m(t) = \text{Re} \{m_1(A)e^{i\omega t} + m_2(A)e^{2i\omega t} + \cdots\} \tag{15.69}$$

$$= f[A \cos \omega t], \tag{15.70}$$

where the square brackets are functional brackets. Thus, from Equations 15.28) and (15.31), find

$$f[A \cos \omega t] = \text{Re} \{AN(A)e^{i\omega t} + \cdots\}$$

$$= \text{Re} \{A(R + iI)e^{i\omega t} + \cdots\}$$

$$= AR(A) \cos (\omega t) - AI(A) \sin (\omega t) + \cdots. \tag{15.71}$$

* This section may be omitted during the first reading of the text.

It follows from the orthogonality of $\{\cos [n\omega t], \sin [n\omega t]\}$ that

$$\int_0^{2\pi} f[A \cos (\omega t)] \cos (\omega t) \, d(\omega t) = \int_0^{2\pi} AR(A) \cos^2 (\omega t) \, d(\omega t)$$

$$= \pi AR(A).$$

Thus,

$$R(A) = \frac{1}{\pi A} \int_0^{2\pi} \cos (\omega t) f[A \cos (\omega t)] \, d(\omega t). \tag{15.72}$$

Similarly,

$$I(A) = -\frac{1}{\pi A} \int_0^{2\pi} \sin (\omega t) f[A \cos (\omega t)] \, d(\omega t). \tag{15.73}$$

Notice that

$$e = A \cos (\omega t),$$
$$de = -\omega A \sin (\omega t) dt \tag{15.74}$$
$$= -A \sin (\omega t) d(\omega t).$$

Also, when ωt varies from 0 to 2π, e varies from $+A$ to $-A$ and then from $-A$ to $+A$. Thus, Equation (15.73) can be rewritten

$$I(A) = +\frac{1}{\pi A^2} \int_{\text{one cycle}} f(e) \, de. \tag{15.75}$$

Notice, however, that

$$\int_{\text{one cycle}} f(e) \, de = \int_A^{-A} f(e) \, de + \int_{-A}^{A} f(e) \, de$$

$$= \frac{-\text{Area under upper loop}}{\text{in Figure (15.49)}} + \frac{\text{Area under lower loop}}{\text{in Figure (15.49)}}.$$

Hence, the imaginary part of the describing function

$$I(A) = -\frac{H}{\pi A^2}, \tag{15.76}$$

where H is the area enclosed by a hysteresis loop in the graph of the non-linearity.

Example 15.10

For the relay in Figure 15.45

$$H = 2K \cdot 2d = 4Kd,$$

so that

$$I = -\frac{4Kd}{\pi A^2}.$$

Check: In the previous section it was shown that

$$I = -\frac{4K}{\pi A} \sin(\phi) = -\frac{4Kd}{\pi A^2}.$$

* * *

Conclusion: Single-valued nonlinearities have *real* describing functions (i.e., $I \equiv 0$).

Notice that in terms of the quantity R, the amplitude locus can be written

$$\boxed{-\frac{1}{N} = \frac{R + iI}{[R^2 + I^2]}.} \qquad (15.77)$$

R is given by Equation (15.72).

References

1. D'AZZO, J., and HOUPIS, C., *Feedback Control System Analysis and Synthesis*, McGraw-Hill, New York, 1966.

2. OGATA, K., *Modern Control Engineering*, Prentice-Hall, Englewood Cliffs, N. J., 1970.

Problems

15.1. Use the Nyquist stability criterion to determine the closed loop stability values of k_c for the following open loop transfer functions:

(a) $\dfrac{k_c}{s[(s/3) + 1]^2}$.

(b) $\dfrac{k_c}{(s + 1)[(s/2) + 1][(s/3) + 1]}$.

(c) $\dfrac{k_c e^{-3\pi s/8}}{(s/2) + 1}$.

(d) $\dfrac{k_c e^{-\pi s/4}}{s[(s/2) + 1]}$.

(e) $\dfrac{k_c}{s - 1}$.

15.2. The positive Nyquist diagram for an uncompensated open loop transfer function is shown in Figure 15.50. Determine the stable ranges of values for a compensating gain k_c. Try to explain the fact that the stable range of k_c is disjoint by using root locus arguments.

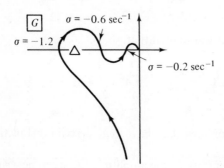

Figure 15.50. Nyquist diagram for Problem 15.2.

15.3. Construct the root locus diagram for the open loop transfer function of Example 15.2 and verify the results obtained using the Nyquist criterion.

15.4. Prove that all points on the semicircular part of the closed path illustrated in Figure 15.13 map to the $+1$ point in the $1 + G$ plane if $m \leq n$ in Equation (15.4).

15.5. Verify the results of Example 15.5 by constructing a root locus diagram.

15.6. Explain why the graph in Figure 15.28(b) might well be associated with the spool valve illustrated in Figure 7.14.

15.7. Is the limit cycle indicated by the intersection point in Figure 15.33 stable?

15.8. The input, e, and the output, m, of a nonlinearity satisfy the equation

$$m(t) = -e^2 \frac{de}{dt}.$$

Determine a describing function for this nonlinearity. [*Hint:* (1) Assume that $e(t) = A \cos \omega t$; (2) using trigonometric identities, reduce $m(t)$ to a superposition of sine and cosine terms; and then (3) assume that the plant is a low-pass filter.]

15.9. Assume that a regulator contains an ideal relay and a linear plant whose Nyquist diagram is shown in Figure 15.50. Determine whether or not the theoretical limit cycles are stable. Is "chatter" one of the theoretical limit cycles?

15.10. Determine the describing function for the nonlinearity shown in Figure 15.51.

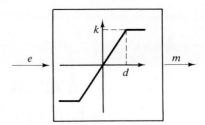

Figure 15.51. Saturating element.

15.11. Determine the describing function for the nonlinearity shown in Figure 15.52. This nonlinearity is commonly associated with spool valves.

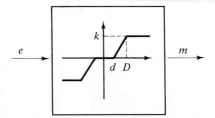

Figure 15.52. Dead zone and saturation in a spool valve.

15.12. Determine the describing function for the nonlinearity in Figure 15.53.

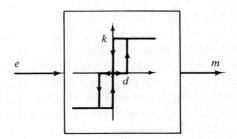

Figure 15.53. Dead zone with hysteresis.

Design of Scalar
Automatic Control
Systems by Frequency
Domain Methods

16

The design of scalar (single input, single output) feedback systems was discussed in Chapter 11. The same topic is discussed in the present chapter; however, the frequency domain formalism is used as the basis for the present discussion.[1, 3] As has been indicated in the last three chapters, the frequency techniques provide the analyst with the ability to deal with open loop systems described by transcendental transfer functions (systems with distributed parameter components and/or dead time). Other differences between the application of root locus and of frequency domain methods will be pointed out.

The discussion centers about the development of frequency domain feedback design criteria, which are, roughly speaking, analogous to the criteria introduced in Chapter 10 for the root locus formalism. These criteria must quantify the four considerations summarized in Section 10.6 and are:

1. Transient state criteria:
   a. Stability.
   b. Speed of response.
   c. Damping.
2. Equilibrium state criteria:
   d. Steady state error.
3. Sensitivity criteria:
   e. Sensitivity to noise (random disturbances).
   f. Sensitivity to parameter variation.

4. Effect of nonlinearities:
   g. Stability.

Frequency domain criteria for considerations a and g were presented in Chapter 15 and will not be discussed further here. As indicated in Chapter 13, bandwidth is a frequency domain quantifier of speed of response; other such quantifiers will be introduced below. Consideration 2 was also discussed in earlier chapters.

As will be shown, frequency domain criteria can be developed for *all* of the above considerations. This fact is another of the important advantages of the use of the frequency domain formalism. As will be seen in the first few sections, the quantification of the speed of response and of damping can be done in only a very approximate manner when frequency domain methods are used. The approximate characterization of transient response may, in some cases, prove to be a disadvantage.

We shall begin the discussion with some preparatory material on the relationship between closed loop frequency response and open loop frequency response.

16.1 Graphical Determination of the Closed Loop Frequency Response from the Open Loop Frequency Response (Nyquist Diagram)

For convenience, consider the unity feedback system illustrated in Figure 16.1. For this system, the closed loop transfer function

$$G_L(s) \triangleq \frac{C(s)}{R(s)} = \frac{G}{1 + G} \tag{16.1}$$

where the open loop transfer function

$$G = G_c G_p. \tag{16.2}$$

Define

$$M \triangleq |G_L(i\omega)|. \tag{16.3}$$

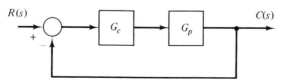

Figure 16.1. Unity feedback system.

It follows from Equations (16.1) and (16.3) that

$$M = \frac{|G(i\omega)|}{|1 + G(i\omega)|},$$ (16.4)

and so $M(\omega)$ may be given the graphical interpretation illustrated in Figure 16.2 since $a = 1 + G$ and $b = G$. Equation (16.4) is a relationship between the closed loop frequency response, $M(\omega) = |G_L(i\omega)|$, and the open loop

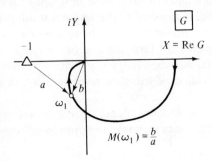

$$M(\omega_1) = \frac{b}{a}$$

Figure 16.2. Graphical interpretation of the closed loop frequency response, $M(\omega)$.

frequency response, $|G(i\omega)|$. Effort will now be made to develop a more complete graphical interpretation of this relationship. More explicitly, Equation (16.4) will be used to determine lines of constant M in the G plane.

It is left as an exercise for the reader to show that the closed loop phase

$$<G_L(i\omega) = <b - <a,$$ (16.5)

where a and b are the complex numbers illustrated in Figure 16.2.

Denote the open loop quantities

$$X(\omega) = \text{Re}\{G(i\omega)\},$$
$$Y(\omega) = \text{Im}\{G(i\omega)\}.$$ (16.6)

Then, the closed loop magnitude

$$M(\omega) = \frac{|X + iY|}{|1 + X + iY|} = \left[\frac{X^2 + Y^2}{(1 + X)^2 + Y^2} \right]^{1/2},$$

or, after squaring both sides of this equation,

$$M^2 = \frac{X^2 + Y^2}{(1 + X)^2 + Y^2}.$$ (16.7)

This equation may be rewritten

$$X^2(M^2 - 1) + 2XM^2 + M^2 = Y^2(1 - M^2)$$

or

$$X^2 + 2X\frac{M^2}{M^2 - 1} + Y^2 = \frac{M^2}{1 - M^2}. \tag{16.8}$$

Complete squares to find that

$$X^2 + 2X\frac{M^2}{M^2 - 1} + \frac{M^4}{(M^2 - 1)^2} + Y^2 = \frac{M^2}{1 - M^2} + \frac{M^4}{(M^2 - 1)^2}$$
$$= \frac{-M^2(M^2 - 1) + M^4}{(M^2 - 1)^2},$$

or, finally,

$$\left[X + \frac{M^2}{M^2 - 1}\right]^2 + Y^2 = \frac{M^2}{(M^2 - 1)^2}. \tag{16.9}$$

The equation of a circle is

$$(X - X_0)^2 + (Y - Y_0)^2 = R^2. \tag{16.10}$$

Thus, it follows from Equations (16.9) and (16.10) that the constant M contours are circles in the G plane with the center at

$$X_0 = -\frac{M^2}{M^2 - 1}, \tag{16.11}$$

$$Y_0 = 0. \tag{16.12}$$

The radius of these circles is given by

$$R = \left|\frac{M}{M^2 - 1}\right|. \tag{16.13}$$

The question not answered by Equations (16.11) and (16.13) is, Where is the $M = 1$ contour? From the interpretation illustrated in Figure 16.2, it follows that the $M = 1$ contour is a vertical line through the $-\frac{1}{2}$ point.
Notice that the limits

$$\lim_{M \to \infty} X_0 = -1 \quad \text{and} \quad \lim_{M \to \infty} R = 0 \tag{16.14}$$

and

$$\lim_{M \to 0} X_0 = 0 \quad \text{and} \quad \lim_{M \to 0} R = 0. \tag{16.15}$$

One could find the closed loop frequency response from the open loop response in the following way: (1) Construct the constant M contours in the

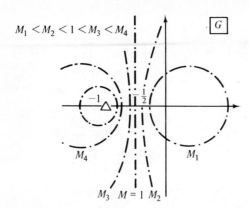

$M_1 < M_2 < 1 < M_3 < M_4$

Figure 16.3. Constant $M(\omega)$ circles for unity feedback systems.

G plane (see Figure 16.3); (2) construct the open loop Nyquist diagram; (3) a point on the closed loop magnitude curve (M, ω) corresponds to each intersection of an M contour with the Nyquist diagram. Of course to obtain the entire closed loop response (i.e., to also obtain the phase curve) one would also have to make use of Equation (16.5) and its graphical interpretation.

The idea developed above is exploited in the next section.

16.2 The Concept of a *Dominant Pole Pair*; Quantifiers of Speed of Response and Damping

The concepts developed in the previous section are applied to three particular cases in Figure 16.4. A remarkable fact is that the sketches of the closed loop magnitude curves have similar qualitative features and are quite similar to the magnitude curve of a quadratic factor. This similarity is exploited below, but first it is important to establish some of the quantitative features summarized in Figure 16.4. The value of M corresponding to the tangency illustrated in the G-plane plots of Figure 16.4 is denoted M_m.

A consequence of relations (16.14) and (16.15) is that M_m may be interpreted as the maximum value of the closed loop frequency response. This fact is illustrated in Figure 16.4. It follows that the closed loop resonant frequency, ω_R, is that frequency where the open loop Nyquist diagram is tangent to the M_m circle. This fact is also illustrated in Figure 16.4. Finally, notice that closed loop static gain (i.e., the magnitude at $\omega = 0.0$ radians/sec) is near 0 db for all cases considered. The reader should verify this result by applying the interpretation of Figure 16.2 to Figure 16.4.

A widely applied approximation for the closed loop transfer function is

$$G_L(s) \simeq \frac{K\omega_n^2}{s^2 + 2\xi\omega_n s + \omega_n^2}, \qquad (16.16)$$

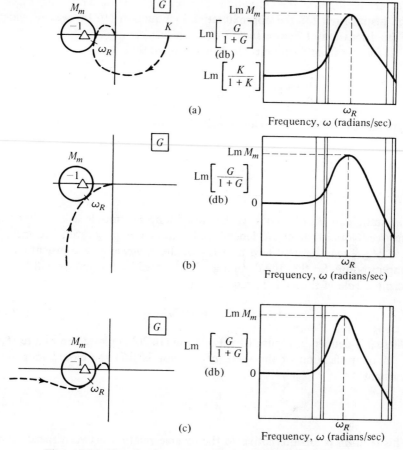

Figure 16.4. The sketches on the left are the Nyquist diagrams for radically different open loop transfer functions. (a) Three poles, (b) two poles, one at the origin, (c) four poles, two at the origin, and one zero. M_m corresponds to that M contour which is just tangent to the Nyquist diagrams. The sketches to the right are the corresponding closed loop Bode magnitude curves obtained by using the interpretation illustrated in Figure 16.2.

where the damped natural frequency

$$\beta = \omega_n \sqrt{1 - \xi^2}. \tag{16.17}$$

Recall that the resonant frequency

$$\omega_R = \omega_n \sqrt{1 - 2\xi^2}. \tag{16.18}$$

ω_n denotes the undamped natural frequency and ξ denotes the damping ratio. The terminology and the derivation of these equations were provided earlier. The justification for the approximation in Equation (16.16) was provided in the discussion of Figure 16.4.

It follows from Equations (16.3) and (16.16) that

$$\frac{M}{K} \simeq \frac{1}{\{[1 - (\omega^2/\omega_n^2)]^2 + 4\xi^2(\omega^2/\omega_n^2)\}^{1/2}}. \tag{16.19}$$

Combine Equations (16.18) and (16.19) to show that

$$\boxed{\frac{M_m}{K} \simeq \frac{1}{2\xi\sqrt{1 - \xi^2}}.} \tag{16.20}$$

The static gain K, of course, is the closed loop magnitude at $\omega = 0.0$ radians/sec. Notice that M_m/K depends only on the damping ratio, ξ. Conclude that M_m/K may be used as a measure of the *damping* of a transient closed loop response. So that the damping of the closed loop response will be sufficient, a rule of thumb is to require that

$$0° < \theta < 70°, \tag{16.21}$$

where θ is defined in Figure 16.5. Equation (16.20) is plotted in Figure 16.6.

A quantification of the *speed of response* is that the closed loop time constant

$$\frac{1}{T} = \xi\omega_n. \tag{16.22}$$

This relation is an alternative to the inverse relation between speed of response and bandwidth derived in Chapter 13.

M_m, K, and ω_R are easily read off the Nyquist diagram. M_m is the value of M whose circle is just tangent to the Nyquist locus in the G plane, ω_R is the

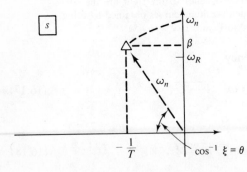

Figure 16.5. Dominant closed loop poles.

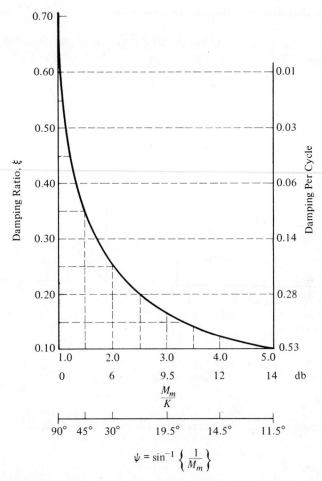

Figure 16.6. Plot of Equation (16.20), which relates approximate values of closed loop damping to M_m/K.

value of frequency which corresponds to this point of tangency, and K is the gain at zero frequency.

The poles of the approximate transfer function (16.16) are called the *dominant* pole pair. On occasion it is appropriate to add some numerator dynamics to approximation (16.16). The reader will be asked, in the problem assignments, to demonstrate that a zero should be added for the system described by Figs 16.4(a) and 16.4(c).

Example 16.1

An open loop system with a distributed parameter component is placed in a feedback system as is illustrated in Figure 16.7.

Notice that if $K_c = 1.0$,

$$G(i\omega) = 10e^{-\pi\sqrt{2i\omega}}. \tag{16.23}$$

Notice that

$$\sqrt{i} = [e^{i(\pi/2)}]^{1/2} = e^{i(\pi/4)},$$

also, in rectangular form,

$$e^{i(\pi/4)} = \frac{1}{\sqrt{2}} + \frac{i}{\sqrt{2}},$$

(refer to Figure 16.8).

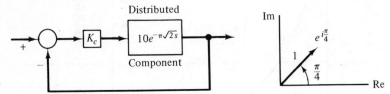

Figure 16.7. Case where the plant is described by a transcendental transfer function.

Figure 16.8. Vector interpretation of $\sqrt{i}$.

Thus,

$$G(i\omega) = 10e^{-\pi\sqrt{\omega}\,(1+i)}$$

$$= 10e^{-\pi\sqrt{\omega}}e^{-i\pi\sqrt{\omega}}. \tag{16.24}$$

The last equation may be interpreted as the polar form of $G(i\omega)$ so that

$$|G| = 10e^{-\pi\sqrt{\omega}} \tag{16.25a}$$

and

$$<G = -\pi\sqrt{\omega}. \tag{16.25b}$$

These equations were used to form Table 16.1. The results summarized

Table 16.1

CALCULATION OF OPEN LOOP FREQUENCY RESPONSE

$<G$	ω	$|G|$
$-\pi/2$	$\frac{1}{4}$	$10e^{-\pi/2} = 2.09$
$-2\pi/3$	$\frac{4}{9}$	$10e^{-2\pi/3} = 1.23$
$-5\pi/6$	$\frac{25}{36}$	$10e^{-5\pi/6} = 0.73$
$-\pi$	1	$10e^{-\pi} = 0.43$
$-3\pi/2$	$\frac{9}{4}$	$10e^{-3\pi/2} = 0.091$

Table 16.2

CLOSED LOOP FREQUENCY RESPONSE

ω	$\dfrac{G}{1+G}$
0	$10/11 = 0.91$
0.25	$2.09/2.3 = 0.91$
0.45	$1.23/1.14 = 1.08$
0.70	$0.73/0.54 = 1.38$
1.0	$0.43/0.57 = 0.83$
2.25	$0.091/1.02 = 0.09$

in Table 16.1 are plotted in Figure 16.9. The results summarized in Table 16.2 were obtained graphically from Figure 16.9 using the interpretation of Figure 16.2. The closed loop response is plotted in Figure 16.10.

As in the cases illustrated in Figure 16.4, the closed loop response reminds one of the response of a quadratic factor for which

$$\omega_n = 0.8 \text{ radians/sec,}$$

and it can be shown, using Figure 16.10, that

$$\frac{M_m}{K} = \frac{1.4}{0.91} = 1.5.$$

From Figure 16.6,

$$\xi = 0.35.$$

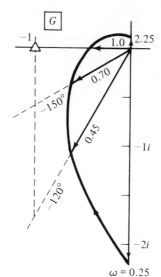

Figure 16.9. Open loop response for Example 16.1. The numbers near the vectors are the frequencies.

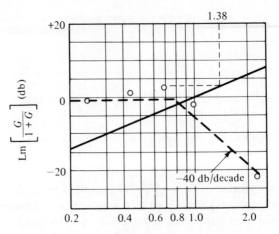

Figure 16.10. Closed loop Bode diagram for the system illustrated
in Figure 16.7. The 0s denote points from Table 16.2. The solid line
is drawn in order to convert decibels to ordinary numbers.

The closed loop time constant is found from Equation (16.22) to be

$$T = \frac{1}{\zeta \omega_n} = 1.9 \text{ sec.}$$

Clearly, the use of transfer functions which are transcendental functions
of s offers no special problems.

*   *   *

A commonly used approximation is

$$\omega_n \simeq \omega_R. \tag{16.26}$$

If $\zeta > 0.707$, then ω_R will not exist, and a sketch similar to that illustrated
in Figure 16.10 will have to be constructed in order to determine ω_n for the
closed loop dominant poles.

16.3 M-Circle Formulas: A Technique
for Obtaining a Desired Value of M_m
with Proportional Feedback Control

In this section, a technique is presented for obtaining a desired value of
damping, M_m, when the compensator, illustrated in Figure 16.1, is a proportional gain, K_c.

It is clear from Figure 16.3 that M_m (and hence ω_R) can be changed by

changing the open loop gain. The question is, How can the appropriate value
of gain be found that will provide a desired value of M_m? The G-domain
relationship that will now be derived greatly aids in the task of finding this
gain. Consider the M circle in Figure 16.11. The symbol b denotes the center
of the M circle. The line $\overline{ac}$ is constructed perpendicular to the line $\overline{0b}$.

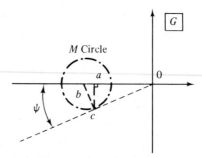

Figure 16.11. M circle and a radial
tangent line in the G plane.

First, it will be proved that a is the -1 point: Begin by noting that

$$\overline{0c}^2 = \overline{0b}^2 - \overline{bc}^2 \tag{16.27a}$$

and

$$\overline{0a}^2 = \overline{0c}^2 - \overline{ac}^2 \tag{16.27b}$$

and

$$\overline{ac} = \overline{0c}\sin(\psi). \tag{16.28}$$

Thus,

$$\overline{0a}^2 = \overline{0c}^2[1 - \sin^2(\psi)]. \tag{16.29}$$

Notice also that

$$\sin(\psi) = \frac{\overline{bc}}{\overline{0b}}. \tag{16.30}$$

Use Equations (16.11), (16.13), and (16.30) to show that

$$\sin(\psi) = \frac{M/(M^2 - 1)}{M^2/(M^2 - 1)}$$

or

$$\boxed{\sin(\psi) = \frac{1}{M}.} \tag{16.31}$$

Substitute Equation (16.31) into Equation (16.29) to show that

$$\overline{0a^2} = [\overline{0b^2} - \overline{bc^2}]\left[1 - \frac{1}{M^2}\right]$$

$$= \left[\frac{M^4}{(M^2-1)^2} - \frac{M^2}{(M^2-1)^2}\right]\left[\frac{M^2-1}{M^2}\right] = 1.0.$$

The last equation demonstrates that

$$a = -1. \tag{16.32}$$

Suppose that a designer wishes to obtain a desired value of M_m for a feedback system. Let ψ_d be the G-plane tangent angle associated with the desired value of M_m. If, as in Figure 16.12, a circle (*not* the M circle) is

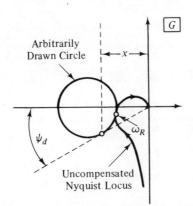

Figure 16.12. Graphical determination of compensating gain. The circle is tangent to both the radial line and the Nyquist diagram.

drawn tangent to both the ψ_d line and the open loop Nyquist locus, the correct value of compensating gain is that gain for which *this circle becomes the M circle*. Since changing the gain causes the Nyquist locus to expand or contract radially, it follows from Equation (16.32) that

$$\boxed{K_c = \frac{1}{x},} \tag{16.33}$$

where x is as defined in Figure 16.12.

Notice that Equation (16.31) is represented in graphical form in Figure 16.6. Notice that one can relate ψ_d directly to desired values of damping ratio or damping per cycle with this figure. Typically,

$$\psi_d = 60° \quad (M_m = 1.15) \tag{16.34}$$

or

$$\psi_d = 45° \quad (M_m = \sqrt{2}) \tag{16.35}$$

or

$$\psi_d = 30° \quad (M_m = 2). \tag{16.36}$$

Example 16.2

Find K_c such that $M_m = \sqrt{2}$ ($\psi_d = 45°$) when

$$G_p(s) = \frac{5}{s(s+1)}.$$

The Nyquist diagram for the uncompensated transfer function is constructed in Figure 16.13 as is the radial line corresponding to $\psi_d = 45°$.

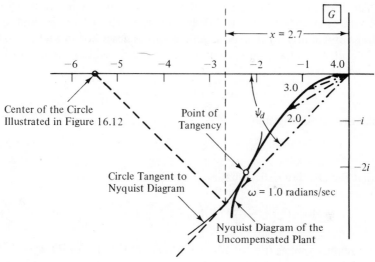

Figure 16.13. Constructions of Figure 16.2. for the particular case of Example 16.2.

A trial and error procedure must be used to find the center of the circle of tangency (refer to Figure 16.12). The procedure begins by projecting lines *perpendicular to the ψ_d radial line* back to the real axis. The point where the projected line and the real axis intersect is a possible center of the tangency circle. The correct center was found by constructing circular arcs until tangency was obtained.

As is illustrated,

$$x = 2.7$$

for this example so that the proper value of K_c is, from Equation (16.33),

$$K_c = \frac{1}{2.7} = 0.37.$$

* * *

Notice that if the scales on the axes of Figure 16.13 are changed by a factor of 2.7 to 1.0, the figure will be the Nyquist diagram *of the compensated system*. In general, the correct scale change is x to 1.0.

16.4 An Alternative Procedure: Modifying the Mapping Concept and the Nyquist Criterion to the Task of Feedback System Synthesis

The discussion in this section will draw heavily on the material in Section 10.2 (see especially Figure 10.11) and on the material in the first several sections of Chapter 15. The basic philosophy is described in Section 10.9.

Consider the characteristic equation for the closed loop system in Figure 16.2, which is

$$1 + G_c(s_p)G_p(s_p)H(s_p) = 0. \tag{16.37}$$

Make the transformation

$$s_p = s_0 - \frac{1}{T_0}, \tag{16.38}$$

so that the characteristic equation

$$1 + G_c\left(s_0 - \frac{1}{T_0}\right)G_p\left(s_0 - \frac{1}{T_0}\right)H\left(s_0 - \frac{1}{T_0}\right) = 0. \tag{16.39}$$

If the analyst applies the Nyquist criterion (Chapter 12) to Equation (16.39), he will then determine the conditions for which all of the closed loop poles lie to the left of $-1/T_0$, where T_0 can be taken to be equal to some specified closed loop time constant. The graphical interpretation of this procedure is illustrated in Figure 16.14. Notice that a vertical line at $-1/T_0$ is mapped to the G plane rather than the imaginary axis.

Notice that for the example illustrated in Figure 16.14 the number of open loop poles in the right half of the s_0 plane is

$$N_0 = 2. \tag{16.40}$$

Thus, two counterclockwise encirclements of the -1 point in the G_0 plane must occur if the closed loop system is to have closed loop time constants less than T_0. Conclude that the gain must be *increased* until the -1 point lies in

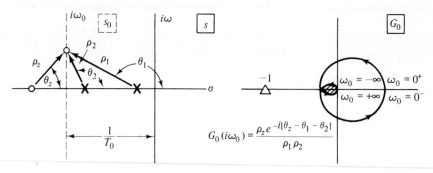

Figure 16.14. Graphical interpretation of the application of the Nyquist criterion to Equation (16.39).

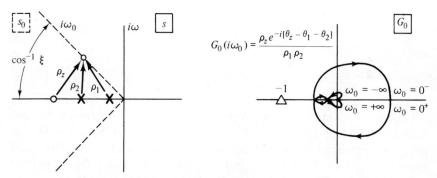

Figure 16.15. Modification of the procedure illustrated in Figure 16.14. in order to deal with specifications on closed loop damping.

the shaded region indicated in Figure 16.14. It is left as an exercise for the reader to compare this conclusion with one obtained using root locus analysis.

The above procedure can be used to design a system to meet specifications on damping. The appropriate modification is illustrated in Figure 16.15. Here the axis to be mapped (i.e., the $i\omega_0$ axis) is taken at an angle $\cos^{-1}\xi$, where ξ is the specified closed loop damping ratio. It is left as an exercise for the reader to verify the form of the Nyquist locus illustrated in Figure 16.15 as well as to compare conclusions obtained from a Nyquist analysis with those obtained using a root locus analysis.

Finally, it is noted that speed of response and damping can be dealt with simultaneously by mapping the combined vertical-radial lines illustrated in Figure 10.11.

Example 16.3

Assume that

$$G_p(s) = \frac{0.5}{(s+1)^2} \tag{16.41}$$

and use the extension of the Nyquist criterion suggested above to find that range of K_c such that for the closed loop poles

$$\xi < 0.707 \quad (\text{i.e., } \cos^{-1}\xi < 45°). \tag{16.42}$$

On the radial line for which $\cos^{-1}\xi = 45°$,

$$s_0 = -\omega + i\omega, \tag{16.43}$$

so that

$$G_p(s_0) = \frac{0.5}{[(1-\omega)+i\omega]^2}.$$

Show that

$$|G_p(s_0)| = \frac{0.5}{[(1-\omega)^2 + \omega^2]} \tag{16.44}$$

and that

$$<G_p(s_0) = 2\tan^{-1}\left[\frac{-\omega}{1-\omega}\right]. \tag{16.45}$$

The last two equations are plotted in the form of an extended Nyquist diagram in Figure 16.16. This diagram could also have been constructed using the graphical procedure indicated in Figure 16.15. The important quantity is $|G_p(s_0)|$ when $<G_p(s_0) = -180°$. Obviously, this important quantity is 0.5. It then follows from the Nyquist criterion that

$$K_c = \frac{1}{0.5}$$

$$= 2.0. \tag{16.46}$$

$$* \quad * \quad *$$

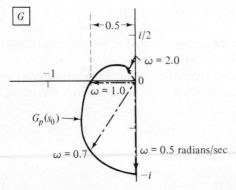

Figure 16.16. Mapping of the radial line at —135° in the s plane by the plant transfer function of Example 16.3.

16.5 Factors of Safety for Parameter Variation: *Gain Margin* and *Phase Margin*

Refer to Section 10.5 where it is emphasized that the chief reason for using feedback is to compensate for parameter variation. The so-called *sensitivity problem* is dealt with in a direct manner in the next two sections. In the present section, the classical concepts of *phase margin* and *gain margin* are introduced. These parameters are used in frequency domain design procedures to guard against the unfortunate effects of parameter variation. Parameter variation manifests itself in errors in frequency response diagrams. This error will be called *parasitic response* and is analogous to the *parasitic poles* of the time domain formalism (refer to Section 11.5).

When the system in Figure 16.1 is at the stability limit, a closed loop pole

$$s_p = i\omega_u \tag{16.47a}$$

for some ω_u so that at the stability limit

$$G(i\omega_u) = -1. \tag{16.47b}$$

This is an equation between complex quantities which can be rewritten by equating moduli and angular parts; i.e.,

$$|G(i\omega_u)| = 1 \tag{16.48}$$

or

$$Lm[G(i\omega_u)] = 0 \text{ db} \tag{16.49}$$

and

$$<G(i\omega_u) = -180°. \tag{16.50}$$

The last two equations motivate the definition of the important design factors phase margin and gain margin, which can be described on the Bode diagram as shown in Figure 16.17 or on the Nyquist diagram as shown in Figure 16.18.

In words, when $G(i\omega)$ satisfies Equation (16.48) for some ω, the phase margin, γ, represents the additional phase lag that (in the form of parasitic response) would be required in order that $G(i\omega)$ satisfy Equation (16.50) at that same frequency. Thus, phase margin is a factor of safety. Similarly, when $G(i\omega)$ satisfies Equation (16.50) for some ω, the gain margin, a, is the multiplicative constant that (in the form of parasitic response) would also allow Equation (16.48) to be satisfied.

So that parasitic response will not cause a previously stable system to become unstable, constraints are placed on the minimum values of these

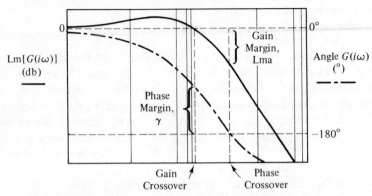

Figure 16.17. Phase and gain margins illustrated on the Bode diagram.

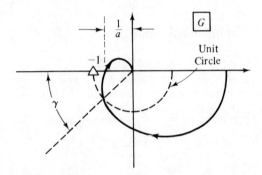

Figure 16.18. Phase and gain margins illustrated on the Nyquist diagram.

margins. These constraints are factors of safety and are based on a good deal of practical experience.

As a rule of thumb, it is often sufficient to require

$$\gamma \geq 30°, \tag{16.51}$$

while, *simultaneously,*

$$\text{Lm}(a) \geq 8 \text{ db.} \tag{16.52}$$

The last constraint is equivalent to

$$a \geq 2.5. \tag{16.53}$$

16.6 The *Sensitivity Measures* of Bode[2]

Bode developed quantitative measures of the ability of feedback systems to compensate for parameter variation. His work was completed shortly after World War II and was the starting point for research which has continued to the present day. Curiously, this topic has not, however, been in the mainstream of system and control theory.

Recall that the closed loop transfer function for the system illustrated in Figure 16.19 is

$$G_L = \frac{C}{R}$$

$$= \frac{G_c G_p}{1 + G_c G_p H}. \qquad (16.54)$$

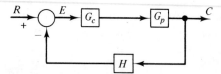

Figure 16.19. Standard scalar variable feedback control system.

If the plant is the source of parasitic response, the Bode sensitivity function

$$S_p^L = \frac{G_p}{G_L} \frac{\partial G_L}{\partial G_p}. \qquad (16.55)$$

This normalized partial derivative is read "the sensitivity of G_L with respect to G_p."

For the system illustrated in Figure 16.19, it follows from Equations (16.54) and (16.55) that

$$S_p^L = \frac{(1 + G_c G_p H)}{G_c} \left[\frac{G_c}{1 + G_c G_p H} - \frac{G_c^2 H G_p}{(1 + G_c G_p H)^2} \right]$$

or

$$S_p^L = \frac{1}{1 + G_c G_p H}. \qquad (16.56)$$

It is interesting to compare this result with that obtained for the system illustrated in Figure 16.20 for which

$$G_L = \frac{G_p}{1 + G_c G_p H}. \qquad (16.57)$$

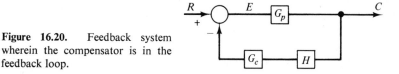

Figure 16.20. Feedback system wherein the compensator is in the feedback loop.

It follows that

$$S_p^L = (1 + G_cG_pH)\left[\frac{1}{1 + G_cG_pH} - \frac{G_cG_pH}{(1 + G_cG_pH)^2}\right]$$

or

$$S_p^L = \frac{1}{1 + G_cG_pH}.$$ (16.58)

Clearly, *if the plant is the source of parasitic response, the placement of the compensator in the feedback path provides the same closed loop sensitivity as does the placement of the compensator in the forward path.*

If, however, the compensator itself is a source of parasitic response, another important sensitivity measure is

$$S_c^L = \frac{G_c}{G_L}\frac{\partial G_L}{\partial G_c}.$$ (16.59)

For the system illustrated in Figure 16.19, it follows from Equations (16.54) and (16.59) that

$$S_c^L = \frac{1 + G_cG_pH}{G_p}\left[\frac{G_p}{1 + G_cG_pH} - \frac{G_cHG_p^2}{1 + G_cG_pH}\right]$$

or

$$S_c^L = \frac{1}{1 + G_cG_pH}.$$ (16.60)

For the system illustrated in Figure 16.20, it follows from Equations (16.57) and (16.59) that

$$S_c^L = \frac{G_c(1 + G_cG_pH)}{1}\left[\frac{-G_p^2H}{(1 + G_cG_pH)^2}\right]$$

or

$$S_c^L = \frac{-G_cG_pH}{1 + G_cG_pH}.$$ (16.61)

Clearly, *if the compensator is the source of the parasitic response, the placement of the compensator is of some importance.* Usually, systems are such that the magnitude of G_cG_pH is greater than one at low frequencies and much less than one at high frequencies. If the parasitic response provided by the compensator is at relatively low frequencies, then a comparison of Equations (16.60) and (16.61) leads to the conclusion that the synthesis illustrated in Figure 16.19 is preferable.

In the remainder of the discussion, it is assumed that the plant is the source of parasitic response and that $H = 1.0$ so that the important sensitivity is

$$S_p^L = \frac{1}{1 + G_c G_p}$$

$$= \frac{1}{1 + G}. \tag{16.62}$$

The plot of $S_p^L(i\omega)$ on the G plane is called the *sensitivity locus*. The sensitivity locus is easily obtained when the Nyquist diagram is given. The manipulations involved are illustrated in Figure 16.21. The angle, θ, and the modulus, r, of $1 + G(i\omega)$ are determined graphically. The modulus and angle of $S_p^L(i\omega)$ are $1/r$ and $-\theta$, respectively. Repeating this procedure for several values of ω leads to the construction of the sensitivity locus. In this text, the -1 point in the G plane will be used for the origin of $S_p^L(i\omega)$.

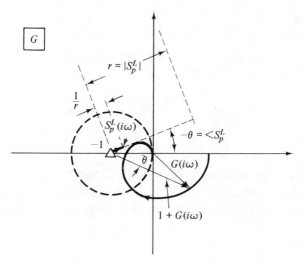

Figure 16.21. Construction of the sensitivity locus from the Nyquist diagram.

Obviously, so that the closed loop will be less sensitive to parameter change than the open loop, it is required that

$$|S_p^L(i\omega)| < 1.0 \tag{16.63a}$$

for ω in the frequency range of the parasitic response. Thus, hopefully, the sensitivity locus will lie inside the unit circle about -1 for that range of frequencies most affected by parameter change.

Example 16.4

The Nyquist diagram and sensitivity locus for a first-order open loop transfer function are sketched in Figure 16.22. For this example $S_p^L(i\omega)$ falls completely inside the unit circle so that the closed loop system will be less sensitive than the open loop to *any* parameter change.

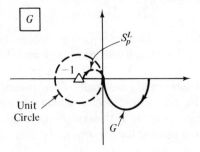

Figure 16.22. Sensitivity locus for a first-order system.

* * *

Example 16.5

The Nyquist diagram and sensitivity locus for a third-order system are sketched in Figure 16.23. Notice that

$$|S_p^L(i\omega_m)| > 1.0$$

if

$$\omega_m > \omega_1.$$

This shows that on a percentage basis the effect of a change in the plant transfer function in this high-frequency range is *magnified* in the closed loop transfer function. As indicated in previous sections, ω_m is a measure of the

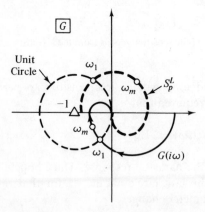

Figure 16.23. Sensitivity locus for a third-order system.

response time and damping in the closed loop system; thus, the magnification effect could be important.

*   *   *

It is interesting to notice that the sensitivity of Equation (16.62) is equal to the closed loop error transfer function

$$\frac{E}{R} = \frac{1}{1 + G}. \tag{16.63b}$$

16.7 The Closed Loop Response to Disturbance (*Noise*) Signals*

A closed loop system will often be subjected to an unwanted input. Such inputs are called *disturbance signals*. One of the purposes of the feedback system will often be to regulate against this disturbance.

Disturbances can enter at many different points in the feedback loop. In this section, attention is focused on the type of disturbance illustrated in Figure 16.24.

To derive the transfer function C/D, note that

$$(R - C)G_c G_p + D = C,$$

so that

$$C = \frac{G_c G_p}{1 + G_c G_p} R + \frac{1}{1 + G_c G_p} D.$$

Thus, the disturbance transfer function

$$G_N(s) = \frac{C(s)}{D(s)}$$

$$= \frac{1}{1 + G(s)}. \tag{16.64}$$

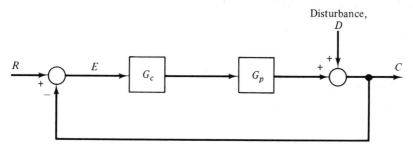

Figure 16.24. Closed loop system subjected to a disturbance signal.

* This section may be omitted during the first reading of the text.

Notice that this transfer function is equal to S_p^L and, therefore, has the same graphical interpretation as that indicated in Figure 16.21 for the sensitivity.

The important frequency range is determined by the spectral content of the disturbance signal. It would be favorable if, for these frequencies, the $1 + G(i\omega)$ vector were to fall outside [$G_N(i\omega)$ falls inside] the unit circle indicated in Figure 16.20, for if this is the case, the important spectral components of the disturbance signal will be attenuated by the closed loop system.

16.8 Selection of a Cascade Compensator

Frequency domain concepts which are used to decide on the form of $G_c(s)$ [see Figure 16.1] are introduced in this section.

First consider Figure 16.25 wherein the Nyquist diagrams of typical plant and compensator transfer functions are indicated. Obviously the magnitude of the open loop transfer function

$$|G(i\omega_*)| = |G_c(i\omega_*)G_p(i\omega_*)|$$
$$= r_c r_p \tag{16.65}$$

and the angle

$$<G(i\omega_*) = \theta_p + \theta_c. \tag{16.66}$$

These equations indicate the manner in which the open loop Nyquist diagram is modified by the inclusion of a *cascade* compensator, $G_c(s)$.

Suppose that the speed of response and the damping of the closed loop system meet specifications but that the steady state error is too large; thus, the designer needs to increase the gain without making too great a change in the Nyquist diagram in the vicinity of the -1 point (i.e., without appreciably altering M_m, ω_R). The compensator that can often be used to accomplish this task is the *lag compensator:*

$$G_c(s) = A\frac{(1 + Ts)}{1 + T\alpha s}, \quad \alpha > 1.0. \tag{16.67}$$

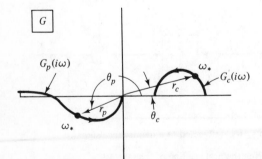

Figure 16.25. Nyquist diagrams for a typical plant and compensator. ω_* is a particular frequency.

In Figure 16.26, it is illustrated that if the gain, A, is taken to be equal to α (which implies that the open loop gain is increased), the Nyquist diagram near the -1 point is not greatly modified.

Denote

$$\phi_c \triangleq <G_c. \tag{16.68}$$

It can be shown that

$$\tan (\phi_c) = \frac{T\omega - \alpha T\omega}{1 + \alpha T^2\omega^2}. \tag{16.69}$$

Thus,

$$T^2 + \frac{\alpha - 1}{\alpha\omega \tan (\phi_c)} T + \frac{1}{\omega^2\alpha} = 0. \tag{16.70}$$

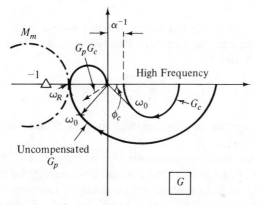

Figure 16.26. Lag compensation to decrease steady state error.

To find the maximum additional phase lag introduced by the compensator, set $d\phi_c/d\omega = 0$ and find from Equation (16.69) that ϕ_c is maximized when

$$\omega_m = \frac{1}{T\sqrt{\alpha}}. \tag{16.71}$$

Then use Equation (16.70) to show that the maximum phase lag is

$$\phi_m = \sin^{-1}\left[\frac{1 - \alpha}{1 + \alpha}\right]. \tag{16.72}$$

The last two quantities are illustrated in Figure 16.27. The value of ω_m is taken such that ϕ_m makes as little change as possible in the value of ω_R for the closed loop system.

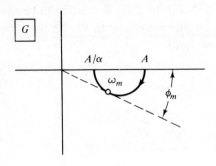

Figure 16.27. Nyquist diagram of the lag compensator.

Notice that the open loop gain

$$= AK_p,$$

where K_p is the plant gain. To decrease steady state error, A should be as large as possible.

Suppose now that it is desired to make a closed loop system respond faster. A compensator that can often be used to achieve this end is the *lead compensator*, defined by

$$G_c = \frac{A}{\alpha}\left(\frac{1 + Ts}{1 + \alpha Ts}\right), \quad \alpha < 1. \tag{16.73}$$

The Bode diagrams in Figures 16.28 and 16.29 illustrate why the lead compensator is used.

The lead compensator increases the gain margin so that the gain, A/α, can be taken greater than 1.0. In effect, the lead compensator increases the bandwidth of the closed loop system, hence reducing the response time of the closed loop system. The difference between ω_R and ω_{RC} is a measure of the increase of the bandwidth.

The same conclusion can be drawn from an examination of the Nyquist diagram in Figure 16.30. Notice that a phase *lead* is contributed by the lead compensator so that the resonant frequency of the closed loop system is

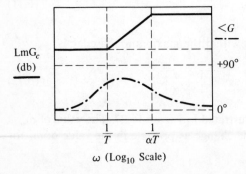

Figure 16.28. Bode diagram of a lead compensator.

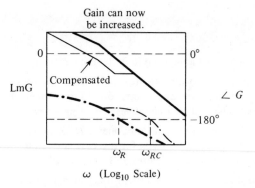

Figure 16.29. Cascade compensation with a lead compensator. ω_R is the crossover frequency for the uncompensated system and ω_{RC} is the crossover frequency for the compensated system.

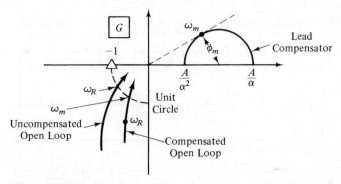

Figure 16.30. Lead compensation. ω_R is the resonant frequency of the uncompensated system and ω_m is the resonant frequency of the compensated system.

increased by compensation. ϕ_m is taken to be the phase difference between ω_R and ω_m on the uncompensated Nyquist diagram. The value of α is found from Equation (16.72) to be

$$\alpha = \frac{1 - \sin \phi_m}{1 + \sin \phi_m}. \qquad (16.74)$$

References

1. D'Azzo, J., and Houpis, C., *Feedback Control System Analysis and Synthesis*, McGraw-Hill, New York, 1966.

2. Horowitz, I., *Synthesis of Feedback Systems*, Academic Press, New York, 1963.

3. OGATA, K., *Modern Control Engineering*, Prentice-Hall, Englewood Cliffs, N.J., 1970.

Problems

16.1. Construct lines of constant closed loop damping ratio in the G plane.

16.2. Consider Figure 16.4(a) and 16.4(c). Show that for these systems, a zero might well be added to the approximation of Equation (16.16). [*Hint:* Sketch the closed loop phase angle diagrams.]

16.3. Suppose that the measurement component, $H(s)$, is the source of parasitic response. Define a Bode sensitivity for this case and relate it to G_p, G_c, and H for the system illustrated in Figures 16.19 and 16.20.

16.4. Consider the thickness control system illustrated in Figure 16.31. Use root locus techniques and choose k_1 so that the dominant roots of the inner loop are critically damped. Let the inner loop be the plant for the outer loop. For $k_c = 1.0$, determine the Nyquist locus of the open loop transfer function of the outer loop.

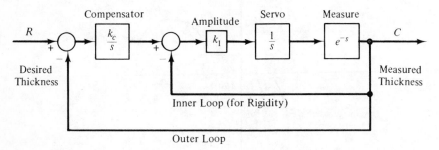

Figure 16.31. Thickness control system with two feedback loops.

16.5. For Problem 16.4, choose k_c so that the gain margin is 8 db for the outer loop. Estimate the response time for the resulting system. If the output measurement is "noisy" (see Figure 16.24), what disturbance frequencies would be the most bothersome?

16.6. For the servo system illustrated in Figure 16.32, it is desired that $M_m = \sqrt{2}$

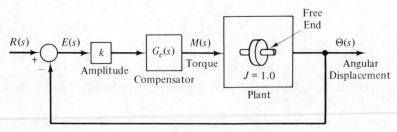

Figure 16.32. Servo system.

and $\omega_R = 10$ radians/sec. Select k and $G_c(s)$ to meet these specifications. What disturbance frequencies would be the most bothersome?

16.7. Determine phase margin, gain margin, closed loop damping ratio, and closed loop response time for the following open loop transfer functions:

(a) $\dfrac{0.2}{s(2s + 1)^2}$.

(b) $\dfrac{e^{-3\pi s/4}}{s + 1}$.

A Short Table of APPENDIX
Laplace Transforms A

Equation	$f(t)$	$F(s)$
(A.1)	$\delta(t)$, unit impulse	1
(A.2)	$u(t)$, unit step	$\dfrac{1}{s}$, $\sigma > 0$
(A.3)	$t \cdot u(t)$, ramp	$\dfrac{1}{s^2}$, $\sigma > 0$
(A.4)	$e^{-\alpha t} \cdot u(t)$	$\dfrac{1}{s + \alpha}$, $\sigma > -\alpha$
(A.5)	$\sin(\omega t) \cdot u(t)$	$\dfrac{\omega}{s^2 + \omega^2}$, $\sigma > 0$
(A.6)	$\cos(\omega t) u(t)$	$\dfrac{s}{s^2 + \omega^2}$, $\sigma > 0$
(A.7)	$\dfrac{1}{\beta} e^{-\alpha t} \sin(\beta t) \cdot u(t)$	$\dfrac{1}{(s + \alpha)^2 + \beta^2}$, $\sigma > -\alpha$
(A.8)	$\left(\dfrac{1}{\omega_n \sqrt{1 - \xi^2}}\right) e^{-\xi \omega_n t}\left[\sin(\omega_n \sqrt{1 - \xi^2}\, t) \cdot u(t)\right]$	$\dfrac{1}{s^2 + 2\xi \omega_n s + \omega_n^2}$
(A.9)	$\dfrac{\sqrt{(\alpha - a)^2 + \beta^2}}{\beta} e^{-\alpha t} \sin(\beta t + \phi) \cdot u(t)$, where $\phi = \tan^{-1} \dfrac{\beta}{a - \alpha}$	$\dfrac{s + a}{(s + \alpha)^2 + \beta^2}$, $\sigma > -\alpha$
(A.10)	$\sinh(\alpha t) \cdot u(t)$	$\dfrac{\alpha}{s^2 - \alpha^2}$, $\sigma > \alpha$
(A.11)	$\cosh(\alpha t) u(t)$	$\dfrac{s}{s^2 - \alpha^2}$, $\sigma > \alpha$
(A.12)	$\dfrac{1}{(s + \alpha)^{n+1}}$	$\dfrac{t^n e^{-\alpha t}}{n!}$, $\sigma > -\alpha$
(A.13)	$\dfrac{1}{\sqrt{\pi t}} e^{-(k^2/4t)} \cdot u(t)$	$\dfrac{1}{\sqrt{s}} e^{-k\sqrt{s}}$, $k \geq 0$

Reference: R. V. Churchill, *Operational Mathematics*, McGraw-Hill, New York, 1958.

Review of Elementary Analysis

This appendix contains a summary of Taylor series theory, the algebra of matrices and vectors, and the algebra of complex numbers.

B.1 Taylor Series,[3] Linearization, and Newton Iteration

The most useful result from all of mathematical analysis is probably Taylor's theorem. For some function, $f(x)$, the statement of the theorem is

$$f(x) = f(a) + f'(a)(x - a) + \frac{f''}{2!}(a)(x - a)^2 + \cdots + \frac{f^{(k)}}{k!}(a)(x - a)^k$$

$$+ \frac{f^{(k+1)}(\xi)}{(k + 1)!}(x - a)^{k+1}, \tag{B.1}$$

where prime and parenthetical superscripts represent differentiation with respect to x, a is some arbitrary but fixed value, and ξ is a value intermediate between x and a. There are restrictions on the range of values of x for which Equation (B.1) applies. These restrictions, which are related to the existence of the derivatives of $f(x)$, will not be of present concern. Examples of Taylor series are

$$e^x = 1 + x + \frac{x^2}{2!} + \frac{x^3}{3!} + \cdots, \tag{B.2}$$

$$\cos (x) = 1 - \frac{x^2}{2!} + \frac{x^4}{4!} - \frac{x^6}{6!} + \cdots, \tag{B.3}$$

$$\sin(x) = x - \frac{x^3}{3!} + \frac{x^5}{5!} - \frac{x^7}{7!} + \cdots. \tag{B.4}$$

For these examples, $a = 0$.

If x is sufficiently close to a, then one may neglect terms of $(x - a)^2$ and higher powers so that

$$f(x) \simeq f(a) + f'(a)(x - a). \tag{B.5}$$

Clearly, the error is

$$\frac{f''(\xi)(\xi - a)^2}{2}.$$

In other words, the linear function of x, on the right-hand side of Equation (B.5), may be used in place of the nonlinear function, $f(x)$, when x is sufficiently close to a. This result and its generalizations are used throughout the text to "linearize" the equations of nonlinear system dynamics.

Still another extremely important application of Taylor's series is to the nonlinear analysis problem: Find x_* such that

$$f(x_*) = 0. \tag{B.6}$$

This problem occurs quite often in the type of analysis discussed in the text. An important numerical scheme for solving this problem is developed as follows: Choose $a = x_*$ and assume that x is close to x_* so that Equation (B.5) may be rewritten

$$f(x) \simeq f(x_*) + f'(x_*)(x - x_*) = f'(x_*)(x - x_*). \tag{B.7}$$

The last equality follows from Equation (B.6). Thus,

$$x_* \simeq x - \frac{f(x)}{f'(x_*)}. \tag{B.8}$$

Since x_* is close to x, $f'(x_*) \simeq f'(x)$ so that

$$x_* \simeq x - \frac{f(x)}{f'(x)}. \tag{B.9}$$

This formula for x_* in terms of x suggests the following iterative solution to problem (B.6):

1. Arbitrarily select an initial guess, $x_{(0)}$.
2. Take the kth iteration, $x_{(k)}$, to be

$$x_{(k)} = x_{(k-1)} - \frac{f[x_{(k-1)}]}{f'[x_{(k-1)}]}, \quad k = 1, 2, \ldots. \tag{B.10}$$

This scheme, known as Newton iteration, should be appended with two remarks.

1. The scheme may not converge.
2. If the scheme does converge, it converges very quickly.

It is useful to note an important generalization of formula (B.1). Consider the function of several variables $f(x_1, x_2, \ldots, x_n)$ and arbitrary values $a_1, a_2, \ldots, a_n$ of the independent variables. The generalization of Equation (B.1) is

$$f(x_1, x_2, \ldots, x_n) = f(a_1, a_2, \ldots, a_n) + \sum_k^n \frac{\partial f}{\partial x_k}(a_1, \ldots, a_n) \cdot (x_k - a_k)$$

$$+ \frac{1}{2!} \sum_l^n \sum_k^n \frac{\partial^2 f}{\partial x_l \partial x_k}(a_1, a_2, \ldots, a_n) \cdot (x_k - a_k)(x_l - a_l)$$

$$+ \text{ higher-order terms.} \tag{B.11}$$

The linearizing approximation

$$f(x_1, x_2, \ldots, x_n) \simeq f(a_1, a_2, \ldots, a_n) + \sum_k^n \frac{\partial f}{\partial x_k}(a_1, \ldots, a_n) \cdot (x_k - a_k). \tag{B.12}$$

is accurate if all of the x_k are sufficiently close to the corresponding values of a_k.

B.2 The Algebra of Matrices and Vectors[2]

A *matrix*, quite simply, is an array of numbers. A matrix is denoted by a boldface script symbol, for instance,

$$\boldsymbol{\mathcal{C}} = \begin{bmatrix} a_{11} & a_{12} & \cdots & a_{1n} \\ a_{21} & a_{22} & \cdots & a_{2n} \\ \cdot & \cdot & & \cdot \\ \cdot & \cdot & & \cdot \\ \cdot & \cdot & & \cdot \\ a_{m1} & a_{m2} & \cdots & a_{mn} \end{bmatrix}. \tag{B.13}$$

This particular matrix has m rows and n columns. An *element* or entry is given two subscripts: The first subscript indicates the row of the element, while the second subscript denotes the element's column. A matrix with a single column is called a *vector*, and the column subscript is not ordinarily used when denoting a vector's elements. In this book boldface symbols are used for vectors, for instance,

$$\mathbf{x} = \begin{bmatrix} x_1 \\ x_2 \\ \cdot \\ \cdot \\ \cdot \\ x_n \end{bmatrix}. \tag{B.14}$$

A vector *is* a matrix, and the use of a special notation is not meant to give the impression that vectors are distinct mathematical entities. Rather, it is useful to keep track of matrices and vectors when writing matrix equations, much as it is useful to keep track of units or dimensions.

The *transpose* of a matrix is denoted with a prime and is a matrix formed by replacing the rows of the original matrix with its columns. For instance, if

$$\mathcal{L} = \begin{bmatrix} 2 & 1 \\ 3 & 0 \\ 1 & 4 \end{bmatrix}, \tag{B.15}$$

then

$$\mathcal{L}' = \begin{bmatrix} 2 & 3 & 1 \\ 1 & 0 & 4 \end{bmatrix}. \tag{B.16}$$

Also, if

$$\mathbf{x} = \begin{bmatrix} x_1 \\ x_2 \\ x_3 \end{bmatrix}, \tag{B.17}$$

then,

$$\mathbf{x}' = [x_1 \quad x_2 \quad x_3]. \tag{B.18}$$

Occasionally, it is useful to consider a matrix partitioned into sub-matrices or vectors. This partitioning will be emphasized with notation similar to that illustrated below:

$$\mathcal{A} = [\mathbf{a}_1 \; \vdots \; \mathbf{a}_2 \; \vdots \; \cdots \; \vdots \; \mathbf{a}_n]. \tag{B.19}$$

Here $\mathcal{A}$ is partitioned into columns (or vectors) denoted $\mathbf{a}_k$. Another example of this notation is

$$\mathcal{B} = \begin{bmatrix} \mathbf{b}'_1 \\ \mathbf{b}'_2 \\ \cdot \\ \cdot \\ \cdot \\ \mathbf{b}'_m \end{bmatrix}, \tag{B.20}$$

where $\mathcal{B}$ is partitioned into rows (or transposed vectors).

The term *dimension* of a vector refers to the number of elements of the vector. For instance, the vector of Equation (B.14) is n-dimensional, while that of Equation (B.18) is three-dimensional.

The algebra of matrices is summarized in the list provided below.

Addition: The sum of two matrices is defined only if the two matrices have the same number of columns *and* rows. Further, if

$$\mathbf{S} = \mathbf{\alpha} + \mathbf{\mathcal{B}}, \tag{B.21}$$

then

$$s_{ik} \stackrel{\triangle}{=} a_{ik} + b_{ik} \tag{B.22}$$

for all i and k; i.e., one merely adds the elements of $\mathbf{\alpha}$ to those of $\mathbf{\mathcal{B}}$ to form the elements of $\mathbf{S}$.

Scalar multiplication: If α is a scalar, then

$$\alpha\mathbf{\alpha} = \begin{bmatrix} \alpha a_{11} & \alpha a_{12} & \cdots & \alpha a_{1n} \\ \alpha a_{21} & \alpha a_{22} & \cdots & \alpha a_{2n} \\ \cdot & \cdot & & \cdot \\ \cdot & \cdot & & \cdot \\ \cdot & \cdot & & \cdot \\ \alpha a_{m1} & \alpha a_{m2} & \cdots & \alpha a_{mn} \end{bmatrix}. \tag{B.23}$$

Vector multiplication: A row vector, $\mathbf{x}'$, may be multiplied by a column vector, $\mathbf{y}$, only if $\mathbf{x}$ and $\mathbf{y}$ are of the same dimension. The product is a scalar and is defined by

$$\mathbf{x}'\mathbf{y} = \sum_{k=1}^{n} x_k y_k; \tag{B.24}$$

n is the common dimension of $\mathbf{x}$ and $\mathbf{y}$. Notice that

$$\mathbf{x}'\mathbf{y} = \mathbf{y}'\mathbf{x}. \tag{B.25}$$

This product is often given the name *scalar* or *dot* or *inner* product.

Matrix multiplication: The multiplication of matrices is a generalization of the vector product. Consider the product $\mathbf{\alpha}\mathbf{\mathcal{B}}$ and assume that $\mathbf{\alpha}$ and $\mathbf{\mathcal{B}}$ are partitioned as shown:

$$\mathbf{\mathcal{P}} = \mathbf{\alpha}\mathbf{\mathcal{B}} = \begin{bmatrix} \mathbf{a}'_1 \\ \mathbf{a}'_2 \\ \cdot \\ \cdot \\ \cdot \\ \mathbf{a}'_m \end{bmatrix} [\mathbf{b}_1 \;\vdots\; \mathbf{b}_2 \;\cdots\; \vdots\; \mathbf{b}_n]. \tag{B.26}$$

The product is a matrix and is defined only if the dimension of the $\mathbf{a}'_k$ equals the dimension of the $\mathbf{b}_k$, i.e., only if the number of columns of $\mathbf{\alpha}$ equals the number of rows of $\mathbf{\mathcal{B}}$. Further, the $i - k$ element is defined by

$$(\mathcal{P}_{ik}) \triangleq \mathbf{a}'_i \mathbf{b}_k = \sum_{\alpha=1}^{r} a_{i\alpha} b_{\alpha k}, \tag{B.27}$$

i.e., the inner product of the ith row of $\mathbf{\alpha}$ with the kth column of $\mathbf{\mathcal{B}}$, where r denotes the number of columns of $\mathbf{\alpha}$ and the number of rows of $\mathbf{\mathcal{B}}$.

For example, if

$$\mathbf{\alpha} = \begin{bmatrix} 1 & 2 \\ 1 & -1 \end{bmatrix}, \quad \mathbf{\mathcal{B}} = \begin{bmatrix} 1 & -1 \\ 1 & 4 \end{bmatrix},$$

then

$$\mathbf{\alpha}\mathbf{\mathcal{B}} = \begin{bmatrix} 3 & 7 \\ 0 & -5 \end{bmatrix} \quad \text{and} \quad \mathbf{\mathcal{B}}\mathbf{\alpha} = \begin{bmatrix} 0 & 3 \\ 5 & -2 \end{bmatrix}.$$

The above example illustrates that, in general,

$$\mathbf{\alpha}\mathbf{\mathcal{B}} \neq \mathbf{\mathcal{B}}\mathbf{\alpha}. \tag{B.28}$$

In fact, $\mathbf{\mathcal{B}}\mathbf{\alpha}$ may not be defined even though $\mathbf{\alpha}\mathbf{\mathcal{B}}$ is defined.

The utility of matrix notation begins to become apparent when one considers the system of algebraic equations

$$\begin{aligned}
a_{11}x_1 + a_{12}x_2 + \cdots + a_{1n}x_n &= b_1, \\
a_{21}x_1 + a_{22}x_2 + \cdots + a_{2n}x_n &= b_2, \\
&\vdots \\
a_{m1}x_1 + a_{m2}x_2 + \cdots + a_{mn}x_n &= b_m.
\end{aligned} \tag{B.29}$$

The reader should verify the fact that this system of equations can be written in the compact matrix equation

$$\mathbf{\alpha}\mathbf{x} = \mathbf{b}. \tag{B.30}$$

As will be seen, the utility of matrix algebra results from more important uses than development of compact notation; however, it is useful to explore the notational advantage a bit further. Notice that a scalar-valued function of several variables can be denoted

$$f(x_1, x_2, \ldots, x_n) = f(\mathbf{x}). \tag{B.31}$$

Denote the *gradient vector* of $f(\mathbf{x})$ by $\partial f/\partial \mathbf{x}$ and define the kth component of this vector by

$$\left(\frac{\partial f}{\partial \mathbf{x}}\right)_k = \frac{\partial f}{\partial x_k}. \tag{B.32}$$

Denote the *Hessian matrix* of $f(\mathbf{x})$ by $\mathcal{H}$ and define the $i - k$ element by

$$(\mathcal{H})_{ik} = \frac{\partial^2 f}{\partial x_i \partial x_k}. \tag{B.33}$$

It is left as an exercise for the reader to show that the first three terms of Equation (B.11) can be written in the compact form

$$f(\mathbf{x}) \simeq f(\mathbf{a}) + [\mathbf{x} - \mathbf{a}]'\frac{\partial f}{\partial \mathbf{x}}(\mathbf{a}) + \frac{1}{2}[\mathbf{x} - \mathbf{a}]'\mathcal{H}(a)[\mathbf{x} - \mathbf{a}]. \tag{B.34}$$

To generalize these notions still further, consider a set of functions of several variables

$$\begin{aligned} f_1 &= f_1(x_1, x_2, \ldots, x_n), \\ f_2 &= f_2(x_1, x_2, \ldots, x_n), \\ &\qquad\vdots \\ f_m &= f_m(x_1, x_2, \ldots, x_n), \end{aligned} \tag{B.35}$$

which could have been written more succinctly as

$$\mathbf{f} = \mathbf{f}(\mathbf{x}). \tag{B.36}$$

The right-hand sides of all m equations could be expanded in the manner indicated in Equation (B.11). Under appropriate conditions, the resulting expansion can be truncated in the manner indicated in Equation (B.12). It is left as an exercise for the reader to show that the entire *set* of linearizations can be written compactly as

$$\mathbf{f}(\mathbf{x}) \simeq \mathbf{f}(\mathbf{a}) + \mathcal{J}(a)[\mathbf{x} - \mathbf{a}], \tag{B.37}$$

where the *Jacobian*, $\mathcal{J}$, is defined by the component equation

$$\mathcal{J}_{ik} = \frac{\partial f_i}{\partial x_k}. \tag{B.38}$$

Matrix manipulation and algebra are even more fascinating than the compactness of matrix-vector notation. Begin the review of matrix manipulation with a discussion of an important scalar quantity called the *determinant*, which is associated with *square* matrices (i.e., with matrices which have the same number of rows as columns). For the 2×2 matrices the determinant is denoted and defined as follows:

$$\det \mathbf{\alpha} = \det \begin{bmatrix} a_{11} & a_{12} \\ a_{21} & a_{22} \end{bmatrix} \triangleq a_{11}a_{22} - a_{12}a_{21}. \tag{B.39}$$

For the $n \times n$ case, the "definition" of the determinant is begun as follows: For the $n \times n$ matrix the $i - k$ *minor* is the determinant of the matrix formed by crossing out the ith row and kth column and is denoted M_{ik}.

Example B.1

For the matrix

$$\mathbf{\alpha} = \begin{bmatrix} 1 & 1 & -10 \\ 1 & 2 & 3 \\ 0 & 1 & 11 \end{bmatrix},$$

it follows that

$$M_{23} = \det \begin{bmatrix} 1 & 1 \\ 0 & 1 \end{bmatrix} = 1 \cdot 1 - 1 \cdot 0 = 1.$$

*   *   *

Define the $i - k$ *cofactor* of the matrix $\mathbf{\alpha}$ by

$$A_{ik} = (-1)^{i+k} M_{ik}. \tag{B.40}$$

Notice the notation used for the cofactor.

*   *   *

Example B.2

For the matrix of Example B.1

$$A_{23} = (-1)^{2+3} M_{23} = -1$$

and

$$A_{33} = (-1)^{3+3} \det \begin{bmatrix} 1 & 1 \\ 1 & 2 \end{bmatrix} = +1(1 \cdot 2 - 1 \cdot 1) = +1.$$

*   *   *

Finally, define

$$\det \mathbf{\alpha} \triangleq \sum_{\alpha=1}^{n} a_{k\alpha} A_{k\alpha}, \quad \textit{expansion by a row}. \tag{B.41}$$

Two remarkable facts, which are the basis of matrix algebra, are summarized in the following two theorems which are provided without proof.

Theorem B.1

The value of det $\mathcal{Q}$ is independent of the row number, k. (B.42)

Theorem B.2

The determinant is also given by

$$\det \mathcal{Q} = \sum_{\alpha=1}^{n} a_{\alpha k} A_{\alpha k}, \quad \textit{expansion by a column}, \qquad (B.43)$$

and is independent of the column number, k.

The validity of these important theorems is illustrated by example.

Example B.3

For the matrix of Example B.1, use Equation (B.41) and $k = 3$ to find that

$$\det \mathcal{Q} = -0(-1)^{3+1} \det \begin{bmatrix} 1 & -10 \\ 2 & 3 \end{bmatrix} + 1(-1)^{3+2} \det \begin{bmatrix} 1 & -10 \\ 1 & 3 \end{bmatrix}$$

$$+ 11(-1)^{3+3} \det \begin{bmatrix} 1 & 1 \\ 1 & 2 \end{bmatrix} = 0 - (3 + 10) + 11(2 - 1) = -2.$$

Alternatively, use $k = 1$ to find that

$$\det \mathcal{Q} = 1(-1)^{1+1} \det \begin{bmatrix} 2 & 3 \\ 1 & 11 \end{bmatrix} + 1(-1)^{1+2} \det \begin{bmatrix} 1 & 3 \\ 0 & 11 \end{bmatrix}$$

$$-10(-1)^{1+3} \det \begin{bmatrix} 1 & 2 \\ 0 & 1 \end{bmatrix} = (22 - 3) - (11) - 10(1) = -2.$$

Theorem B.1 seems to apply. Finally, use Equation (B.43) and $k = 1$ to find that

$$\det \mathcal{Q} = 1(-1)^{1+1} \det \begin{bmatrix} 2 & 3 \\ 1 & 11 \end{bmatrix} + 1(-1)^{2+1} \det \begin{bmatrix} 1 & -10 \\ 1 & 11 \end{bmatrix}$$

$$+ 0(-)^{3+1} \det \begin{bmatrix} 1 & -10 \\ 2 & 3 \end{bmatrix} = (22 - 3) - (11 + 10) + 0(3 + 20)$$

$$= -2.$$

* * *

The following theorem, provided without proof, is important to the succeeding discussion:

Theorem B.3

If $\mathcal{B}$ is formed from $\mathcal{C}$ by interchanging two rows of $\mathcal{C}$ or two columns of $\mathcal{C}$, then

$$\det \mathcal{B} = -\det \mathcal{C}. \tag{B.44}$$

Example B.4

Take

$$\mathcal{C} = \begin{bmatrix} 1 & 2 \\ -3 & 4 \end{bmatrix}, \qquad \mathcal{B} = \begin{bmatrix} -3 & 4 \\ 1 & 2 \end{bmatrix}, \qquad \mathcal{C} = \begin{bmatrix} 2 & 1 \\ 4 & -3 \end{bmatrix}.$$

Then

$$\det \mathcal{C} = +10, \qquad \det \mathcal{B} = -10, \qquad \det \mathcal{C} = -10.$$

* * *

The following theorems lead to the concept of an inverse matrix.

Theorem B.4

If two rows or two columns of $\mathcal{C}$ are equal, then

$$\det \mathcal{C} = 0. \tag{B.45}$$

Proof: Interchange the two equal rows or columns; then, from Equation (B.44) it follows that

$$\det \mathcal{C} = -\det \mathcal{C}. \quad \text{Q.E.D.}$$

* * *

Theorem B.5

$$\sum_{\alpha=1}^{n} a_{i\alpha} A_{k\alpha} = \begin{cases} \det \mathcal{C}, & \text{if } i = k, \\ 0, & \text{if } i \neq k. \end{cases} \tag{B.46}$$

A similar statement can be made about column expansions.

Proof: The equality for $i = k$ follows immediately from definition (B.41). To prove the equality for $i \neq k$, form the matrix $\mathcal{B}$ from the matrix $\mathcal{C}$ by replacing the kth row of $\mathcal{C}$ with the ith row of $\mathcal{C}$. The left-hand side of Equation (B.46) is $\det \mathcal{B}$ which is zero by Theorem B.4.   Q.E.D.

* * *

The *adjoint* of a matrix $\mathcal{A}$ is a matrix denoted adj $\mathcal{A}$ and is defined by the component equation

$$(\text{adj } \mathcal{A})_{ik} \triangleq A_{ki}, \tag{B.47}$$

where A_{ki} is a cofactor of the matrix $\mathcal{A}$. Note the reversal in subscripts.

Theorem B.6

$$\mathcal{A}(\text{adj } \mathcal{A}) = (\text{adj } \mathcal{A})\mathcal{A} = \begin{bmatrix} \det \mathcal{A} & 0 & \cdots & 0 \\ 0 & \det \mathcal{A} & \cdots & 0 \\ \cdot & \cdot & & \cdot \\ \cdot & \cdot & & \cdot \\ \cdot & \cdot & & \cdot \\ 0 & 0 & \cdots & \det \mathcal{A} \end{bmatrix}. \tag{B.48}$$

Proof: From Equation (B.27)

$$(\mathcal{A}[\text{adj } \mathcal{A}])_{ik} = \sum_{\alpha=1}^{n} a_{i\alpha} \text{ adj } A_{\alpha k} = \sum_{\alpha=1}^{n} a_{i\alpha} A_{k\alpha}.$$

From Equation (B.46)

$$(\mathcal{A}[\text{adj } \mathcal{A}])_{ik} = \begin{cases} \det \mathcal{A}, & i = k, \\ 0, & i \neq k. \end{cases} \quad \text{Q.E.D.}$$

$$* \quad * \quad *$$

Define the *unit matrix*, denoted $\mathcal{I}$, by

$$\mathcal{I} \triangleq \begin{bmatrix} 1 & 0 & \cdots & 0 \\ 0 & 1 & \cdots & 0 \\ \cdot & \cdot & & \cdot \\ \cdot & \cdot & & \cdot \\ \cdot & \cdot & & \cdot \\ 0 & 0 & \cdots & 1 \end{bmatrix}. \tag{B.49}$$

It is easy to show that

$$\mathcal{I}\mathcal{M} = \mathcal{M} \tag{B.50}$$

for any matrix $\mathcal{M}$.

Define the *inverse* of a matrix $\mathcal{A}$, which is denoted $\mathcal{A}^{-1}$, by

$$\mathcal{A}^{-1} = \frac{1}{\det \mathcal{A}} \text{adj } \mathcal{A}. \tag{B.51}$$

It follows from Equation (B.48) that

$$\mathcal{Q}^{-1}\mathcal{Q} = \mathcal{Q}\mathcal{Q}^{-1} = \mathcal{J} \qquad (B.52)$$

if the inverse exists. Clearly, one condition for the existence of $\mathcal{Q}^{-1}$ is that

$$\boxed{\det \mathcal{Q} \neq 0.} \qquad (B.53)$$

An application of Equation (B.52) is as follows: Suppose that the x_k in the system of equations (B.29) are unknowns. Suppose further that $m = n$ and that inequality (B.53) is true. Multiply Equation (B.30) by $\mathcal{Q}^{-1}$ to find

$$\mathcal{Q}^{-1}\mathcal{Q}\mathbf{x} = \mathcal{Q}^{-1}\mathbf{b}.$$

From Equations (B.52) and (B.50) it follows that

$$\boxed{\mathbf{x} = \mathcal{Q}^{-1}\mathbf{b}.} \qquad (B.54)$$

Example B.5

Solve the system of equations

$$
\begin{aligned}
x_1 + 2x_3 \qquad\;\; &= 5, \\
x_1 + 2x_2 + x_3 &= 4, \\
-x_1 + \; x_2 - x_3 &= 2,
\end{aligned}
$$

which may be rewritten in matrix form as

$$
\begin{bmatrix} 1 & 0 & 2 \\ 1 & 2 & 1 \\ -1 & 1 & -1 \end{bmatrix}
\begin{bmatrix} x_1 \\ x_2 \\ x_3 \end{bmatrix} =
\begin{bmatrix} 5 \\ 4 \\ 2 \end{bmatrix}.
$$

To calculate the inverse, find the cofactors

$$
\begin{aligned}
A_{11} &= -3, & A_{21} &= \;\;2, & A_{31} &= -4, \\
A_{12} &= \;\;0, & A_{22} &= \;\;1, & A_{32} &= \;\;1, \\
A_{13} &= \;\;3, & A_{23} &= -1, & A_{33} &= \;\;2.
\end{aligned}
$$

Expanding by the first row of $\mathcal{Q}$, it follows that

$$\det = 1(-3) + 0(0) + 2(3) = +3.$$

Thus, from Equations (B.51) and (B.54), it follows that

$$\begin{bmatrix} x_1 \\ x_2 \\ x_3 \end{bmatrix} = \begin{bmatrix} -1 & \frac{2}{3} & -\frac{4}{3} \\ 0 & \frac{1}{3} & \frac{1}{3} \\ 1 & -\frac{1}{3} & \frac{2}{3} \end{bmatrix} \begin{bmatrix} 5 \\ 4 \\ 2 \end{bmatrix} = \begin{bmatrix} -5 \\ 2 \\ 5 \end{bmatrix}.$$

* * *

An important application of Equation (B.54) to functional analysis will now be demonstrated. The generalization of problem (B.6) is, Find the vector $\mathbf{x}_*$ such that

$$\mathbf{f}(\mathbf{x}_*) = \mathbf{0}, \tag{B.55}$$

where $\mathbf{f}$ is the vector representation of the set of functions in Equation (B.35) and $\mathbf{0}$ is the vector whose elements are all zero. An iterative solution to this problem is derived by setting $\mathbf{a} = \mathbf{x}_*$ in Equation (B.37) so that

$$\mathbf{f}(\mathbf{x}) \simeq \mathbf{f}(\mathbf{x}_*) + \mathcal{J}[\mathbf{x} - \mathbf{x}_*] = \mathcal{J}[\mathbf{x} - \mathbf{x}_*], \tag{B.56}$$

so that, from Equation (B.54), it follows that

$$\mathbf{x}_* \simeq \mathbf{x} - \mathcal{J}^{-1}\mathbf{f}(\mathbf{x}).$$

Make the additional approximation

$$\mathcal{J}(\mathbf{x}_*) \simeq \mathcal{J}(\mathbf{x}), \tag{B.57}$$

so that, finally,

$$\mathbf{x}_* \simeq \mathbf{x} - \mathcal{J}^{-1}(\mathbf{x})\mathbf{f}(\mathbf{x}). \tag{B.58}$$

This equation is the generalization of Equation (B.9) and suggests the following iteration scheme for solving problem (B.55).

1. Select an initial guess, $\mathbf{x}_{(0)}$, arbitrarily.
2. Take the kth iteration, $\mathbf{x}_{(k)}$, to be

$$\mathbf{x}_{(k)} = \mathbf{x}_{(k-1)} - \mathcal{J}^{-1}(\mathbf{x}_{(k-1)})\mathbf{f}(\mathbf{x}_{(k-1)}). \tag{B.59}$$

This method is obviously the generalization of Newton iteration.

Still another important application of the concept of a matrix inverse is to the solution of the *homogeneous* equation

$$\mathcal{A}\mathbf{x} = \mathbf{0}, \tag{B.60}$$

where $\mathbf{0}$ is the vector whose elements are all zero. It follows from Equation

(B.54) that if $\boldsymbol{\alpha}^{-1}$ exists, then

$$\mathbf{x} = \boldsymbol{\alpha}^{-1}\mathbf{0} = \mathbf{0}. \tag{B.61}$$

This solution to the homogeneous equation is known as the *trivial* solution. Very often one is interested in the conditions that must be satisfied in order that *non*trivial solutions to the homogeneous equation exist. It follows from Equation (B.61) that nontrivial solutions can exist only if $\boldsymbol{\alpha}^{-1}$ does not exist. It then follows from Equation (B.51) that a necessary condition for the existence of nontrivial solutions to the homogeneous equation is

$$\det \boldsymbol{\alpha} = 0. \tag{B.62}$$

Nontrivial solutions are not unique.

The final topic to be summarized in this section is the concept of functions of matrices. A function of a matrix is often a matrix and defined by analogy with the Taylor series expansion of the function of a scalar variable. For instance, use Equation (B.2) to define

$$\exp \boldsymbol{\alpha} \triangleq \boldsymbol{\jmath} + \boldsymbol{\alpha} + \frac{1}{2!}\boldsymbol{\alpha}^2 + \frac{1}{3!}\boldsymbol{\alpha}^3 + \frac{1}{4!}\boldsymbol{\alpha}^4 + \cdots. \tag{B.63}$$

The important fact to remember about functions of matrices is that one cannot simply assume that the algebra of these functions is analogous to the algebra of the corresponding functions of scalar variables. For instance, the reader should verify for himself that, because of inequality (B.28),

$$\exp(\boldsymbol{\alpha}\boldsymbol{\beta}) \neq \exp \boldsymbol{\alpha} \exp \boldsymbol{\beta}. \tag{B.64}$$

B.3   The Algebra of Complex Numbers[1]

In a trivial way, complex numbers have something of the character of two component vectors. In fact, complex numbers are often given the vector-like graphical representation indicated in Figure B.1. The first major difference between complex numbers and two-dimensional vectors is the representation of these entities. The two components, x and y, of a complex number, z, are not distinguished by their location in an array [as in Equation (B.14)] but rather by their placement in the expression

$$z = x + iy, \tag{B.65}$$

where

$$i = \sqrt{-1}. \tag{B.66}$$

x is called the *real* part of z, and y is called the *imaginary* part of z and one often sees the notations

$$x = \text{Re}(z), \qquad (B.67)$$

$$y = \text{Im}(z). \qquad (B.68)$$

The chief motivation for the use of this notational difference is related to the fact that it is useful to define a product of complex numbers which has no counterpart in vector algebra. Define the product of two complex numbers, $z_1 = x_1 + iy_1$ and $z_2 = x_2 + iy_2$, by

$$z_1 z_2 \triangleq x_1 x_2 - y_1 y_2 + i(x_1 y_2 + x_2 y_1), \qquad (B.69)$$

which is precisely the result that one would obtain if he represented z_1 and z_2 as in Equation (B.65) and applied the normal algebra of scalar variables while observing Equation (B.66). Thus, by adopting the notation indicated in Equation (B.65), one reduces the algebra of complex numbers to the algebra of scalar variables.

The polar coordinate quantities, ρ and θ, indicated in Figure B.1 are called the *modulus* and *angle*, respectively, of the complex number z. Notice that

$$\rho = \sqrt{x^2 + y^2} \qquad (B.70)$$

and that

$$\theta = \tan^{-1} \frac{y}{x}. \qquad (B.71)$$

Also notice that Equation (B.65) can be rewritten

$$z = \rho(\cos \theta + i \sin \theta). \qquad (B.72)$$

The *complex conjugate* of a complex number, $z = x + iy$, is denoted $\bar{z}$ and defined by

$$\bar{z} = x - iy = \rho(\cos \theta - i \sin \theta). \qquad (B.73)$$

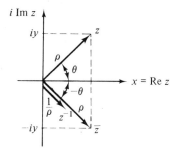

Figure B.1. Graphical representation of z, $\bar{z}$, and z^{-1}.

The graphical interpretation of $\bar{z}$ is indicated in Figure B.1. It is left as an exercise for the reader to show that

$$z\bar{z} = \rho^2 + i \cdot 0; \tag{B.74}$$

i.e., the product of a number with its conjugate is a real number and equal to the square of the modulus.

Example B.6

Consider the reciprocal of a complex number

$$\frac{1}{z} = \frac{1}{x + iy}. \tag{B.75}$$

Multiply numerator and denominator by $\bar{z}$ to find

$$\frac{1}{z} = \frac{\bar{z}}{\rho^2} = \frac{x - iy}{\rho^2}$$

or

$$\frac{1}{z} = \frac{1}{\rho}(\cos \theta - i \sin \theta). \tag{B.76}$$

The reciprocal, z^{-1}, is given the graphical interpretation indicated in Figure B.1.

* * *

As in the case of matrices, one defines the function of a complex number by an infinite series which is analogous to the corresponding Taylor series of a scalar variable. For instance, define

$$e^z = 1 + z + \frac{1}{2!}z^2 + \frac{1}{3!}z^3 + \frac{1}{4!}z^4 + \cdots. \tag{B.77}$$

Unlike the matrix case, the rules of algebra of functions of scalar variables carry over to the case of functions of complex variables.

An extremely important special case of Equation (B.77) is the case when z is pure imaginary, i.e., for the case $z = iy$. Then

$$e^{iy} = 1 + iy + \frac{1}{2!}(iy)^2 + \frac{1}{3!}(iy)^3 + \frac{1}{4!}(iy)^4 + \cdots$$

$$= 1 - \frac{1}{2!}y^2 + \frac{1}{4!}y^4 - \cdots + i\left(y - \frac{1}{3!}y^3 + \frac{1}{5!}y^5 - \cdots\right).$$

It follows from Equations (B.3) and (B.4) that

$$e^{iy} = \cos y + i \sin y. \tag{B.78}$$

It follows that Equations (B.71), (B.72), and (B.75) can be rewritten

$$z = \rho e^{i\theta}, \tag{B.79}$$

$$\bar{z} = \rho e^{-i\theta}, \tag{B.80}$$

and

$$z^{-1} = \frac{1}{\rho} e^{-i\theta}. \tag{B.81}$$

The significance of these formulas is that the algebraic manipulations of these quantities reduce to the use of the standard rules of manipulation of exponential functions. The usefulness of this fact is illustrated many times in the text.

Example B.7

Equation (B.68) can be rewritten

$$z_1 z_2 = \rho_1 e^{i\theta_1} \rho_2 e^{i\theta_2}$$
$$= \rho_1 \rho_2 e^{i(\theta_1 + \theta_2)}. \tag{B.82}$$

Thus the modulus of $z_1 z_2$ is $\rho_1 \rho_2$ and the angle is $\theta_1 + \theta_2$. Similarly, the modulus of z_1 / z_2 is ρ_1 / ρ_2 and the angle is $\theta_1 - \theta_2$.

* * *

An often-used concept is the graphical interpretation of the complex quantity $z - z_0$. The addition or subtraction of complex numbers is defined in the same manner as in the case of matrices and vectors [Equation (B.22)], i.e.,

$$z_1 \pm z_2 = (x_1 + x_2) \pm i(y_1 + y_2). \tag{B.83}$$

This definition of addition leads to the well-known *parallelogram law* for vectors and complex numbers. This "law" is illustrated in Figure B.2. $z - z_0$

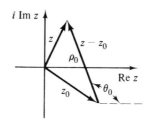

Figure B.2. Graphical interpretation of $z - z_0$.

must be the complex number indicated in this figure because when it is added, graphically, to z_0, the result is z, which is consistent with

$$(z - z_0) + z_0 = z. \tag{B.84}$$

The important point to notice is that

$$z - z_0 = \rho_0 e^{i\theta_0}, \tag{B.85}$$

where ρ_0 and θ_0 are defined in Figure B.2.

References

1. CHURCHILL, R. V., *Complex Variables and Applications*, McGraw-Hill, New York, 1960.

2. OGATA, K., *State Space Analysis of Control Systems*, Prentice-Hall, Englewood Cliffs, N. J., 1967.

3. SOKOLNIKOFF, I. S., and REDHEFFER, R.M., *Mathematics of Physics and Modern Engineering*, McGraw-Hill, New York, 1966.

Computer Programs of Support Subroutines for Computer-Aided Design of Automatic Control Systems

This appendix contains listings of programs useful in the computer-aided analysis and design of dynamic systems.

C.1 Modifying the Q-R Algorithm to Find Roots of Polynomials[1,2]

The ability to quickly calculate the poles of an unfactored transfer function is the basis of many of the programs listed in this appendix. For this purpose the program of the Q-R algorithm written by Van Ness and Imad[2] for SHARE has been modified to provide such a capability. The SHARE program finds the eigenvalues of a matrix in *upper Hessenburg form*, i.e., a matrix of the form

$$\begin{bmatrix} x & x & x & \cdots & x & x & x \\ x & x & x & \cdots & x & x & x \\ 0 & x & x & \cdots & x & x & x \\ 0 & 0 & x & \cdots & x & x & x \\ 0 & 0 & 0 & \cdots & x & x & x \\ \cdot & & \cdot & & & \cdot \\ \cdot & & \cdot & & & \cdot \\ \cdot & & \cdot & & & \cdot \\ 0 & 0 & 0 & \cdots & 0 & x & x \end{bmatrix}, \qquad (C.1)$$

where x denotes a possible nonzero matrix element.

For present purposes, it is desired to find the roots of the polynomial equation

$$b_0 + b_1 s + b_2 s^2 + \cdots + b_n s^n = 0. \tag{C.2}$$

It is clear that in order to use the Q-R algorithm for finding the roots of Equation (C.2), one must find a matrix which is in upper Hessenburg form and whose characteristic equation is Equation (C.2). Consider the $n \times n$ matrix

$$\begin{bmatrix} \dfrac{-b_{n-1}}{b_n} & \dfrac{-b_{n-2}}{b_n} & \cdots & \dfrac{-b_0}{b_n} \\ 1 & 0 & \cdots & 0 \\ 0 & 1 & \cdots & 0 \\ \cdot & \cdot & \cdot & \cdot \\ \cdot & \cdot & \cdot & \cdot \\ \cdot & \cdot & \cdot & \cdot \\ 0 & 0 & \cdots & 0 \end{bmatrix}. \tag{C.3}$$

This matrix is in upper Hessenburg form, and it will now be shown that its characteristic equation is, indeed, Equation (C.2). The characteristic equation of matrix (C.3) is

$$\det \begin{bmatrix} \dfrac{-b_{n-1}}{b_n} - \lambda & \dfrac{-b_{n-2}}{b_n} & \cdots & \dfrac{-b_0}{b_n} \\ 1 & -\lambda & \cdots & 0 \\ 0 & 1 & \cdots & 0 \\ \cdot & \cdot & \cdot & \cdot \\ \cdot & \cdot & \cdot & \cdot \\ \cdot & \cdot & \cdot & \cdot \\ 0 & 0 & \cdots & -\lambda \end{bmatrix} = 0. \tag{C.4}$$

Expand this determinant in terms of the minors of the first row and obtain

$$\left(\frac{-b_{n-1}}{b_n} - \lambda \right) c_1 + \sum_{i=2}^{n} \left(\frac{-b_{n-i}}{b_n} \right) (-1)^{i-1} c_i = 0, \tag{C.5}$$

where the c_i are the minors formed by deleting the first row and the ith column from the determinant of Equation (C.4). By inspection

$$c_i = \det \begin{bmatrix} 1 & -\lambda & \cdots & 0 & 0 & 0 & \cdots & 0 \\ 0 & 1 & \cdots & 0 & 0 & 0 & \cdots & 0 \\ \vdots & \vdots & \vdots & & & & & \\ 0 & 0 & \cdots & 1 & 0 & 0 & \cdots & 0 \\ 0 & 0 & \cdots & 0 & -\lambda & 0 & \cdots & 0 \\ 0 & 0 & \cdots & 0 & 1 & -\lambda & \cdots & 0 \\ \vdots & \vdots & & & & \ddots & & \\ \vdots & \vdots & & & & & \ddots & \\ 0 & 0 & \cdots & 0 & 0 & 0 & \cdots & -\lambda \end{bmatrix} \longleftarrow (i-1)\text{st row.}$$

$$[n-1]\,[n-1]$$

$$(C.6)$$

Successively expand this determinant in terms of the minors of the first column until the $(i-1)$st column is reached; then expand in terms of the minors of the first row and find that

$$c_i = (-\lambda)^{(n-1)-(i-1)} = \lambda^{n-i}(-1)^{n-i},$$

so that Equation (C.5) becomes

$$\left(\frac{-b_{n-1}}{b_n} - \lambda\right)\lambda^{n-1} + \sum_{i=2}^{n}\left(\frac{-b_{n-i}}{b_n}\right)\lambda^{n-i} = 0,$$

which may be rewritten

$$-\lambda^n - \frac{b_{n-1}}{b_n}\lambda^{n-1} - \frac{b_{n-2}}{b_n}\lambda^{n-2} - \cdots - \frac{b_0}{b_n} = 0$$

or

$$b_n\lambda^n + b_{n-1}\lambda^{n-1} + \cdots + b_0 = 0,$$

which, indeed, is Equation (C.2) and the assertion has been proved.

C.2 Representation of a Polynomial in a FORTRAN Program

In the computer subroutines, which will now be described, it has been found convenient to represent a polynomial by a vector. The first component of the vector corresponds to the zeroth-order coefficient of the polynomial, the second component corresponds to the first-order coefficient, and so on. Transfer functions are represented by matrix arrays with two rows; the first row represents the numerator polynomial, and the second row represents the denominator polynomial.

C.3   The Subroutines EIGRAD and RESIDU

At the end of this appendix are listed support programs for computer-aided design of automatic control systems. The first two of these programs are related to the application of the Q-R algorithm to factoring polynomials.

EIGRAD (NN, CHARAC, INDEX) is the program of the Q-R algorithms modified in the manner indicated in Section C.1. CHARAC (22) is a vector representing the characteristic polynomial. NN is the order of the polynomial plus 1. INDEX is a control signal discussed below. Upon completion of the algorithm the roots of the characteristic polynomial will be returned in the array, ROOT (2,22): The real parts of the roots are returned in the first row of this array, and the imaginary parts are returned in the second row. If INDEX is not set equal to integer 0 by the user, then the powers of the roots are calculated as well as the e_α, where the e_α are the partial derivatives defined by Equation (10.49). The real parts of the e_α are returned in the first row of SMALR (2,22) and the imaginary parts are returned in the second row. The real parts of the powers of the roots are returned in EVECR (22,22) and the imaginary parts are returned in EVECI (22,22). The first component of these arrays corresponds to the number of the root and the second component is the power minus 1 of the root. The e_α and the root powers are used in computer-aided sensitivity analysis.

RESIDU (MN,N) calculates the transient state residues and/or variations of poles with changes in parameters [Equation (10.49)]. N is the order of the transfer function plus 1 and MN is the number of sets of residues and/or sets of pole variations to be calculated. It is assumed that SMALR (2,22), EVECR (22,22), and EVECI (22,22) have previously been loaded in the subroutine EIGRAD. Whether residues or pole variations are calculated is determined by the manner in which the matrix BN (10,10) is loaded; if the numerator coefficients of the transform are loaded into BN, then the real and imaginary parts of the ith residue are returned in the ith row of the arrays RESIDR (10,10), RESIDI (10,10). The pole variations will be returned in the kth row of these latter arrays if, before calling the routine, the kth row of BN was loaded with the negative of the partial derivative of the denominator of the transfer function with respect to some parameter.

C.4   The Function IROUTH

IROUTH (L,EQN,FRSTCN,ARRAY) returns the number of sign changes in the first column of the Routh array for the polynomial represented by EQN (22). The order of this polynomial is $L - 1$. The first column of the array is returned in FRSTCN (22). This function is quite handy for fast stability checks and for checking for nonminimum phase conditions.

C.5 Manipulating Polynomials: The Subroutine SERMUL

The following program is useful when developing programs for performing block diagram algebra.

The subroutine SERMUL (C,A,B,M,N) determines the product of the $(M - 1)$st-order polynomial represented by A(22) and the $(N - 1)$st-order polynomial represented by B(22) and returns the result in C(22). The order of C(22) is $M + N - 1$.

C.6 Further Important Uses of the Q-R Algorithm; Numerical Laplace Transform Inversion and Analytical Root Locus

The final two listings at the end of this appendix are programs which call the above routines to perform numerical Laplace inversions and to determine, numerically, root locus diagrams. The importance of these programs cannot be overemphasized.

These routines are very fast. For instance, eight eighth-order transforms have been inverted in 15 sec of execution time using a moderate-sized computer. As of now the Laplace inverter is restricted to transforms with non-repeated roots. *This is a restriction imposed by the method of calculating residues* and is not a restriction on the Q-R algorithm.

The COMMENT and READ statements in these programs provide sufficient explanation of the operation of the programs. Discussion here is limited to the last line of print-out provided after numerical Laplace inversion. DET is the lowest-order nonzero term of the denominator of the transform divided by the highest-order term. TRACE is the negative of the second highest-order term of the denominator divided by the highest-order term and multiplied by -1 if the order of the denominator is odd. These numbers should equal, respectively, the product and the sum of the poles of the transform, which are the first two numbers in the last line of print-out.

The third number of the print-out is the sum of the residues. This number should equal the initial value of the time function minus the final value, which is the last term in this line of print-out and which is determined by applying the initial and final value theorems to the transform being inverted. Thus, three checks are provided for evaluating the accuracy of the numerical inversion. To date, no error has been indicated by these checks for any transform inverted by the author.

```
      SUBROUTINE EIGRAD(NN,CHARAC,INDEX)
C         THIS ROUTINE IS AN ADAPTATION OF A SHARE
C     PROGRAM OF THE Q-R SCHEME TO THE FINDING
C     OF THE ROOTS OF THE POLYNOMIAL REPRESENTED
C     BY CHARAC .
      DIMENSION CHARAC(22)
      COMMON PSI(22),O(22),ST2(22),ST3(22),A(22,22)
     1,ET(22)/EIG/ROOT(2,22),SMALR(2,22),EVECR(22,22)
     1,EVECI(22,22)
      N=NN-1
      NL1=N-1
      NS=N
      NP=NN+2
      K=1
      EVECR(1,1)=1.
      EVECI(1,1)=0.
      DN=CHARAC(N+1)
C         THE POLYNOMIAL IS LOADED INTO THE MATRIX
C     IN UPPER HESSENBERG FORM .
      DO 2 I=2,N
      EVECR(I,1)=1.
      EVECR(I,I)=0.
      NI=N-I+1
      DO 3 J=1,N
    3 A(I,J)=0.
      A(I,I-1)=1.
    2 A(1,I)=-CHARAC(NI)/DN
      A(1,1)=-CHARAC(N)/DN
      IF(N.LT.3) GO TO 7
      ITER=1
      Z1=0.
      Z2=0.
      X=0.
      S=0.
      SQ=0.
      F=0.
      H=0.
      D=0.
      E=0.
      G=0.
      ZERO = 0.0
      JJ=1
      XNN=0.0
      XN2=0.0
      AA = 0.0
      B = 0.0
      C = 0.0
      DD = 0.0
      R=0.0
      SIG=0.0
      GO TO 60
```

```
    7 XNN=0.0
      XN2=0.0
      AA = 0.0
      B = 0.0
      C = 0.0
      DD = 0.0
      R=0.0
      SIG=0.0
      ITER = 0
      IF(N.GT.2) GO TO 12
14    JJ=-1
  12 X = (A(N-1,N-1) - A(N,N))**2
      S = 4.0*A(N,N-1)*A(N-1,N)
      ITER = ITER + 1
      IF(X .EQ. 0.0 .OR. ABS(S/X) .GT. 1.0E-8) GO TO 5
      IF(ABS(A(N-1,N-1)).LE.ABS(A(N,N))) GO TO 32
      E = A(N-1,N-1)
      G = A(N,N)
      GO TO 9
  32 G = A(N-1,N-1)
      E = A(N,N)
   9 F=0.
      H = 0.
      GO TO 24
   5 S = X + S
      X = A(N-1,N-1) + A(N,N)
      IF(S.LT.0.) GO TO 8
      SQ=SQRT(S)
      F=0.0
      H=0.0
      IF (X.GT.0.) GO TO 22
      E=(X-SQ)/2.0
      G=(X+SQ)/2.0
      GO TO 24
  22  G=(X-SQ)/2.0
      E=(X+SQ)/2.0
      GO TO 24
   8 F = SQRT(-S)/2.0
      E=X/2.0
      G=E
      H=-F
24    IF(JJ)  84,70,70
  70 D = 1.0E-10*(ABS(G) + F)
      IF(ABS(A(N-1,N-2)) .GT. D)  GO TO 26
  84  ROOT(1,K)= E
      ROOT(2,K)= F
      ROOT(1,K+1)= G
      ROOT(2,K+1)= H
      IF(INDEX.EQ.0) GO TO 207
      EVECR(K,2)=E
      EVECR(K+1,2)=G
```

```
      EVECI(K,2)=F
      EVECI(K+1,2)=H
      DO 205 J=3,NP
      REAL=EVECR(K,J-1)
      REALK=EVECR(K+1,J-1)
      EMAG=EVECI(K,J-1)
      EMAGK=EVECI(K+1,J-1)
      EVECR(K,J)=E*REAL- F*EMAG
      EVECR(K+1,J)=G*REALK-H*EMAGK
      EVECI(K+1,J)=G*EMAGK+H*REALK
  205 EVECI(K,J)=E*EMAG+ F*REAL
      RN=NS
      PRODR=EVECR(K,NS)*RN*DN
      PRODRK=EVECR(K+1,NS)*RN*DN
      PRODI=EVECI(K,NS)*RN*DN
      PRODIK=EVECI(K+1,NS)*RN*DN
      DO 204 J=2,NS
      RN=J-1
      PRODR=PRODR+RN*CHARAC(J)*EVECR(K,J-1)
      PRODRK=PRODRK+RN*CHARAC(J)*EVECR (K+1,J-1)
      PRODIK=PRODIK+RN*CHARAC(J)*EVECI(K+1,J-1)
  204 PRODI=PRODI+RN*CHARAC(J)*EVECI(K,J-1)
      REAL=PRODR**2+PRODI**2
      REALK=PRODRK**2+PRODIK**2
      SMALR(1,K)=PRODR/REAL
      SMALR(1,K+1)=PRODRK/REALK
      SMALR(2,K)=-PRODI/REAL
      SMALR(2,K+1)=-PRODIK/REALK
  207 K=K+2
      N=N-2
      IF(N-1) 1,86,7
   26 IF(ABS(A(N,N-1)) .GT. 1.0E-10*ABS(A(N,N)))
     1GO TO 50
   86 ROOT(1,K)=A(N,N)
      ROOT(2,K)=0.
      IF(INDEX.EQ.0) GO TO 301
      REAL=A(N,N)
      DO 302 I=2,NP
      EVECI(K,I)=0.
  302 EVECR(K,I)=REAL**(I-1)
      RN=NS
      PRODR=DN*RN*REAL**NL1+CHARAC(2)
      DO 300 I=2,NL1
      RN=I
  300 PRODR=PRODR+RN*CHARAC(I+1)*REAL**(I-1)
      SMALR(2,K)=0.
      SMALR(1,K)=1./PRODR
  301 K=K+1
      N=N-1
      IF(N-1) 1,86,7
```

```
 50    IF(ABS(ABS(XNN/A(N,N-1))-1.0)-1.0E-6)  63,63,62
 62    IF(ABS(ABS(XN2/A(N-1,N-2))-1.0)-1.0E-6)  63,63,700
   63 VQ=ABS(A(N,N-1))-ABS(A(N-1,N-2))
       IF (ITER-15) 53,164,64
  164 IF(VQ) 165,165,166
  165 R = A(N-1,N-2)**2
       SIG = 2.0*A(N-1,N-2)
       GO TO 60
  166 R = A(N,N-1)**2
       SIG = 2.0*A(N,N-1)
       GO TO 60
   64 IF(VQ) 86,86,84
  700 IF(ITER .GT. 50) GO TO 63
       IF(ITER .GT. 5 ) GO TO 53
       Z1=          ((E-AA)**2+(F-B)**2)/(E*E+F*F)
       Z2=        ((G-C)**2+(H-DD)**2)/(G*G+H*H)
       IF(Z1-0.25)  51,51,52
   51    IF(Z2-0.25)    53,53,54
   53    R=E*G-F*H
       SIG=E+G
       GO TO 60
   54    R=E*E
       SIG=E+E
       GO TO 60
   52    IF(Z2-0.25) 55,55,601
   55    R=G*G
       SIG=G+G
       GO TO 60
  601 R = 0.0
       SIG = 0.0
   60    XNN=A(N,N-1)
       XN2=A(N-1,N-2)
       N1 = N - 1
       IA = N - 2
       IP = IA
       IF(N-3) 101,10,20
   20 DO 17 J = 3,N1
       J1 = N - J
       IF(ABS(A(J1+1,J1)).LE.D) GO TO 10
       DEN = A(J1+1,J1+1)*(A(J1+1,J1+1)-SIG)+A(J1+1,J1+2)
      1*A(J1+2,J1+1)+R
       IF(DEN.EQ.0.) GO TO 17
       IF(ABS(A(J1+1,J1)*A(J1+2,J1+1)*(ABS(A(J1+1,J1+1)
      1+A(J1+2,J1+2)-SIG)+ABS(A(J1+3,J1+2)))/DEN)
      1-D)   10,10,17
   17 IP=J1
   10    DO 11  J=1,IP
       J1=IP-J+1
       IF(ABS(A(J1+1,J1)).LE.D) GO TO 13
   11 IQ=J1
   13    DO 100 I=IP,N1
```

```
          IF(I-IP)     16,15,16
 15       O(1)=A(IP,IP)*(A(IP,IP)-SIG)+A(IP,IP+1)*A(IP+1,IP)*
          O(2)=A(IP+1,IP)*(A(IP,IP)+A(IP+1,IP+1)-SIG)
          O(3)=A(IP+1,IP)*A(IP+2,IP+1)
          A(IP+2,IP)=0.0
          GO TO 19
 16       O(1)=A(I,I-1)
          O(2)=A(I+1,I-1)
          IF(I.GT.IA) GO TO 18
          O(3)=A(I+2,I-1)
          GO TO 19
 18       O(3)=0.0
    19    XK = SIGN(SQRT(O(1)**2 + O(2)**2 + O(3)**2), O(1))
          IF(XK.EQ.0.) GO TO 28
          AL=O(1)/XK+1.0
          PSI(1)=O(2)/(O(1)+XK)
          PSI(2)=O(3)/(O(1)+XK)
          GO TO 25
    28 AL=2.0
       PSI(1)=0.0
       PSI(2)=0.0
    25 IF(I.EQ.IQ) GO TO 27
       IF(I.NE.IP) GO TO 29
       A(I,I-1)=-A(I,I-1)
       GO TO 27
 29    A(I,I-1)=-XK
 27    DO 30   J=I,N
       IF(I.GT.IA) GO TO 34
       Z=PSI(2)*A(I+2,J)
       GO TO 33
    34 Z=0.
 33    Y=AL*(A(I,J)+PSI(1)*A(I+1,J)+Z)
       A(I,J)=A(I,J)-Y
       A(I+1,J)=A(I+1,J)-PSI(1)*Y
       IF(I.GT.IA) GO TO 30
       A(I+2,J)=A(I+2,J)-PSI(2)*Y
 30    CONTINUE
       IF(I-IA)     35,35,36
 35    L=I+2
       GO TO  37
 36    L=N
 37    DO  40  J=IQ,L
       IF(I-IA)   38,38,39
 38    Z=PSI(2)*A(J,I+2)
       GO TO   41
    39 Z=0.
 41    Y=AL*(A(J,I)+PSI(1)*A(J,I+1)+Z)
       A(J,I)=A(J,I)-Y
       A(J,I+1)=A(J,I+1)-PSI(1)*Y
       IF(I-IA)   42,42,40
 42    A(J,I+2)=A(J,I+2)-PSI(2)*Y
```

```
   40     CONTINUE
          IF(I-N+3)      43,43,100
   43     Y=AL*PSI(2)*A(I+3,I+2)
          A(I+3,I)=-Y
          A(I+3,I+1)=-PSI(1)*Y
          A(I+3,I+2)=A(I+3,I+2)-PSI(2)*Y
  100     CONTINUE
  101     AA=E
          B=F
          C=G
          DD=H
          GO TO 12
    1     RETURN
          END


          SUBROUTINE RESIDU(MN,N)
C     THIS ROUTINE CALCULATES RESIDUES, POLE VARIATIONS
          COMMON /RESID/ A(10,10),BN(10,10),RESIDR(10,10)
         1,RESIDI(10,10),FR(10,10),FI(10,10)/EIG/ROOT(2,22)
         1,SMALR(2,22),EVECR(22,22),EVECI(22,22)
          NL1=N-1
          DO 1 I=1,MN
          DO 2 K=1,NL1
          IF(ROOT(2,K).LT.0.) GO TO 2
          RX=0.
          IF(ROOT(2,K).EQ.0.)GO TO 4
          RY=0.
          DO 5 L=1,N
          RX=RX+BN(I,L)*EVECR(K,L)
    5     RY=RY+BN(I,L)*EVECI(K,L)
          RESIDR(I,K)=RX*SMALR(1,K)-RY*SMALR(2,K)
          RESIDI(I,K)=RX*SMALR(2,K)+RY*SMALR(1,K)
          RESIDR(I,K+1)=RESIDR(I,K)
          RESIDI(I,K+1)=-RESIDI(I,K)
          GO TO 2
    4     REAL=ROOT(1,K)
          DO 3 L=1,N
    3     RX=RX+BN(I,L)*REAL**(L-1)
          RESIDR(I,K)=RX*SMALR(1,K)
          RESIDI(I,K)=0.
    2     CONTINUE
    1     CONTINUE
          RETURN
          END


          FUNCTION IROUTH(L,EQN,FRSTCN,ARRAY)
C         THIS ROUTINE CHECKS THE ROUTH ARRAY OF EQN .
          DIMENSION EQN(22),ARRAY(22,22),FRSTCN(22),
          N=L
```

```
      1 IF (EQN(N).NE.0.) GO TO 2
        N=N-1
        GO TO 1
C          LOAD THE ARRAY .
      2 I=2
        J=0
        K=N+1
    100 K=K-1
        I=I-1
        J=J+1
        ARRAY(I,J)=EQN(K)
        IF(K.EQ.1) GO TO    150
        K=K-1
        I=I+1
        ARRAY(I,J)=EQN(K)
        IF(K.GE.2) GO TO 100
    150 F=N
        KK=F/2.
        IF (N.EQ.2*KK) GO TO 200
        KK=KK+1
        ARRAY(2,KK)=0.0
    200 I=3
        FRSTCN(1)=ARRAY(1,1)
        FRSTCN(2)=ARRAY(2,1)
    250 F=N-I
        FRSTCN(I)=1.
        KK=F/2.
        KK=KK+1
        KKK=KK+1
        IL1=I-1
        IL2=I-2
C          PERFORM THE ALGORITHM .
        DO 300 J=1,KK
        JP1=J+1
        TEM=ARRAY(IL2,1)*ARRAY(IL1,JP1)/ARRAY(IL1,1)
    300 ARRAY(I,J)=ARRAY(IL2,JP1)-TEM
        ARRAY(I,KKK)=0.0
        IF(ARRAY(I,1).NE.0.0) GO TO 325
        ARRAY(I,1)=SIGN(1.,ARRAY(IL1,1))*10.**(-12)
        FRSTCN(I)=0.
    325 IF(FRSTCN(I).NE.0.) FRSTCN(I)=ARRAY(I,1)
        I=I+1
        IF (I.LE.N) GO TO 250
        IND=0
        STORE=+1.
C          CHECK THE FIRST COLUMN .
        STORE=SIGN(1.,FRSTCN(1))
        DO 400 I=1,N
        IF(FRSTCN(I).EQ.0.) IND=IND+1
        TEM=SIGN(1.,FRSTCN(I))
        IF (STORE*TEM.EQ.-1.0) IND=IND+1
```

```
  400 STORE=TEM
      IROUTH=IND
      RETURN
      END



      SUBROUTINE SERMUL(C,A,B,M,N)
C         POLYNOMIAL MULTIPLICATION ROUTINE
      DIMENSION A(22),B(22),C(22)
      K=M+N-1
      DO 20 J=1,K
      C(J)=0.0
      KK=J-N+1
      JP1=J+1
      IF(KK.GT.1) GO TO 6
      MAX=J
      IF (M.LE.MAX) MAX=M
      DO 5 I=1,MAX
      JP1LI=JP1-I
    5 C(J)=C(J)+A(I)*B(JP1LI)
      GO TO 20
    6 MAX=M
      IF(J.LT.MAX) MAX=J
      DO 7 I=KK,MAX
      JP1LI=JP1-I
    7 C(J)=C(J)+A(I)*B(JP1LI)
   20 CONTINUE
      RETURN
      END



      SUBROUTINE LPLACE(TF,M,N,SCALE)
C    ROUTINE FOR INVERTING AN UNFACTORED LAPLACE TRANSFORM
C    THE METHOD USED TO DETERMINE RESIDUES LIMITS USE OF
C    ROUTINE TO TRANSFORMS WITH DISTINCT POLES .
      COMMON /RESID/ A(10,10),BN(10,10),RESIDR(10,10)
     1,RESIDI(10,10),FR(10,10),FI(10,10)/EIG/ROOT(2,22)
     1,SMALR(2,22),EVECR(22,22),EVECI(22,22)
      DIMENSION CHARAC(22),TF(2,22)
C    TF IS THE LAPLACE TRANSFORM
      PRINT 90
      PRINT 91
      IF (SCALE.EQ.0.) SCALE=1.
      DO 1000 I=1,N
      K=I+M
      TF(1,K)=0.
      BN(1,I)=TF(1,I)
 1000 CHARAC(I)=TF(2,I)
      PRINT 50,(BN(1,I),I=1,M)
      PRINT51,(CHARAC(I),I=1,N)
```

```
      NN=0
      IF(CHARAC(1).EQ.O.) NN=1
      SS=0.
      K=N
      IF(NN.EQ.0) GO TO 400
      K=N-1
      SS=BN(1,1)/CHARAC(2)
      DO 301 L=1,K
  301 CHARAC(L)=CHARAC(L+1)
C    DETERMINE THE TRANSFORM'S POLES
  400 CALLEIGRAD(K,CHARAC,1)
      DET=CHARAC(1)/CHARAC(K)
      TRACE=CHARAC(K-1)/CHARAC(K)
      MN=1
C    DETERMINE THE RESIDUES
      CALLRESIDU(MN,K)
      KK=K-1
      IF(NN.EQ.0) GO TO 410
C    MAKE NECESSARY MODIFICATIONS IF TRANSFORM HAS POLF
C    AT ORIGIN
      DO 405 L=1,KK
      REAL=RESIDR(1,L)
      EMAG=RESIDI(1,L)
      E=ROOT(1,L)
      F=ROOT(2,L)
      R=E**2+F**2
      RESIDR(1,L)=(E*REAL+EMAG*F)/R
  405 RESIDI(1,L)=(E*EMAG-F*REAL)/R
  410 RT=1.E30
      Y=0.
      DO 411 I=1,KK
      IF(ABS(ROOT(1,I)).LT.RT) RT=ROOT(1,I)
  411 IF(ROOT(2,I).GT.Y) Y=ROOT(2,I)
C    DETERMINE TIME INCREMENT
      IF(Y.NE.O.) DEL=5.*3.1416/(200.*Y)
      IF(RT.EQ.O.) GO TO 412
      DEL=-5./RT
      DEL=DEL/200.
      DEL=DEL*SCALE
      DEL=ABS(DEL)
  412 PRINT 53
      CHARAC(1)=0.
      CHARAC(2)=0.
      IF(M.EQ.N-1) CHARAC(2)=BN(1,M)/CHARAC(K)
      SI=CHARAC(2)
C    DETERMINE TIME FUNCTION
      DO 500 I=2,5
      R=I-1
      T=DEL*R*40.
      L=2*I
      LL=L-1
```

```
          CHARAC(LL)=T
          RES=SS
          DO 499 LK=1,KK
          F=ROOT(2,LK)
          IF(F.LT.0.) GO TO 499
          E=ROOT(1,LK)
          IF(F.EQ.0.) GO TO 488
          RES=RES+2.*EXP(E*T)*(RESIDR(1,LK)*COS(F*T)-
         1 RESIDI(1,LK)*SIN(F*T))
          GO TO 499
      488 RES=RES+RESIDR(1,LK)*EXP(E*T)
      499 CONTINUE
      500 CHARAC(L)=RES
          PRINT 54,(CHARAC(I),I=1,10)
          DO 601 LLL=2,40
          RR=LLL-1
          DO 600 I=1,5
          R=I-1
          T=RR*DEL+R*DEL*40.
          L=2*I
          LL=L-1
          CHARAC(LL)=T
          RES=SS
          DO 599 LK=1,KK
          F=ROOT(2,LK)
          IF(F.LT.0.) GO TO 599
          E=ROOT(1,LK)
          IF(F.EQ.0.) GO TO 588
          RES=RES+2.*EXP(E*T)*(RESIDR(1,LK)*COS(F*T)-
         1 RESIDI(1,LK)*SIN(F*T))
          GO TO 599
      588 RES=RES+RESIDR(1,LK)*EXP(E*T)
      599 CONTINUE
      600 CHARAC(L)=RES
      601 PRINT 54,(CHARAC(I),I=1,10)
          SUM=0.
          SUMR=0.
          PROD=1.
          RT=-1./RT
          PRINT 93,RT
C     MAKE CALCULATIONS FOR ERROR CHECKS
          DO 700 I=1,KK
          F=ROOT(2,I)
          E=ROOT(1,I)
          SUM=SUM-E
          SUMR=SUMR+RESIDR(1,I)
          IF(F.EQ.0.) PROD=PROD*E
          IF(F.GT.0.) PROD=PROD*(E**2+F**2)
          CHARAC(1)=ROOT(1,I)
          CHARAC(2)=ROOT(2,I)
          CHARAC(3)=RESIDR(1,I)
```

```
      CHARAC(4)=RESIDI(1,I)
  700 PRINT 94,(CHARAC(L),L=1,4)
      SS=SI-SS
      PROD=PROD*(-1.)**KK
      PRINT 95,SUM,PROD,SUMR,TRACE,DET,SS
  100 CONTINUE
   10 FORMAT(16F5.0)
   50 FORMAT(1X,10HNUMERATOR ,2X,10F11.5)
   51 FORMAT(1X,12HDENOMINATOR ,10F11.5)
   52 FORMAT(2I5,F5.0,I5)
   53 FORMAT(1X,5(4X,4HTIME,16X)/1X,5(1X,
     120H(SEC.S)        RESPONSE,3X))
   54 FORMAT(1X,5(E10.3,1X,E11.4,1X,1H*))
   56 FORMAT(I5)
   90 FORMAT(1H1)
   91 FORMAT(41HLAPLACE TRANSFORM (INCREASING ORDER OF S))
   92 FORMAT(1X,F12.3)
   93 FORMAT(5X,17HTRANSFORM POLES..,20X,10HRESIDUES..,
     115X,15HCHARAC. TIME = E10.3)
   94 FORMAT(1X,E12.5,2X,E12.5,5X,E12.5,2X,E12.5)
   95 FORMAT(1X,6HSUM = E10.3,2X,7HPROD = E10.3,6X,
     16HSUM = E10.3,7X,8HTRACE = E10.3,2X,6HDET = E10.3,
     22X,5HSS = E10.3)
      RETURN
      END
```

ROOT LOCUS ROUTINE

```
C    THIS ROUTINE EMPLOYS THE Q-R ALGORITHM TO DETERMINE
C    ROOT LOCUS DIAGRAMS .
     COMMON S(22),S1(22),S2(22),S3(22),A(22,22),E(22)
     COMMON /EIG/RT(2,22),SR(2,22),EVR(22,22),EVI(22,22)
     COMMON /ORDER/N(10),M(10),J(10)
     DIMENSION CHARAC(22),PLANT(2,22),COMPEN(2,22),TF(2,22
    1TF(2,22)
     READ 75,NUMB
     DO 20 KL=1,NUMB
     NXT=0
     READ 75,MC,NC,MP,NP,NXT
     PRINT 76
     READ 73,(COMPEN(1,I),I=1,MC)
     PRINT 74,(COMPEN(1,I),I=1,MC)
     READ 73,(COMPEN(2,I),I=1,NC)
     PRINT 74,(COMPEN(2,I),I=1,NC)
     M(1)=MC
     N(1)=NC
     PRINT 77
     IF(MP.EQ.0) GO TO 10
C    IF MP = 0 ,THE NUMERATOR FROM LAST CLOSED LOOP
C    TF OF PREVIOUS PROBLEM IS USED FOR PLANT
```

```
      M(2)=MP
      READ 73,(PLANT(1,I),I=1,MP)
   10 MP=M(2)
      PRINT 74,(PLANT(1,I),I=1,MP)
      IF(NP.EQ.0) GO TO 11
C    IF NP = 0 ,THE DENOMINATOR OF LAST CLOSED LOOP
C    TF OF PREVIOUS PROBLEM IS USED FOR PLANT
      READ 73,(PLANT(2,I),I=1,NP)
      N(2)=NP
   11 NP=N(2)
      PRINT 74,(PLANT(2,I),I=1,NP)
      READ 73,SK,FK,SGN
      PRINT 78
      LN=NXT+10
      R=LN
      DEL=(FK-SK)/R
      J(1)=1
      J(2)=2
      J(3)=3
      CALL CASCDE(COMPEN,PLANT,TF)
C    TF IS NOW THE OPEN LOOP TRANSFER FUNCTION
      KK=N(3)
      LL=KK-1
C    DETERMINE THE ROOT LOCUS
      DO 6 I=1,LN
      G=I
      G=SK+G*SGN*DEL
      DO 1 L=1,KK
    1 CHARAC(L)=TF(2,L)+G*TF(1,L)
      CALL EIGRAD(KK,CHARAC,0)
      PRINT 80,G
      DO 2 L=1,LL
    2 PRINT 79,RT(1,L),RT(2,L)
    6 CONTINUE
      N(2)=N(3)
      M(2)=N(3)
      LL=N(3)
      DO 7 I=1,LL
      PLANT(2,I)=CHARAC(I)
    7 PLANT(1,I)=TF(2,I)
   20 CONTINUE
   73 FORMAT(12F7.0)
   74 FORMAT(1X,10(E10.3,1X))
   75 FORMAT(6I5)
   76 FORMAT(1H1,1X,13HCOMPENSATOR..)
   77 FORMAT(1X,7HPLANT..)
   78 FORMAT(5X,12HROOT LOCUS../5X,4HGAIN,6X,4HREAL,7X,
     14HIMAG)
   79 FORMAT(11X,E10.3,1X,E10.3)
   80 FORMAT(1X,E10.3)
      STOP
      END
```

```
      SUBROUTINE CASCDE(ERROR,COMPEN,VMANIP)
C         VMANIP IS THE TRANSFER FUNCTION PRODUCT
C     OF ERROR AND COMPEN .
      DIMENSION ERROR(2,22),VMANIP(2,22),COMPEN(2,22)
      COMMON ST(22),ST1(22),ST2(22)/ORDER/ N(10),M(10),J(10
      K=J(1)
      KK=J(2)
      KKK=J(3)
      DO 1 I=1,22
      VMANIP(1,I)=0.
      VMANIP(2,I)=0.
      ST(I)=ERROR(1,I)
      ST1(I)=COMPEN(1,I)
    1 CONTINUE
      CALL SERMUL(ST2,ST,ST1,M(K),M(KK))
      DO 2 I=1,22
      VMANIP(1,I)=ST2(I)
      ST(I)=ERROR(2,I)
      ST1(I)=COMPEN(2,I)
    2 CONTINUE
      CALL SERMUL(ST2,ST,ST1,N(K),N(KK))
      KKKK=N(K)+N(KK)-1
      DO 3 I=1,KKKK
      VMANIP(2,I)=ST2(I)
    3 CONTINUE
      N(KKK)=KKKK
      M(KKK)=M(K)+M(KK)-1
      RETURN
      END
```

Subroutine CASCDE is used in the root locus routine.

References

1. BREWER, J. W., *The Application of Numerical Methods to System Design*, Ph.D. Dissertation, Div. of Mechanical Design, University of California, Berkeley, 1966.

2. VAN NESS, J. E., and IMAD, F. P., *An I.B.M. SHARE Program for the Q-R Algorithm.*

Index

A

Accelerators *74, 422*
  piezoelectric *74*
Accumulators *123*
Active bond *174, 180, 188, 195*
Adiabatic expansion *138*
Adjoint matrix *517*
Amplidyne *175*
Amplitude locus *459*
Analyzer matrix *230, 237*
Angle (complex number) *521*
Angle criterion *317*
Angle of approach (departure) *331*
Anti-friction bearing *92*
Approximation *424, 432*
Armature *173*
Automatic control *275*
Axiom *53*

B

Backlash *453*
Bandwidth *402, 410, 414*
Bellows *187*
Bernoulli force *185*

B (second column)

Bilateral coupling *137, 174*
Black box problem *425*
Block diagrams *194*
Bode *3*
Bode diagram *387, 407, 416, 426*
Bonemain *5*
Breakaway (breakin) point *329*
Bulk modulus *122, 124*
Buntenbach *68*

C

Capacitance (compliance):
  carrier *148*
  electric *53*
  hydraulic *113, 122*
  magnetic *70*
  mechanical *78, 91*
  pneumatic *126*
  thermal *141*
Carrier approximation *146*
Carrier fluid *146*
Cascade system *137, 500*
Causal function *12, 19*
Center of gravity *327*